Probability and Statistics for Engineering and the Sciences with Modeling using R

Probability and statistics courses are more popular than ever. Regardless of your major or your profession, you will most likely use concepts from probability and statistics often in your career.

The primary goal behind this book is offering the flexibility for instructors to build most undergraduate courses upon it. This book is designed for a one-semester course in introductory probability and statistics (not calculus based) or a one-semester course in a calculus-based probability and statistics course.

The book focuses on engineering examples and applications, while also including social sciences and more examples. Depending on the chapter flows, a course can be tailored for students at all levels and from all backgrounds.

Over many years of teaching this course, the authors created problems based on real data, student projects, and labs. Students have suggested these enhance their experience and learning. The authors hope to share projects and labs with other instructors and students to make the course more interesting for both.

R is an excellent platform to use. This book uses R with real data sets. The labs can be used for group work, in class, or for self-directed study. These project labs have been class-tested for many years with good results and encourage students to apply the key concepts and the use of technology to analyze and present results.

Dr. William P. Fox is a visiting professor of Computational Operations Research in the Mathematics Department at the College of William and Mary. He is an emeritus professor in the Department of Defense Analysis at the Naval Postgraduate School. He earned his BS degree from the United States Military Academy, MS in operations research from the Naval Postgraduate School, and his PhD in Industrial Engineering from Clemson University. He has taught at the United States Military Academy and at Francis Marion University. He has many publications and scholarly activities including 16 books, 21 book chapters and technical reports, 150 journal articles, and more than 150 conference presentations and mathematical modeling workshops.

Rodney X. Sturdivant, PhD, is director of the Statistical Consulting Center and an associate professor in the Department of Statistical Science at Baylor University. He has been senior research biostatistician with the Henry M. Jackson Foundation for the Advancement of Military Medicine supporting the Uniformed Services University of Health Science. Previously, he was professor of Applied Statistics at Azusa Pacific University. He was associate professor of Clinical Public Health in the Biostatistics Division of the College of Public Health at The Ohio State University. He retired as a Colonel after 27-year career in the U.S. Army. He completed his military service as an Academy Professor and Professor of Applied Statistics in the Department of Mathematical Sciences at the United States Military Academy, West Point. He earned a B.S. from West Point, an M.S. in statistics and an M.S. in operations research from Stanford, and a PhD in biostatistics from the University of Massachusetts - Amherst.

Textbooks in Mathematics

Series editors:

Al Boggess, Kenneth H. Rosen

Geometry and Its Applications, Third Edition
Walter J. Meyer

Transition to Advanced Mathematics
Danilo R. Diedrichs and Stephen Lovett

Modeling Change and Uncertainty
Machine Learning and Other Techniques
William P. Fox and Robert E. Burks

Abstract Algebra
A First Course, Second Edition
Stephen Lovett

Multiplicative Differential Calculus
Svetlin Georgiev, Khaled Zennir

Applied Differential Equations
The Primary Course
Vladimir A. Dobrushkin

Introduction to Computational Mathematics: An Outline
William C. Bauldry

Mathematical Modeling the Life Sciences
Numerical Recipes in Python and MATLAB™
N. G. Cogan

Classical Analysis
An Approach through Problems
Hongwei Chen

Classical Vector Algebra
Vladimir Lepetic

Introduction to Number Theory
Mark Hunacek

Probability and Statistics for Engineering and the Sciences with Modeling using R
William P. Fox and Rodney X. Sturdivant

For more information on this series, please visit www.routledge.com/Textbooks-in-Mathematics/book-series/CANDHTEXBOOMTH

Probability and Statistics for Engineering and the Sciences with Modeling using R

William P. Fox and Rodney X. Sturdivant

CRC Press is an imprint of the
Taylor & Francis Group, an **informa** business
A CHAPMAN & HALL BOOK

First edition published 2023
by CRC Press
6000 Broken Sound Parkway NW, Suite 300, Boca Raton, FL 33487–2742

and by CRC Press
4 Park Square, Milton Park, Abingdon, Oxon, OX14 4RN

CRC Press is an imprint of Taylor & Francis Group, LLC

ISBN: 9781032330471 (hbk)
ISBN: 9781032330501 (pbk)
ISBN: 9781003317906 (ebk)

DOI: 10.1201/9781003317906

Typeset in Palatino
by Apex CoVantage, LLC

To our wives, Hamilton and Mandy. To Frank R. Giordano, 1942–2022, our mentor and friend, who helped us in many ways along our careers.

Contents

Preface

Probability and statistics courses are more popular than ever. Regardless of your major, you will most likely use concepts from probability and statistics often in your lives. I have taught probability and statistics at all levels, freshman (non-calculus based), advanced as a course for engineering and sciences, and as a graduate course. The content in the first two was very similar except for the integration methods as a few distributions in the advanced statistics course versus the freshman course.

Audience

This book is designed for a one-semester course in introductory probability and statistics (not calculus based) or a one-semester course in calculus-based probability and statistics. The concept of engineering is obvious, but sciences including political sciences do need quality statistical analysis courses. Depending on the chapter flows, a course can be tailored for students at all levels and from all backgrounds. The course is based upon my teaching of probability and statistics for over 30 years at all the levels, including graduate courses.

Technology

I have also taught using MINITAB, EXCEL, the TI-83/84 calculator, and R. I found R to be an excellent platform to do the work in probability and statistics.

This book uses R as the technology of choice. For many years I have used other technologies including calculators. I decided that R, as free downloadable software, appears best to obtain results that can be presented in a quality form.

Problems/Exercises/Labs/Projects

I have found the use of "real" problems such as labs or projects provides relevance to students. I have included in this book labs and projects that I have used for years. These labs and projects force students to apply the key concepts and the use of technology to analyze and present results.

Organization and Flow

The organization of the book is as follows: Chapter 1 discusses a modeling approach to probability and statistics, which depending on your coverage can be omitted without loss of concepts. Chapters 2–5 (Section 5.2) provide coverage of basic statistics, displays, random variables, and basic discrete distributions. Sections 5.3–5.5 cover more discrete distribution for more advanced students. Chapter 6 covers continuous distribution as typically is covered for elementary statistics. Chapter 7 covers a myriad of continuous distributions for the more advanced students. Chapter 8 covers the Central Limit Theorem and its importance. Chapter 9 covers estimation, both point and interval estimation. Chapter 10 covers single sample hypothesis testing. A table/flow diagram is provided to assist students. Chapter 11 then addresses two sample hypothesis testing. In both chapters 10–11 we present testing for means, proportions, and variances using R as the technology to assist. Chapter 12 covers reliability of systems for engineers. Chapter 13 covers correlation and simple linear regression from a standpoint of understanding the procedure and adequacy of the resulting model. Chapters 14–16 are more advanced topics covering advanced regression, ANOVA, and ANCOVA. We also provide labs and projects in the appendices that are tried and tested at all levels of probability and statistics.

Introduction to Probability and Statistics

Chapter flow for elementary P&S: Chapter 1 (possibly); Chapters 2–5 (sections 5.1 and 5.2 only); Chapter 6 (omitting Section 6.3); Chapter 8; Chapter 9 (omitting variances); Chapter 10 (omitting variances); Chapter 11, only the material for matched pairs; and Chapter 13.

Probability and Statistics for Engineering and Sciences

Suggested chapter flow for a more advanced P&S: Chapters 1–13, and pick and choose other chapters as time permits.

Acknowledgments

Thanks to all who helped in the development of the material for this textbook. The list includes the faculty and course directors of Math 208 and Math 206 at USMA from 1990 to 1998. Kaitlyn Crowley and VyVy Vu, students at William and Mary, assisted most recently in the exercises and solutions.

William P. Fox
Visiting Professor of Computational Operations Research
Department of Mathematics
College of William and Mary

Rodney X. Sturdivant
Baylor University

1

Introduction to Statistical Modeling and Models and R

Objectives

1. **Understand the basic flow of statistical modeling as a process.**
2. **Understand the concept of model building as a process.**
3. **Begin the process of using R.**

1.1 What Is Modeling?

Consider the importance of decision-making in such areas as business (B), industry (I), and government (G). BIG decision-making is essential to success at all levels. We do not encourage "shooting from the hip." We recommend good analysis for the decision-maker to examine and question in order to find the best alternative to choose or decision to make. In many cases, we are dealing with uncertainty. In those cases, statistical modeling should be used. So why mathematical or statistical modeling?

In simple terms, **statistical modeling** is a simplified, mathematically formalized way to approximate reality (i.e., what generates your data) and optionally to make predictions from this approximation. The **statistical model** is the mathematical equation that is used.

> We will use the following to express a **statistical model. A statistical model** is a mathematical model that uses a set of statistical assumptions for generating a random sample of data (and similar data from a larger population). A statistical model is usually specified as a mathematical relationship between one or more random variables and other non-random variables. As such, a statistical model is "a formal representation of a probabilistic or statistical theory." All statistical hypothesis tests and all statistical estimators are derived via statistical models. More generally, statistical models are part of the foundation of statistical inference.

A **mathematical model** is a description of a system using mathematical concepts and language. The process of developing a mathematical model is termed **mathematical modeling**. Mathematical models are used not only in the natural sciences (such as physics, biology, earth science, meteorology) and engineering disciplines (e.g., computer science, artificial intelligence), but also in the social sciences (such as business, economics, psychology, sociology, and political science); physicists, engineers, statisticians, operations

DOI: 10.1201/9781003317906-1

research analysts, and economists use mathematical models most extensively. A model may help to explain a system and to study the effects of different components and to make *predictions* about behavior.

Mathematical models can take many forms, including but not limited to dynamical systems, statistical models, differential equations, or game theoretic models. These and other types of models can overlap, with a given model involving a variety of abstract structures. In general, mathematical models may include logical models, as far as logic is taken as a part of mathematics. In many cases, the quality of a scientific field depends on how well the mathematical models developed on the theoretical side agree with the results of repeatable experiments. Lack of agreement between theoretical mathematical models and experimental measurements often leads to important advances as better theories are developed. We are concerned in this book with statistical models.

1.2 Overview and the Modeling Process

Consider sampling as used for inspection of the quality of products. A lot contains N products with defective rate θ. Suppose we take a sample without replacement of n products and get x defective products. What are the defective rates?

We have possible sample outcomes (with G for good and D for defective) that are a set such as GGDGGGDD, as one realization of outcomes. How do we connect a given sample of this sort with the population of all possible samples and use this to make decisions and inferences?

For statistical modeling we think of data as a realization of the random experiment.

Some other examples that may require a statistical model:

- Consider drug evaluation or vaccine evaluation for use with COVID-19. How do you conduct the trial and evaluate success or failure? Statistical models with hypothesis tests might be your best option.
- Consider having data on sales for this year and wanting to predict sales forecasts for the next year. Is simple regression or a more advanced regression technique the model to use?
- You are a new city manager in California. You are worried about the earthquake survivability of your city's water tower. You need to analyze the effects of an earthquake on your water tower and see if any design improvements are necessary. You want to prevent catastrophic failure.
- Maybe you have flown lately. Most airplanes are full these days. As a matter of fact, many times an announcement is made that the plane is overbooked and the airlines are looking for volunteers to take a later flight. Why do airlines overbook? Should they overbook? What impact does this have on the passengers? What impact does it have on the airlines?

These are all events that we can model using probability and statistical mathematics. This textbook will help you understand what a statistical modeler might do for you to become a confident problem-solver using the techniques of statistical modeling. As a decision-maker, understanding the possibilities and asking the key questions will enable better decisions to be made.

1.3 The Modeling Process

In this section, we turn our attention to the process of modeling and examine many different scenarios in which mathematical modeling can play a role. Since statistical modeling is a form of mathematical modeling, the principles apply to the topics covered in this book.

Mathematical modeling requires as much art as it does science. Thus, modeling is more of an *art* than a science. Modelers must be creative—willing to be more artistic or original in their approach to the problem. They must be inquisitive—questioning their assumptions, variables, and hypothesized relationships. Modelers must also think outside the box in order to analyze the model and its results. Modelers must ensure their model and results pass the "common sense" test. Science is very important, and understanding science enables one to be more creative in viewing and modeling a problem. Creativity is extremely advantageous in problem-solving with mathematical modeling.

To gain insight, we should consider a single framework that will enable the modeler to address the largest number of problems. The key is that there is something *changing for which we want to know the effects*. We call this the **system** under analysis. The real-world system can be very complicated or very simplistic. This requires a process that allows for both types of real-world systems to be modeled within the same process.

Figure 1.1 provides a closed loop process for modeling. Given a real-world situation like the ones above, we collect data in order to formulate a mathematical model.

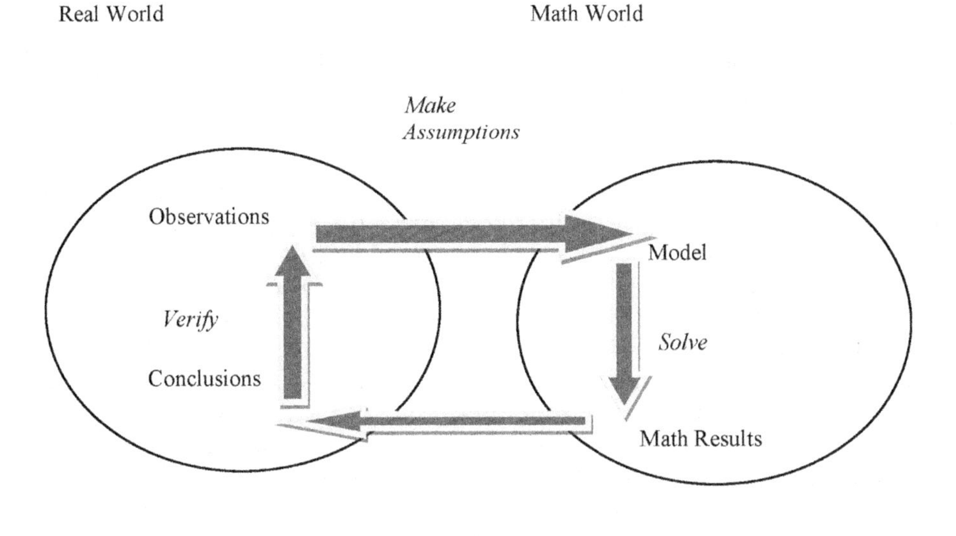

FIGURE 1.1
Modeling real-world systems with mathematics (see Albright & Fox, 2020).

This mathematical model can be one we derive or select from a collection of already-built mathematical models depending on the level of sophistication required. Then we analyze the model that we used and reach mathematical conclusions about it. Next, we interpret the model and either make predictions about what has occurred or offer an explanation as to why something has occurred. Finally, we test our conclusion about the real-world system with new data. We may refine the model to improve its ability to predict or explain the phenomena. We might even reformulate a new mathematical model.

1.3.1 Mathematical Modeling

We will build some mathematical models describing change in the real world. We will solve these models and analyze how good our resulting mathematical explanations and predictions are. The solution techniques that we employ in subsequent chapters take advantage of certain characteristics that the various models enjoy. Consequently, after building the models, we will **classify** the models based on their mathematical structure.

When we observe change, we are often interested in understanding why change occurs the way it does, perhaps to analyze the effects of different conditions or perhaps to predict what will happen in the future. Often, a mathematical model can help us understand a behavior better, while allowing us to experiment mathematically with different conditions. For our purposes, we will consider a mathematical model to be a mathematical construct designed to study a particular real-world system or behavior. The model allows us to use mathematical operations to reach mathematical conclusions about the model as illustrated in Figure 1.1.

1.3.2 Models and Real-World Systems

A system is an assemblage of objects joined by some regular interaction or interdependence: sending a module to Mars, handling the U.S. debt, a fish population living in a lake, a TV satellite orbiting the earth, delivering mail, locations of service facilities, all are examples of a system. The person modeling is interested in understanding not only how a system works but also what interactions cause change and how sensitive the system is to changes in these inputs. Perhaps the person modeling is also interested in predicting or explaining what changes will occur in the system as well as when these changes might occur.

A possible basic technique used in constructing a mathematical model of some system is a combined mathematical-physical analysis. In this approach, we start with some known physical principles or reasonable assumptions about the system. Then we reason logically to obtain conclusions. Sometimes we have data to help us come up with a reasonable model. Modelers must be open to many avenues to solve problems.

Figure 1.2 suggests how we can obtain real-world conclusions from a mathematical model. First, observations identify the factors that seem to be involved in the behavior of interest. Often, we cannot consider, or even identify, all the relevant factors, so we make simplifying assumptions, excluding some of them. Next, we conjecture tentative relationships among the identified factors we have retained, thereby creating a rough "model" of the behavior. We then apply mathematical reasoning that leads to conclusions about the model. These conclusions apply only to the model and may or may not apply to the actual real-world system in question. Simplifications were made in constructing the model, and the observations upon which the model is based invariably contain errors and limitations. Thus, we must carefully account for these anomalies and test the conclusions of the model against real-world observations. If the model is

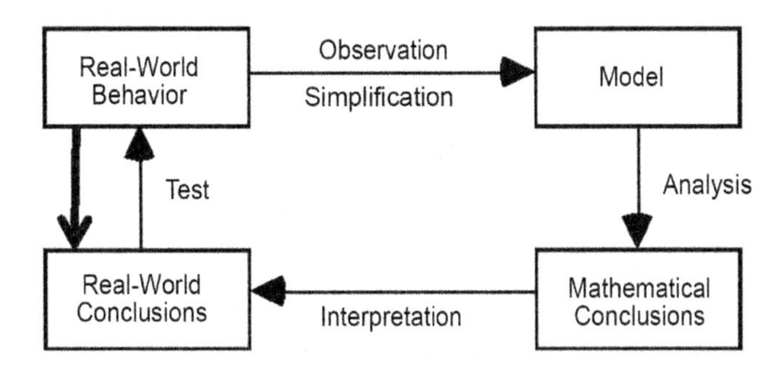

FIGURE 1.2
In reaching conclusions about a real-world behavior, the modeling process is a closed system (adapted from Giordano et al., 2014).

reasonably valid, we can then draw inferences about the real-world behavior from the conclusions drawn from the model. In summary, we present the following procedure for investigating real-world behavior:

Step 1. Observation of the system or hypothesize the system (if one does yet exist), identify the key factors involved in the real-world behavior, simplify initially, and refine later as necessary.

Step 2. Conjecture or guess the possible relationships or interrelationships among the factors and variables identified in step 1.

Step 3. Solve the model.

Step 4. Interpret the mathematical conclusions in terms of the real-world system.

Step 5. Test the model conclusions against real-world observations—the common sense rule.

Step 6. Perform model testing or sensitivity analysis.

There are various kinds of models that we will introduce as well as methods or techniques to solve these models in the subsequent chapters. An efficient process would be to build a library of models and then be able to recognize various real-world situations to which they apply. Another task is to formulate and analyze new models. Still another task is to learn to solve an equation or system in order to find more revealing or useful expressions relating to the variables. Through these activities, we hope to develop a strong sense of the mathematical aspects of the problem, its physical underpinnings, and the powerful interplay between them.

Most models do simplify reality. Generally, models can only approximate real-world behavior. Next, let's summarize a ***process*** for formulating a model.

1.3.3 Model Construction

Let's focus our attention on the process of model construction. An outline is presented as a procedure to help construct mathematical models. In the next section, we will illustrate this procedure with a few examples.

These nine steps are summarized in Figure 1.3 as modified from a six-step approach by Giordano (Giordano et al., 2014). These steps act as a guide for thinking about the problem and getting started in the modeling process.

Step 1. Understand the problem or the question asked.

Step 2. Make simplifying assumptions. Justify your assumptions.

Step 3. Define all variables and provide units.

Step 4. Construct a model.

Step 5. Solve and interpret the model.

Step 6. Verify the model.

Step 7. Identify the strengths and weaknesses of your model.

Step 8. Sensitivity Analysis or Model Testing of the model. Do the results pass the "common sense" test?

Step 9. Implement and maintain the model for future use.

FIGURE 1.3
Mathematical modeling process.

Let's discuss each step in more depth.

Step 1. Understand the problem or the question asked.

Identifying the problem to study is usually difficult. In real life, no one walks up to you and hands you an equation to be solved. Usually, it is a comment like "We need to make more money" or "We need to improve our efficiency." We need to be precise in our formulation of the mathematics to describe the situation.

Step 2. Make simplifying assumptions:

Start by brainstorming the situation. Make a list of as many factors, or variables, as you can. Realize we usually cannot capture all the factors influencing a problem. The task is simplified by reducing the number of factors under consideration. We do this by making simplifying assumptions about the factors, such as holding certain factors as constants. We might then examine to see if relationships exist between the remaining factors (or variables). Assuming simple relationships might reduce the complexity of the problem. Once you have a shorter list of variables, classify them as independent variables, dependent variables, or neither.

Step 3. Define all variables.

It is critical to define all your variables and provide the mathematical notation to be used for each.

Step 4. Select the modeling approach and formulate the model.

Using the tools in this text and your own creativity, build a model that describes the situation and whose solution helps to answer important questions.

Step 5. Solve and interpret the model.

We take the model we constructed in steps 1–4 and solve it. Often, this model might be too complex or unwieldy, so we cannot solve it or interpret it. If this happens, we return to steps 2–4 and simplify the model further.

Step 6. Verify the model.

Before we use the model, we should test it out. There are several questions we must ask. Does the model directly answer the question, or does the model allow for the answer to the questions to be answered? Is the model usable in a practical sense (can we obtain data to use the model)? Does the model pass the "common sense" test?

We like to say that we corroborate the reasonableness of our model rather than verify or validate the model.

Step 7. Strengths and weaknesses.

No model is complete without self-reflection of the modeling process. We need to consider not only what we did right but what we did that might be suspect as well, including what we could do better. This reflection also helps in refining models.

Step 8. Sensitivity analysis and model testing.

A modeler wants to know how the inputs affect the ultimate output for any system. Passing the "common sense" test is essential. One of us once had a class model Hooke's law with springs and weights. The students were then asked to use their model to see how far the spring would stretch using their weight. They all provided the numerical answers, but none said that the spring could break under their weight.

Step 9. Refine, implement, and maintain the model.

A model is pointless if we do not use it. The more user-friendly the model, the more it will be used. Sometimes the ease of obtaining data for the model can dictate its success or failure. The model must also remain current. Often, this entails updating the parameters used in the model.

1.4 Making Assumptions

In its simplest form, we say we need to:

Make assumptions

Do some "math"

Derive and interpret conclusions

Usually, one cannot question the mathematics (unless an obvious error is made), but one can and should challenge the assumptions used to get to the model used. Assumptions drive the modeling as well as the analysis. Every model is based upon some set assumptions. These can be trivial or more complex depending on what we know or can observe about the problem. The assumptions also might be affected by the available data. That is why we say we can question the assumptions and ensure they are justified.

1.5 Illustrative Modeling Examples

Example 1: Ship disaster at sea.

A terrorist bomber explodes a bomb on a cruise ship in the Mediterranean Sea. The results are shown in Table 1.1.

Problem Identification: Determine from the data if one of the "rules of the sea" was followed. One rule of a disaster at sea is rescue women and children first. Was this rule followed?

Assumptions: The table of data is accurate, and we can use it gain insights into the disaster.

Some basic calculations, for which we will show formulas later, reveal that only 19.6% of the men (332 out of 1,692 survived) and 70.4% of the women and children survived (374 out of 531). Such simple calculations are powerful tools in analyzing information and providing insights into results.

Example 2: Probability.

Consider knowledge that manufacturing quality control follows a binomial distribution (discussed in a later chapter). We have an ending contract to produce and deliver 100 items next week. We'd like to know the probability that 100 of the next 120 products are acceptable.

Problem Identification: Build a model that can be used to determine the probability that 100 out of the next 120 items produced are acceptable.

Assumptions: We assume that each product is produced independently but using identical methods and procedures. We assume we can identify a success from a failure. Past quality control states that if we inspect every other item, we can achieve a 95% success rate.

Example 3: Claim we can teach school via the Internet.

Due to COVID-19, all the schools in our state closed; however, education continued over the Internet.

Problem Identification: The governors and the Board of Education can reach all students over the Internet.

Assumptions: Everyone has access to a computer and the Internet.

We can collect data via a survey and determine the proportion of students that has access to the Internet. We can test the claim using hypothesis testing.

TABLE 1.1

Disaster at Sea

	Men	Women	Boys	Girls	Total
Survived	332	318	29	27	706
Died	1,360	104	35	18	1,517
Total	1,692	422	64	45	2,223

Example 4: Bank service problem.

The bank manager is trying to improve customer satisfaction by offering better service. The management wants the average customer to wait less than 2 minutes, and the average length of the queue (length of the line waiting) to be 2 minutes or fewer. The bank estimates about 150 customers per day. The existing arrival and service times are given in Tables 1.2 and 1.3.

Problem Identification: Build a mathematical model to determine if the bank is meeting its goals. Determine if the current customer service is satisfactory according to the manager guidelines. If not, determine through modeling the minimal changes for servers required to accomplish the manager's goal.

Assumptions: Current service time and arrival time rates will remain unchanged.

Example 5: Regression

We have shipping data that we plotted in Figure 1.4, and we want a model to predict future shipping requirements.

Problem Identification: Build a mathematical model to predict the next few months of shipping based upon the data available.

Assumptions: We have accurate and timely data. Is the data linear or nonlinear?

Regression models can be used to build a mathematical model and used to make predictions.

Example 6: Best baseball hitter of all time.

Problem Identification: Using either current or Hall of Fame data, build a mathematical model to rank order the "best" hitters of all time.

TABLE 1.2

Arrival Times

Time between Arrival (minutes)	Probability
0	0.10
1	0.15
2	0.10
3	0.35
4	0.25
5	0.05

TABLE 1.3

Service Times

Service Time (minutes)	Probability
1	0.25
2	0.20
3	0.40
4	0.15

FIGURE 1.4
Scatterplot for shipping data.

Assumptions: We have to decide what "best" hitter implies. Is it batting average? Is it home runs? We have to decide if we look at just one season or a career.

We can use multi-attribute decision-making to assist us.

1.6 Technology

In mathematical modeling, it is generally impossible to proceed through all the steps without technology. The partnering of technology with modeling is both key and essential to good modeling principles and practices. In this book, we illustrate statistical modeling using the software package R.

1.6.1 What Is R?

1.6.1.1 Introduction to R

R is a language and environment for statistical computing and graphics. It is a GNU project that is similar to the S language and environment, which was developed at Bell Laboratories (formerly AT&T, now Lucent Technologies) by John Chambers and colleagues. R can be considered as a different implementation of S. There are some important differences, but much code written for S runs unaltered under R.

R provides a wide variety of statistical (linear and nonlinear modeling, classical statistical tests, time-series analysis, classification, clustering, etc.) and graphical techniques and is highly extensible. The S language was often the vehicle of choice for research in statistical methodology, and R provided an open source route to participation in that activity.

One of R's strengths is the ease with which well-designed publication-quality plots can be produced, including mathematical symbols and formulae where needed. Great care has been taken over the defaults for the minor design choices in graphics, but the user retains full control.

R is available as free software under the terms of the Free Software Foundation's GNU General Public License in source code form. It compiles and runs on a wide variety of UNIX platforms and similar systems (including FreeBSD and Linux), Windows, and macOS.

1.6.2 The R Environment

R is an integrated suite of software facilities for data manipulation, calculation, and graphical display. It includes the following:

- an effective data handling and storage facility;
- a suite of operators for calculations on arrays, in particular matrices;
- a large, coherent, integrated collection of intermediate tools for data analysis;
- graphical facilities for data analysis and display either on-screen or on hardcopy; and
- a well-developed, simple and effective programming language that includes conditionals, loops, user-defined recursive functions, and input and output facilities.

The term "environment" is intended to characterize it as a fully planned and coherent system rather than an incremental accretion of very specific and inflexible tools, as is frequently the case with other data analysis software.

R, like S, is designed around a true computer language, and it allows users to add additional functionality by defining new functions. Much of the system is itself written in the R dialect of S, which makes it easy for users to follow the algorithmic choices made. For computationally intensive tasks, C, C++, and Fortran code can be linked and called at run time. Advanced users can write C code to manipulate R objects directly.

Many users think of R as a statistics system. We prefer to think of it as an environment within which statistical techniques are implemented. R can be extended (easily) via *packages*. There are about eight packages supplied with the R distribution, and many more are available through the CRAN family of Internet sites, covering a very wide range of modern statistics.

R has its own LaTeX-like documentation format, which is used to supply comprehensive documentation, both online in a number of formats and in hardcopy.

We suggest downloading RStudio when you start this course. RStudio offers a user interface to R, which makes certain tasks much simpler.

R and RStudio must both be downloaded and are available from the websites:

R: https://cran.r-project.org

RStudio: www.rstudio.com

Visit the R home page (link on the R CRAN site above) for resources including introductory tutorials and help. Simply searching the web will also produce numerous introductory tutorials and videos for RStudio.

1.6.3 R, Data, and Manipulating Data

We conclude this chapter with a very brief introduction to the important task of getting data into R using RStudio. We will give a few examples of ways to interact with the data. Later chapters will do much more.

1.6.3.1 Importing Data from EXCEL (as a csv file) to RStudio

Perhaps the most generic file format for passing data between programs is a "csv" (Comma separated) file. The file "dataclass.csv" is an example from the text data repository. Save this to your directory where you keep documents for this course on your computer.

Next, open RStudio. There are several ways to import a data file in RStudio. One is to write a command (or commands) to do so. The commands can be entered at the command prompt (">") in the "console" window or typed into an R script file and submitted from there to the console using the "Run" button. The advantage of the latter is the commands are saved, so they can be rerun or copied and modified for other problems.

Remember, commands and syntax, particularly for data import, can be tedious. Thus, we will first examine one of the advantages of RStudio: built in "wizards" for some operations such as data import. We will then return to the commands.

The wizard to import data is in the environment window in RStudio as shown in Figure 1.5. Click on "Import Dataset" to begin the wizard, and from the drop-down choose "From Text (base)". Note that there are wizards to import files from other formats as well.

Navigate to the folder where you saved the data, and select the file. You should see the "raw data" in the upper-right window and the data as it will appear once read into RStudio in the lower left, as pictured in Figure 1.6. On the left are options. Generally, RStudio will pick the correct choices. For example, here the "Heading" is checked "Yes," so the first row of the file is treated as the variable names. This is depicted in the "Data Frame" window in bold. In the upper area is a field to input a name for the data set if desired. Here, we will just use the RStudio default, the filename "dataclass." Click the "Import" button at the bottom right to read the data into RStudio.

The import wizard will automatically open the data set for viewing in a spreadsheet-like form. This is useful for ensuring the import was correctly executed.

You will notice that there is also output in the "Console" window:

```
> dataclass <- read.csv("dataclass.csv")
> View(dataclass)
```

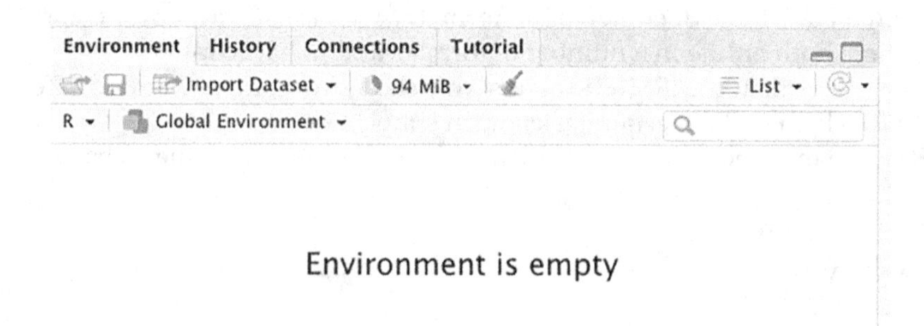

FIGURE 1.5
"Environment" window in RStudio .

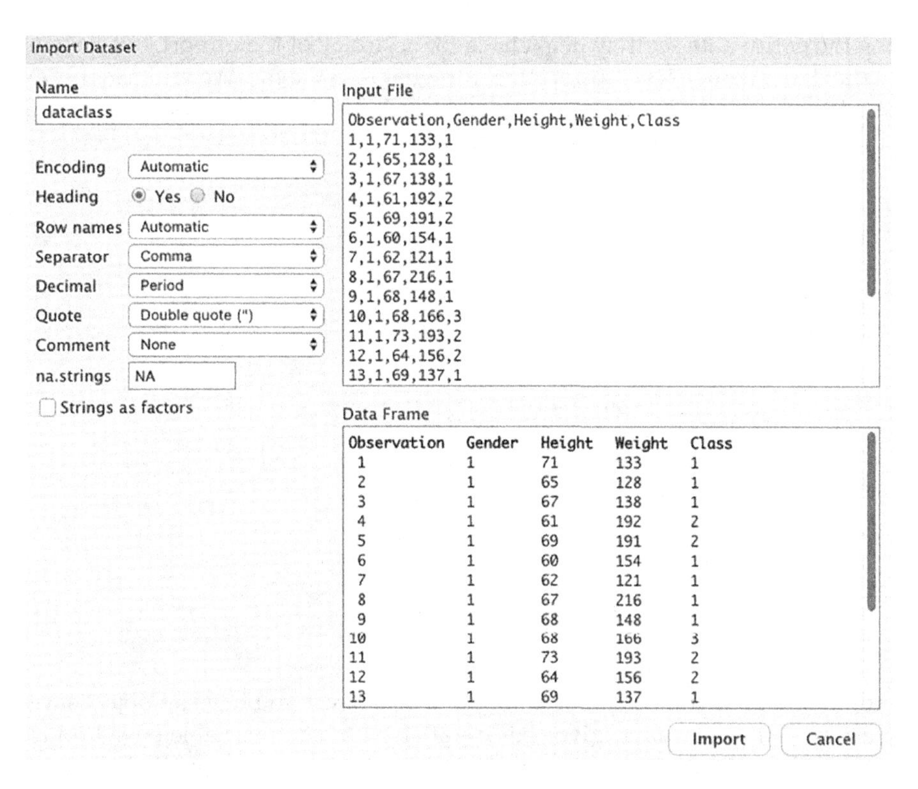

FIGURE 1.6
Import wizard window for "dataclass.csv" .

The first is the command to read in a "csv" file. The second is the command that opened the imported data set for viewing. These commands are useful, and you can copy them to an R script file (click the tab for a new script in the window where the data is viewed; be sure to remove the ">" as it is a prompt not needed in running a script) for later use. Click the "Save" button to keep the script.

Note that your command may include a path to the data set. This too is useful, as you may need to use paths again. A very useful thing to do is to set the "working directory" to the folder where you keep the data sets and will want to save your scripts. This is easily done from the RStudio menu using the following sequence of menu choices:

Session > Set Working Directory > Choose Directory . . .

You will see that RStudio also reports the commands that the menu uses! For example:

```
> setwd("~/probstatsdirectoryname")
```

Again, this is very useful since the working directory needs to be reset each time RStudio is opened. We recommend saving this command in your R script as well.

Note that once the data is in RStudio, the data frame (R language for a data set) will appear in the environment window. Clicking on objects in that window will also open them in the view window.

Viewing the entire data set may help check for accuracy of the import, but we may want to look at the data in other ways to get a feel for what is in the file. We will use this example data set to demonstrate a few useful commands in R. The functions "head" and "tail" will display the first and last observations in a data frame:

```
> head(dataclass)
  Observation Gender Height Weight Class
1           1      1     71    133     1
2           2      1     65    128     1
3           3      1     67    138     1
4           4      1     61    192     2
5           5      1     69    191     2
6           6      1     60    154     1
> tail(dataclass)
   Observation Gender Height Weight Class
45          45      2     59    139     1
46          46      2     71    159     1
47          47      2     69    177     2
48          48      2     72    190     1
49          49      2     60    172     2
50          50      2     69    131     1
```

The "summary" function is a great first option whenever you have an object saved in R. For a data frame, the command gives basic statistics for each variable based on the type of data. Since all the data in this file is numeric, the results are things like percentiles and the mean:

```
> summary(dataclass)
  Observation        Gender          Height           Weight           Class
 Min.    : 1.00   Min.    :1.00   Min.    :59.00   Min.    : 98.0   Min.    :1.00
 1st Qu.:13.25   1st Qu. :1.00   1st Qu. :62.00   1st Qu. :131.2   1st Qu. :1.00
 Median :25.50   Median  :1.00   Median  :67.50   Median  :155.5   Median  :1.00
 Mean   :25.50   Mean    :1.46   Mean    :66.66   Mean    :158.5   Mean    :1.56
 3rd Qu.:37.75   3rd Qu. :2.00   3rd Qu. :71.00   3rd Qu. :190.8   3rd Qu. :2.00
 Max.   :50.00   Max.    :2.00   Max.    :75.00   Max.    :223.0   Max.    :3.00
```

A second command often worth trying on R objects is the "plot" function. Running this function with the data frame as an input produces a "scatterplot matrix" as shown in Figure 1.7. The plot appears in another of the RStudio windows in the plot tab. Note the options in this window (icons at the top) include an export feature. The "Zoom" button will open the plot in a separate window for better viewing. Right-clicking on the plot also allows you to copy the image and then paste into documents if desired. We will discuss various plots throughout the text. For now, notice that the y-axis for each plot is the variable named in that row of plots. So, the y-axis for the second row is the "Gender" variable. The x-axis is then the variable named in the given column of plots. So the third plot in the second row has the "Height" on the x-axis. Note that since Gender has only two values, 1 or 2 (male and female most likely), the plot only has those two y values.

Finally, we may wish to perform operations on just one variable in a data set. In R, there are several ways to "extract" from an object. The simplest is the "$." Suppose we want the average height for those in the data set. The "mean" command produces the average. To do so for just the Height variable, we "extract it" from the data frame using dataclass$Height

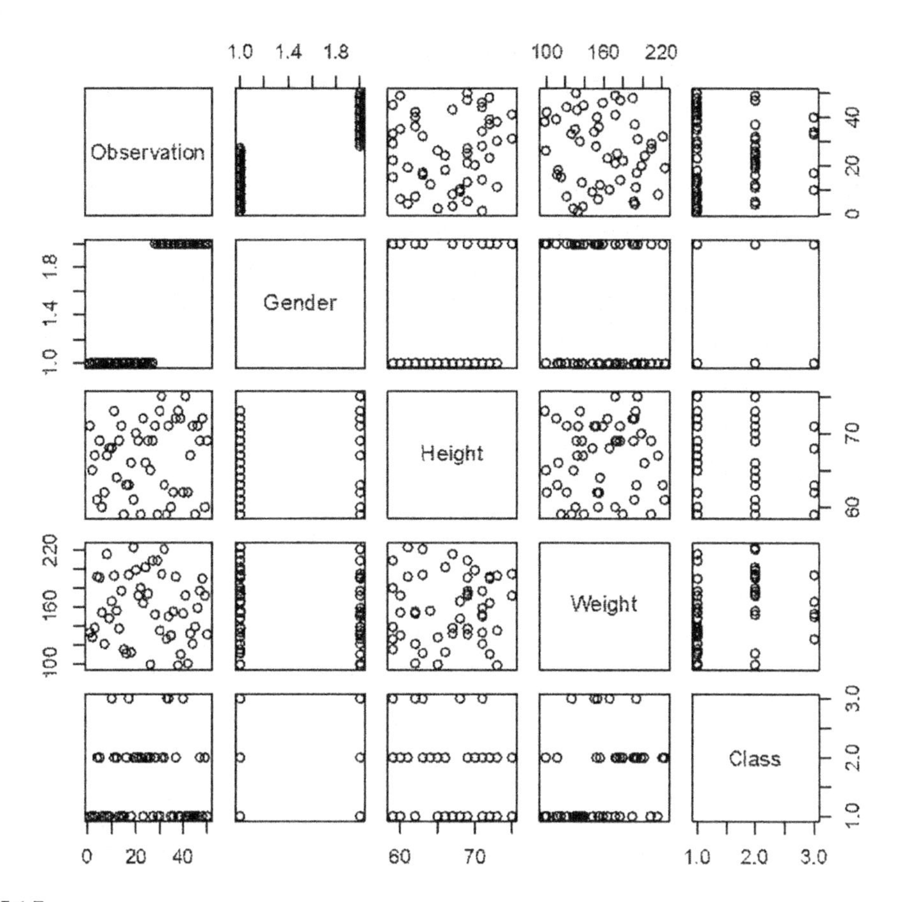

FIGURE 1.7
Output from command plot(dataclass).

as shown below. As a second example, we notice in the scatterplots that the Class variable has just three possible values. We can produce a table showing how many observations are in each class by extracting that variable in the "table" command.

```
> mean(dataclass$Height)
[1] 66.66

> table(dataclass$Class)
 1  2  3
27 18  5
```

1.7 Chapter 1 Exercises

1.1 In modeling the probability that some number of light bulbs among many light bulbs are not operative, what information would you like to be able to obtain? What assumptions might be required?

1.2 How would you approach a problem concerning a test to determine proper drug dosage? Do you always assume the doctor is right? What are other assumptions you would make?

1.3 For the model in Example 3, how would you collect data to build the model? What variables would you consider?

1.4 In the disaster at sea example, what is the impact of the location of the vessel?

1.5 In the bank queue problem, discuss the criticality of the assumptions? Do you feel that more training is as valuable as adding another server?

1.6 In the supply data example, how important is the trend of the data to the model being built?

In exercises 1.7–1.11, use R to compute the following:

1.7 $\frac{211}{312}(99)$ (Hint: the * is used for multiplication.)

1.8 $e^{1.2}$ (Hint: in the RStudio pane where the plot appears, select the "Help" tab and type "exp" in the search bar or input ?(exp) in the RStudio console.)

1.9 $1.2\left(2.5+\frac{12}{4}\right)$

1.10 $\frac{4}{5}\left(\frac{6}{2+3}\right)$

1.11 Obtain a scatterplot for the following supply data. In R, a list of data is created using the "c" function: x = c(1, 2, 3, 4). Look at the help for the plot function to determine how to use it to create a scatterplot of two variables.

Y	17	15	10	18	14	11	20	16	12	22
t	1	2	3	4	5	6	7	8	9	10

1.8 Chapter 1 Projects

1.1 Is Michael Jordan the greatest basketball player of the century? What variables and factors need to be considered?

1.2 What kind of car should you buy when you graduate from college? What factors should be in your decision? Are car companies meeting your needs?

1.3 Consider domestic decaffeinated coffee brewing. Suggest some objectives that could be used if you wanted to market your new brew. What variables and data would be useful?

1.4 Replacing a coaching legend at a school is a difficult task. How would you model this? What factors and data would you consider? Would you equally weigh all factors?

1.5 How would you go about building a model for the "best pro football player of all time"?

1.6 Rumors abound in major league baseball about steroid use. How would you go about creating a model that could imply the use of steroids? Relate baseball steroid rules to the Yankee's Alex Rodriquez case.

1.7 After the 2009 All-Star baseball game, the American League has won 12 straight games dating back to 1997. This is the longest winning streak by either side. Help the National League prepare to win by designing a model for players or a lineup that could help them change their outcome.

1.9 References and Additional Readings

Albright, B. and W. Fox (2020). *Mathematical Modeling with Excel*, 2nd ed., Taylor and Francis Publishers, Boca Raton, FL.

Burden, R. and D. Faires (1997). *Numerical Analysis*, Brooks-Cole, Pacific Grove, CA.

COMAP. *Modeling Competition Sites*, at www.comap.com/contests.

Fox, W. (2018). *Mathematical Modeling for Business Analytics*, Taylor and Francis Publishers, Boca Raton, FL.

Fox, W. and W. Bauldry (2019). *Problem Solving in Maple*, volume I, Taylor and Francis Publishers, Boca Raton, FL.

Fox, W. and W. Bauldry (2020). *Problem Solving in Maple*, volume II, Taylor and Francis Publishers, Boca Raton, FL.

Giordano, F., W. Fox and S. Horton (2014). *A First Course in Mathematical Modeling*, 5th ed., Cengage Publishers, Boston, MA.

2

Introduction to Data

Objectives

1. **Know the different types of data (categorical (qualitative) and quantitative) and which displays are used for the type of data that you are using and displaying.**
2. **Know what constitutes a bad display. Avoid them.**
3. **Know how to use Excel R to create "good" and useful displays of data for both categorical and quantitative data.**
4. **Know how to obtain and use descriptive statistics.**
5. **Know how to interpret key measures mean, median, mode, range, standard deviation, variance, and skewness.**
6. **Distinguish measures of location from measure of spread.**

2.1 Finding Basic Statistics

Statistics is the ***science of reasoning from data,*** so a natural place to begin your study is by examining what is meant by the term "data." The most fundamental principle in statistics is that of variability. If the world were perfectly predictable and showed no variability, there would be no need to study statistics. You will need to find the notion of a **variable** and then first learn how to classify variables.

Any characteristic of a person or thing that can be expressed as a number is called a variable. A value of that variable is the actual number that describes that person or thing. Think of the variables that might be used to describe you: height, weight, income, rank, branch of service, year in school, occupation, and gender.

Data can be quantitative or categorical (qualitative). Let's explain each of these.

Quantitative means that the data are numerical, where the number itself has relative meaning. Examples could be a list of heights and weights of students in your class, weights of soldiers in your unit, or batting averages of the starting lineup for the 1998 New York Yankees.

Heights of Students:

5′10″	6′2″	5′5″	5′2″	6′	5′9″

DOI: 10.1201/9781003317906-2

Weights of Students:

135	155	215	192	173	170	165	142

New York Yankees Batting Averages for Their Starting Lineup:

.276	.320	.345	.354	.269	.275	.300	.254	.309

Traffic Deaths over a 15-Year Period

Period	Deaths	Injuries	Percentages
2001	0	4	0.0
2002	4	25	16.0
2003	3	26	11.54
2004	12	27	44.44
2005	2	7	27.4
2006	41	10	31.54
2007	78	184	42.39
2008	152	263	57.79
2009	275	451	60.98
2010	368	630	58.41
2011	252	49	51.22
2012	132	312	42.31
2013	52	117	44.44
2014	3	13	23.08
2015	2	10	20.0

These data elements provide numerical information. We can determine from the data which height is the greatest or smallest or which batting average is the greatest or the smallest. We can also compare and contrast "mathematically" these values by computing averages for example.

Quantitative data can be either discrete (counting data, 1, 2, 3, 4 . . .) or continuous (1.347, 10.2, . . .). These types become important as we analyze them and use them in models later in the course. Quantitative data allows us to "do meaningful mathematics." (+, −, *, /).

Categorical (qualitative) data can describe objects, such as recording the people with a particular hair color as follows: blonde = 1 or brunette = 0. If we had four colors of hair (blonde, brunette, black, and red), we could use as codes: brunette = 0, blonde = 1, black = 2, and red = 3. We certainly cannot have an average hair color from these numbers. It would not make sense. Another example is categories by gender: male = 0 and female = 1. In general, it may not make sense to do any arithmetic using categorical variables. Eye color as blue, green, brown, or hazel is a category. Student classification as freshman, sophomore, junior, senior, or graduate is a categorical variable.

An important distinction for categorical variables is between nominal and ordinal. Nominal variables are categorical with no order to the categories. Eye and hair color as well as gender are nominal. Ordinal data has categories, but there is a clear order. Student classification is ordinal since you progress in order from freshman to sophomore to junior and so on.

Once you have learned to distinguish between quantitative and categorical data, we need to move on to a fundamental principle of data analysis: "begin by looking at a visual display of the data set."

2.2 Chapter 2 Exercises

2.1 Determine whether the following variables would be quantitative or categorical. If the variable is categorical, determine if it is ordinal or nominal. Provide an example of the value of such a variable, and include the units, if any exist.

a. Flip a dime that lands as a "head" or "tail"
b. The color of peanut M&M's
c. The number of calories in the local fast food selections from McDonald's
d. The life expectancy for males in the United States
e. The life expectancy for females in the United States
f. The number of babies born on New Year's Eve
g. The dollars spent each month out of the allocated supply budget
h. The number of hours that a woman works per week
i. Amount of car insurance paid per year
j. Whether the bride is older, younger, or the same age as the groom
k. The difference in ages of a couple at a wedding
l. Average low temperature in your hometown in January
m. The eye color of a student
n. The gender of a student
o. The number of intramural sports a person plays per year
p. The distance a bullet travels from a specific weapon
q. The number of roommates in three years
r. The size of your immediate family

2.2 The following table represents the number of sports-related injuries treated in U.S. hospital emergency rooms in the past, along with an estimate of the number of participants in that sport. For each variable, determine the type (continuous/discrete or categorical). Are the categorical variables ordinal?

Sport	Injuries	Participants	Sport	Injuries	Participants
Basketball	646,678	26,200,000	Fishing	84,115	47,000,000
Bicycling	600,649	54,000,000	Skateboarding	56,435	8,000,000
Baseball	459,542	36,100,000	Hockey	54,601	1,800,000
Football	453,684	13,300,000	Golf	38,626	24,700,000
Soccer	150,449	10,000,000	Tennis	29,936	16,700,000
Swimming	130,362	66,200,000	Water skiing	26,663	9,000,000
Weightlifting	86,398	39,200,000	Bowling	25,417	40,400,000

(a) If we want to use the number of injuries as a measure of the hazardousness of a sport, which sport is more hazardous between bicycling and football? Between soccer and hockey?

(b) Use either a calculator or a computer to calculate the rate of injuries per thousand participants. Rate is defined as the average number of injuries out of the total participants.

(c) Rank order this new measure for the sports.

(d) How do your answers in part (a) compare if we do the hazardous analysis using the *rates* in (b). If different, why are the results different?

2.3 Displaying the Data

Why do we want to display data? Visual displays such as bar graphs, pie charts, and histograms are very useful because they provide a quick and efficient way to present the data, and reveal characteristics of the data. These displays allow our eyes to take in the overall pattern and see if there are unusual observations of data elements. Graphs and numerical summaries that we will introduce to describe the data are not ends in themselves, but merely aids to our overall understanding of the data.

Displaying data badly is a problem. What is meant by displaying data badly? We will quickly illustrate and explain. The three fundamental elements of bad graphical displays are these: **data ambiguity, data distortion, and data distraction.**

2.3.1 Data Ambiguity

Data ambiguity arises from the failure to precisely define just what the data represent. Every dot on a scatterplot, every point on a time-series line, every bar on a bar chart represents a number (actually, in the case of a scatterplot, two numbers). It is the job of the legend and labels (text) on the chart to tell us just what each of those numbers represents. If a number represented in a chart is, say, 33½, the text in the graph—in the title, the axes labels, the data labels, the legend, and sometimes the footnote—must answer the question: "Thirty-three and a half what?"

2.3.2 Data Distortion

Before the development of spreadsheet graphing, the most common graphical mistake was the use of artist-drawn 3D images with the height of 3D objects representing the magnitude of the data points. In these charts, both the height and the width of the drawn object increase proportionate to the magnitude of the data points. The effect is to exaggerate the differences in magnitude, as the viewer tends to perceive the area of the figures rather than just the height as representing the magnitude. The incredible shrinking family doctor (shown in Tufte, 2001, 69) is a classic example. In this chart, the 1990 doctor is a bit less than half the height of the 1964 doctor. Each doctor has the same relative shape. Imagine two doctors with the same average physical shape, one less than 4 feet tall, the other 8 feet tall. If the 4 ft. doctor weighed 100 lbs., how much would the 8 ft. doctor weigh? Certainly much more than 200 lbs.

With the development of spreadsheet graphics, such visual distortions are no longer common, and the art of lying with graphics has become a technology rather than an art. Today, altogether new forms of bad graphical design predominate.

Most of the bad charting described thus far has the redeeming feature that it does not, for the most part, distort the data being represented by exaggerating or understating the values of some of the data points. We will consider now some of the more complicated ways of using graphical display to mislead.

2.3.3 Data Distraction

Edward Tufte's (2001) fundamental rule of efficient graphical design is to **minimize the ratio of ink-to-data**. This is essentially the same advice offered by Strunk and White to would be writers:

> A sentence should contain no unnecessary words, a paragraph no unnecessary sentences for the same reason that a drawing should contain no unnecessary lines and a machine no unnecessary parts.

The primary source of extraneous lines in charting graphics today is the 3D options offered by conventional spreadsheet graphics. These 3D options serve no useful purpose; they add only ink to the chart, and more often than not, make it more difficult to estimate the values represented. Even worse are the spreadsheet options that allow one to rotate the perspective. For those who would take bad graphical display to even higher levels, the Excel spreadsheet program offers the option of doughnut, radar, cylinder, cone, bubble charts. An example of various displays of the same data, the percentage of active duty military in each branch of service in 1998, is shown in Figure 2.1. Just by glancing at the charts, do you get the same impression about relative percentages from each?

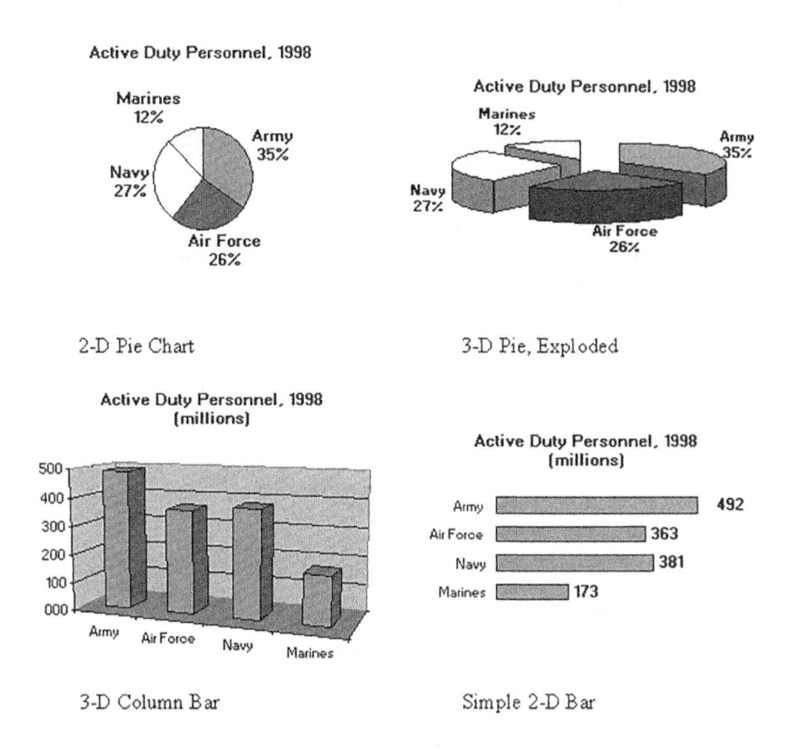

FIGURE 2.1
Pie and bar charts for active duty personnel by branch, 1998.

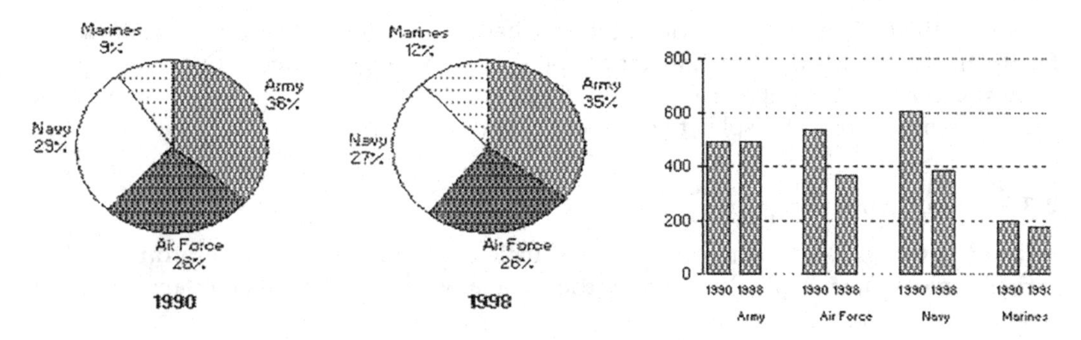

FIGURE 2.2
Pie and bar charts comparing years.

Pie charts (upper two plots in Figure 2.1) should rarely be used. It is more difficult for the eye to discern the relative size of pie slices than it is to assess relative bar length. With the pie chart, without looking at the numbers, it is difficult to figure out whether the navy or air force is larger; from the bar charts, it is obvious. 3D pie charts are even worse, as they also add a visual distortion (in this case, making the air force appear much larger).

Bar charts are shown in the bottom panels of the figure. Note how much less ink the 2D bar chart uses compared to the 3D bar. Using data labels rather than a y-axis scale in this case, the 2D reduces the number of numbers displayed from 6 to 4 and adds precision as well. Normally, we would have sorted the data here so that the navy would be between the army and air force, but since the marines are a part of the navy (and the air force, originally, a part of the army), this order made more sense. A strict application of the ink-to-data in this case, however, would eliminate the bars altogether and simply present the data as a table.

Pies are even less effective when an additional variable is added and comparisons between pies are required (sometimes by adjusting the relative size of the pies). Consider the plots in Figure 2.2. Which makes it easier to identify differences in the percentages between years in each service?

In addition to the distractions and distortions made possible by the use of 3D effects, adding other elements to a graph may actually make it less understandable.

Not content with the distractions and distortions made possible by the use of 3D effects, charters sometimes feel the need to add all sorts of other chart junk to a graph.

Note the extraneous features of this in this graphic.

A completely irrelevant map of the world.

Two entirely different kinds of 3D charts displayed at two different perspectives.

Country names are repeated three times.

To display 24 numeric data points, 28 numbers are used to define the scales.

The countries are sorted in no apparent order (not even alphabetically).

Note the use of the letter "I "to separate the countries on the bottom chart.

While it might be possible to display these data better graphically, a table does the job quite nicely:

Pre-tax Income Distribution in Industrial Nations

	Share of Pre-tax Household Income		
	Top Income Quintile	**Bottom Income Quintile**	**Ratio: Top to Bottom shares**
United States	45	4	12
Canada	42	4	9
France	47	5	9
Britain	45	6	8
W. Germany	39	8	5
Sweden	38	8	5
Netherlands	37	7	5
Japan	36	9	4

*Data estimated from chart.

2.3.4 Two Chart Types That Should Always Be Avoided

Two common charts easily produced by spreadsheet programs that should almost always be avoided are the stacked bar chart and the pie chart. The stacked bar chart, made even worse by the use of 3D effects which makes it very difficult to estimate the values of the variables represented on the top of the bars. Similar "stacking" can also been done with time-series area charts and should be avoided as well.

Pie charts are fun to look at but generally involve using a great deal of ink to display very little data. In addition, the charts often make it difficult to discern the exact magnitude of the size of the pie slices. Using multiple pie charts to display more than one variable is also a bad idea. All this is made even worse by exploiting the power of spreadsheet technology to produce 3D pie charts and "exploding" 3D pie charts. If you think that you really must use a pie chart, make sure it is for data that does indeed at add up to a total (i.e., the percentages for the slices add up to 100), and stay away from the "fancy stuff."

2.3.5 Good Graphical Displays

We will discuss five methods of displaying data: pie chart, bar chart, stem and leaf, histogram, and boxplot. The choice depends on the type of variable as shown in Table 2.1. The displays should supply visual information to the viewer without a struggle.

TABLE 2.1

Graphical Displays Based on Type of Variable

Data	Categorical (Nominal or Ordinal)	Quantitative (Continuous or Discrete)	Comment
Displays	Pie chart	Stem and leaf	
	Bar chart	Dot plot	
		Histogram	May overlay distribution
Concern	Comparisons	Shape and skewness	

2.4 Chapter 2 Exercises

2.3 Search the Internet for an example and a "bad" display of data. Explain briefly why it is a bad display.

2.4 Below are displays from Fox News from two different news programs. Are they good or bad displays? Why? What would you do to improve these graphics?

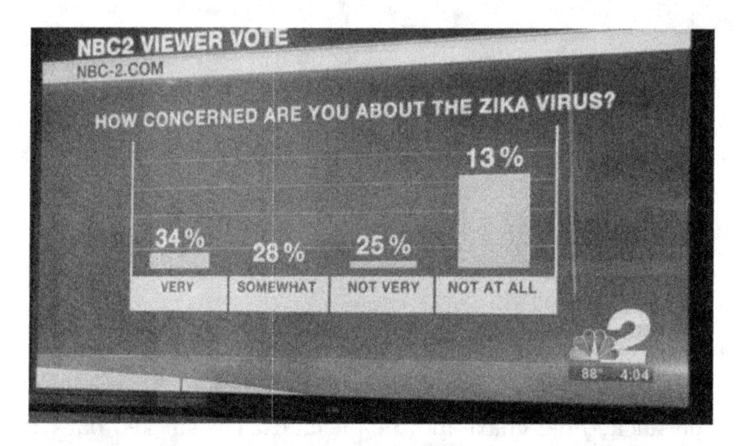

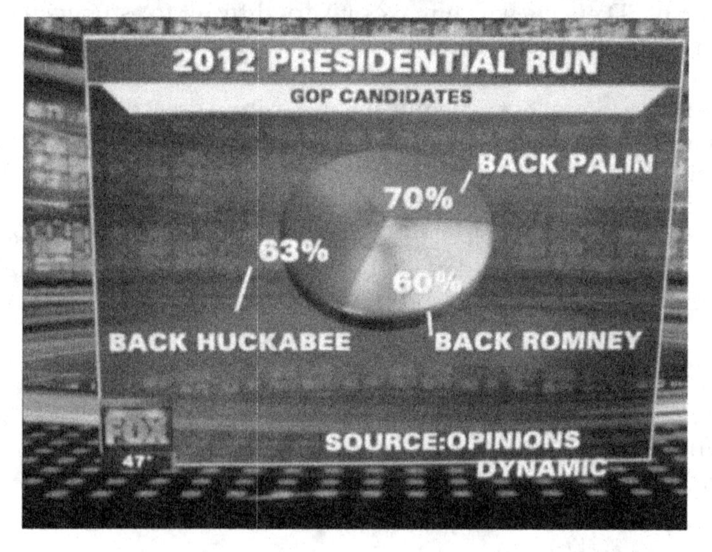

2.5 References and Suggested Readings

Cochran, Clarke et al. (1999). *American Public Policy: An Introduction*, St. Martin's Press, New York.
Phillips, Kevin (1991). *The Politics of Rich and Poor*, Harper Perennial, New York.
Putnam, Robert D. (2000). *Bowling Alone*, Simon and Schuster, New York.
Strunk, William Jr. and E. B. White (1976). *The Elements of Style*, 3rd ed., Macmillan Publishing, New York.
Tufte, E. (2001). *The Visual Display of Quantitative Information*, 2nd ed., Self-published.

2.6 Good Displays of Categorical Data

We will discuss five methods of displaying data: pie chart, bar chart, stem and leaf (by hand only), histogram, and boxplot (by hand only). The displays should supply visual information to the viewer without a struggle.

Data Display	Categorical	Quantitative (Continuous or Discrete)	Comment
	Pie chart	Stem and leaf	
	Bar chart	Dot plot	
		Histogram	
Concern	Comparisons	Shape and skewness	Often overlay the distribution of interest over the histogram

Pie Chart

The *pie chart* is useful to show the division of a total quantity into component parts. A pie chart, if done correctly, is usually safe from misinterpretation. The total quantity, or 100%, is shown as the entire circle. Each wedge of the circle represents a component part of the total. These parts are usually labeled with *percentages* of the total. Thus, a pie chart helps us see what part of the whole each group forms.

Let's review percentages. Let a represent the partial amount and b represent the total amount. Then P represents a percentage calculated by $P = a/b\ (100)$.

A percentage is thus a part of a whole. For example, $0.25 is what part of $1.00? We let $a = 25$ and $b = 100$. Then, $P = 25/100\ (100) = 25\%$.

Now, let's see how R would create a pie chart for us in the following scenario. Suppose five countries contribute aid to a region struck by a natural disaster. We have the amounts (in millions of dollars) from each country. The data is input to R using the code below. The first variable, aid, is the amount, and the second is the country.

```
> aid <- c(10, 11, 4, 16, 8)
> countries <- c("US", "UK", "Australia", "Germany", "France")
```

The most basic pie chart in R is produced using the "pie" function with input as the aid and the country names as labels, an option in the command after a comma:

```
> pie(aid, labels = countries)
```

We do not display the output of this command. Run this command yourself; do you see why we are not satisfied with this plot? For pie charts, we should also include the percentages for reference. To do so in R, we first compute the percentages and, in this case, round to one decimal place (the "1" after the comma):

```
> pct <- round(aid/sum(aid)*100, 1)
> pct
[1] 20.4 22.4 8.2 32.7 16.3
```

You should recognize the formula for a percent in this code. R makes this easy using the sum function to add up the total aid and automatically divided each element in the "aid" variable by that amount.

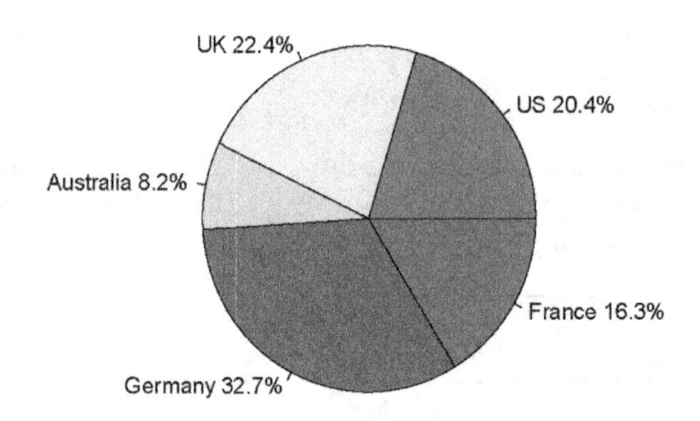

FIGURE 2.3
Pie chart for country aid.

We next use the "paste0" command in R to combine the country name with the percentages. The "0" in the command name refers to no spaces between items listed to combine, so we manually insert a space between the country name and the percent. We also add a percentage symbol:

```
> lbls <- paste0(countries, " ", pct, "%")
> lbls
[1] "US 20.4%" "UK 22.4%" "Australia 8.2%" "Germany 32.7%" "France 16.3%"
```

The final pie chart, shown in Figure 2.3, is produced with the command below. We added options to improve the colors (there are many more palettes available in R) for each of the pie slices (the length(aid) lets R know how many of the rainbow colors are needed) and added a title to the chart.

```
> pie(aid, labels = lbls,
 col = rainbow(length(aid)),
 main = "Pie Chart of Aid from Countries")
```

```
# Pie Chart with Percentages
slices <- c(10, 12, 4, 16, 8)
lbls <- c("US", "UK", "Australia", "Germany", "France")
pct <- round(slices/sum(slices)*100)
lbls <- paste(lbls, pct) # add percents to labels
lbls <- paste(lbls, "%", sep = "") # ad % to labels
pie(slices, labels = lbls, col = rainbow(length(lbls)),
   main = "Pie Chart of Countries")
```

Pie Chart with Annotated Percentages

```
# Pie Chart with Percentages
slices <- c(10, 12, 4, 16, 8)
lbls <- c("US", "UK", "Australia", "Germany", "France")
pct <- round(slices/sum(slices)*100)
lbls <- paste(lbls, pct) # add percents to labels
lbls <- paste(lbls,"%",sep="") # ad % to labels
pie(slices,labels = lbls, col=rainbow(length(lbls)),
   main="Pie Chart of Countries")
```

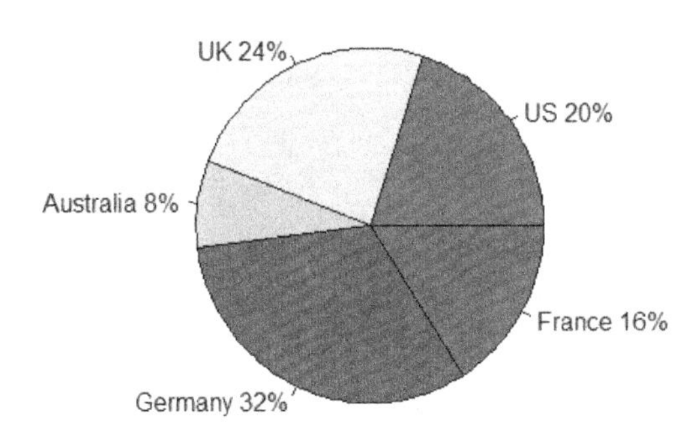

Infantry	250	0.3955
Armor	53	0.0838
Artillery	35	0.0554
Air defense	41	0.0649
Aviation	125	0.1978
Signal	45	0.0712
Logistics	83	0.1313
Total	632	1.00

```
# Pie Chart with Percentages
slices <- c(250,53,35,41,125,45,83)
lbls  <-  c("Inf",  "Armor",  "Artillery",  "Air  Def",  "Aviation",  "Signal",
"Logistics")
pct <- round(slices/sum(slices)*100)
lbls <- paste(lbls, pct) # add percents to labels
lbls <- paste(lbls, "%", sep = "") # ad % to labels
pie(slices, labels = lbls, col = rainbow(length(lbls)),
   main = "Pie Chart of MOS")
```

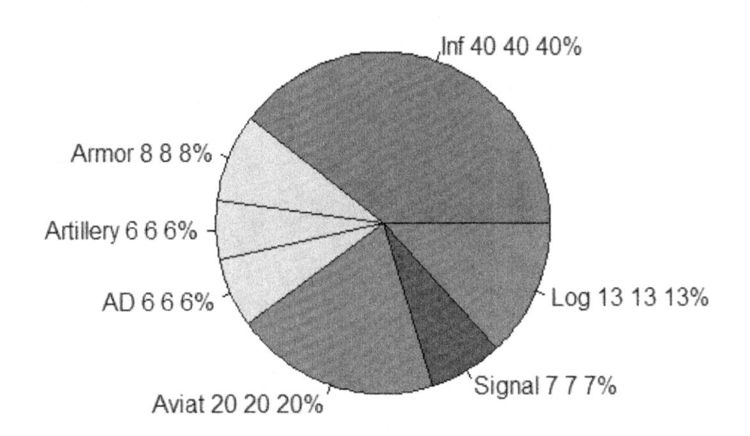

2.6.1 Bar Charts

Let's view a bar chart:

> > bar(slices)

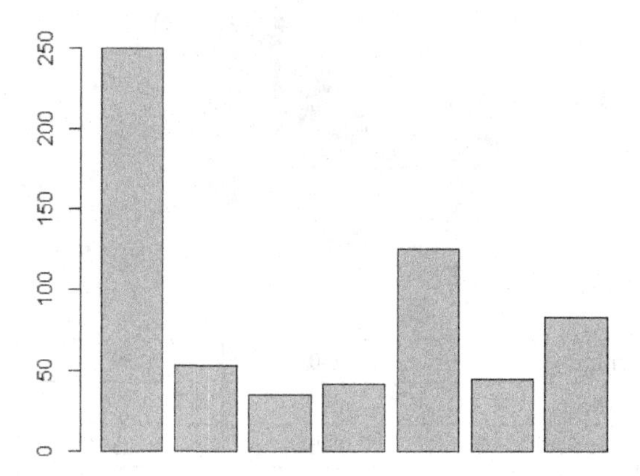

Bar charts are useful when comparing relative sizes of data groups, especially when they come from categorical variables. For example, consider the eye color from patients visiting the local eye clinic last year. The bar chart for the country aid example in Figure 2.4 was produced with the "barplot" command in R:

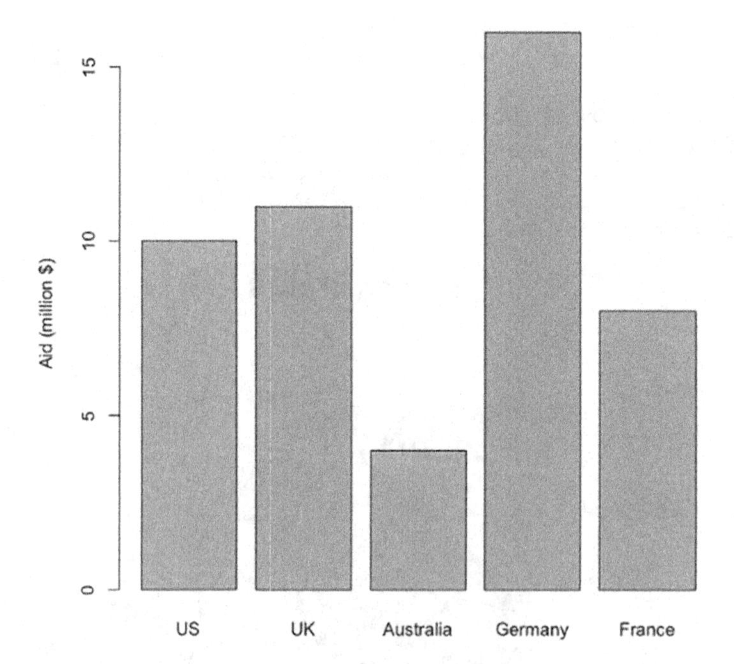

FIGURE 2.4
Bar chart for country aid.

Eye Color	Count	Percent
Blue	113	61.7486
Green	13	7.10383
Brown	41	22.4044
Mixed	16	8.74317
Total	183	100

You can quickly and clearly compare the relative sizes of the color groups. From the bar chart, which eye color occurs the most frequently? Blue eyes occur the most frequently. It appears twice as large as the next most frequent: brown eyes. Again, bar charts might be very useful to display categorical data.

2.7 Chapter 2 Exercises

2.5 Consider soldiers choosing their military occupational specialty (MOS). Out of the 632 new soldiers recruited in South Carolina that actually choose a MOS, the breakdown of selection is as follows.

Infantry	250
Armor	53
Artillery	35
Air defense	41
Aviation	125
Signal	45
Logistics	83
Total	632

Produce a pie chart that displays the percentage (%) of soldiers out of 632 that chose that MOS. Which MOS appears to have the least and most recruits? What advantages and disadvantages can you see with using pie charts? For this data, would your point be clearer with the pie chart or bar chart?

2.6 Counts of the eye colors for a group of students are shown in the table below. Produce a bar chart of the data. What are advantages and disadvantages of this plot?

Eye Color	Count
Blue	113
Green	13
Brown	41
Mixed	16
Total	183

2.7 Given the following 1990 data from the U.S. Bureau of Census, Current Population Reports, p. 385, in the *World Almanac* 1993, construct a pie chart. From the pie chart, rank order the age distribution of the U.S. population in 1990. Repeat using a bar chart. Is this all the information that you would need to discuss population trends? What other information would you like to have been given?

Category	Category #	Population
Under 5	1	18,354,443
5–17	2	45,249,989
18–20	3	11,726,668
21–24	4	15,010,898
25–44	5	80,754,835
45–54	6	25,223,066
55–59	7	10,531,756
60–64	8	10,616,167
65 and up	9	31,241,831

2.8 Displaying Quantitative Data

In this section, we discuss displaying quantitative data using stem and leaf plots and histograms.

In quantitative data, we are concerned with the shape of the data. Shape is primarily concerned with the symmetry of the data. Is the distribution of the data symmetric? Is it skewed? These are questions we ask and answer using visualizations of quantitative data.

2.8.1 Symmetry

We look at these shapes as symmetric or skewed. A symmetric distribution is one in which the shape to the right and left of center is reasonably the same. One example of a common symmetric distribution looks like a bell-shaped curve as depicted in graph A of Figure 2.5.

Three generic risk distribution shapes (symmetric, skewed right, and skewed left)

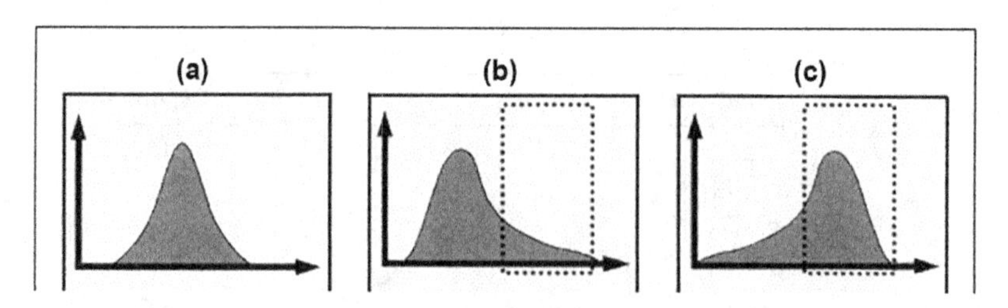

FIGURE 2.5
Examples of shapes of distributions: (a) Symmetric, (b) Skewed right, (c) Skewed left.

A skewed distribution, therefore, means that the plot appears "lopsided." Examples are shown in parts (b) and (c) of the figure. The direction of the skew is often very important. An easy way to remember the direction is that it is the direction the longer "tail" of the distribution points. In plot (b), the longer tail is to the right side of the distribution "pointing" to the right. The distribution of plot (b) is said to be "right skewed." In similar fashion, the plot in (c) is of a distribution that is left skewed.

Note: The shape of the distribution has important implications for many statistical methods, which we discuss in later chapters. There are also practical implications. For example, suppose the variable of interest measures risk, with a larger value meaning higher risk. From a risk management standpoint, in part (a) of Figure 2.5, the risk distribution is symmetric, and as a result, there are an equal number of people experiencing high risk as there are low risk. In part (b), the risk distribution is skewed to the right, with most people experiencing low risk and a few people experiencing high risk, compared to part (c), where the distribution is skewed to the left, translating to many people experiencing high risk and only a few people experiencing low risk. From a risk management or policy perspective, each of these situations would need to be assessed differently in light of the following considerations: the population (children, elderly, etc.) experiencing high risk; the actual magnitude of high risk ("high risk," as defined in this context, may not be very high when compared to other competing risks); if the high risk is being borne as a result of voluntary or involuntary actions; whether the people bearing high risk are in control of the risk situation; and so on.

Plots presented in this section can assist with assessing the symmetry of the underlying data distribution. A key aspect of the risk characterization stage is providing insight not only into the risk estimates, but also into our confidence in the generated assessment. Such insights include the following:

- the steps that could be taken to reduce the risk;
- points in the process about which we have uncertainty and could benefit from more information;
- points that have a significant impact on the risk and as such would be ideal areas to focus more attention on so as to ensure they are under control.

In general, quantitative risk assessment models can be considered as contributing toward risk management decision-making by providing input along four avenues:

- focusing attention on risk-reducing areas;
- focusing attention on research areas;
- helping in the formulation of risk-reduction strategies;
- providing a tool to test out formulated risk-reduction strategies prior to implementation.

2.8.2 Stem and Leaf Plots

A stem and leaf plot uses the actual data points in making a plot. The plot will appear strange because your plot is typically presented "sideways." The rules are as follows: plot is best produced with software and understood by looking at an example.

Step 1: Order the data.
Step 2: Separate according to the one or more leading digits. List stems in a vertical column.
Step 3: Leading digit is the stem and trailing digit is the leaf. In Example 1, 5 is the stem and 3 is the leaf. Separate the stem from the leaves by a vertical line.
Step 4: Indicate the units for stems and leaves in the display.

You will probably begin creating these plots by hand (***Excel*** will not produce a stem and leaf plot) or in R.

Example 1: Grades for 20 students in a course on a 100-point exam are provided. We wish to examine the distribution of the grades. In R, the command stem produces the plot:

```
> grades <- c(53,55,66,69,71,78,75,79,77,
  75,76,73,82,83,85,74,90,92,95,99)
> stem(grades)

 The decimal point is 1 digit(s) to the right of the |

 5 | 35
 6 | 69
 7 | 134556789
 8 | 235
 9 | 0259
```

Can you tell how the plot is produced? The "stems" are the leading digit for each grade: 5, 6, 7, 8, and 9 in this case. These values are placed to the left of a vertical line. They can be thought of as scores in the 50s, 60s, 70s, 80s, and 90s:

5
6
7
8
9

Standing for 50s, 60s, 70s, 80s, and 90s.

If there had been a score of 100, then the leading digit would be 10, and we would add a row to the chart in 100s. So we would need:

05
06
07
08
09
10
for 50s, 60s, 70s, 80s, 90s, and 100s
Draw a vertical line after each stem.

5|
6|
7|
8|
9|

Now add the "leaves," which are the trailing digits, placed to the left of the vertical line for each observation with scores in that group. For example, there are two scores in the 50s in the data set: 53 and 55. The "leaves," values trailing the "stem" of 5, are then "3" and "5" for these two grades, and we see a 3 and 5 placed on the row for the 5 stem. In other words:

5| 3, 5 is read as data elements 53 and 55.

How many scores were in the 70s? You can figure this out by counting "leaves" on the 7 row of the plot. There are nine such scores. Notice that two students scored a 75, so there are two "leaves" with a value of 5. It is customary to put the leaves in order as R has done.

53, 55, 66, 69, 71, 73, 74, 75, 75, 76, 77, 78,79, 82, 83, 85, 90, 92, 95, 99

5| 3, 5
6| 6, 9
7| 1, 3, 4, 5, 5, 6, 7, 8, 9
8| 2, 3, 5
9| 0, 2, 5, 9

We can determine the shape of the distribution based on the plot. With a small data set such as our example, it is not easy to make a determination using the stem and leaf plot. The shape appears as almost *symmetric*. The center of the data is in the 70s, and there are roughly similar amounts of data in two "branches" to each side. Note how we read the values from the stem and leaf.

2.9 Chapter 2 Exercises

Make a stem and leaf of the following data sets, and comment about the shape.

2.8 100, 105, 111, 115, 121, 129, 131, 131, 133, 135, 137, 145, 146, 150, 160, 180

2.9 0.10, 0.15, 0.22, 0.23, 0.50, 0.62, 0.62, 0.65, 0.66, 0.69, 0.72

2.10 63, 65, 72, 81, 83, 85, 92, 93, 94, 105, 106, 121, 135

2.9.1 Displaying Quantitative Data with Histograms

We begin by stating that while they look similar, there is an important difference between bar charts and histograms. Bar charts have discrete (or categorical) values as their horizontal axis. Thus, bars are centered at certain discrete values. A histogram has continuous values as its horizontal axis, and the bars represent a range of values. Thus, there are no spaces between the bars unless there are no data in that particular range. Since most of the data that you will use are large, we will again go directly to displays with technology using R to explore histograms.

Example: Grades continued with R.

The histogram for the grades example is produced in R very simply:

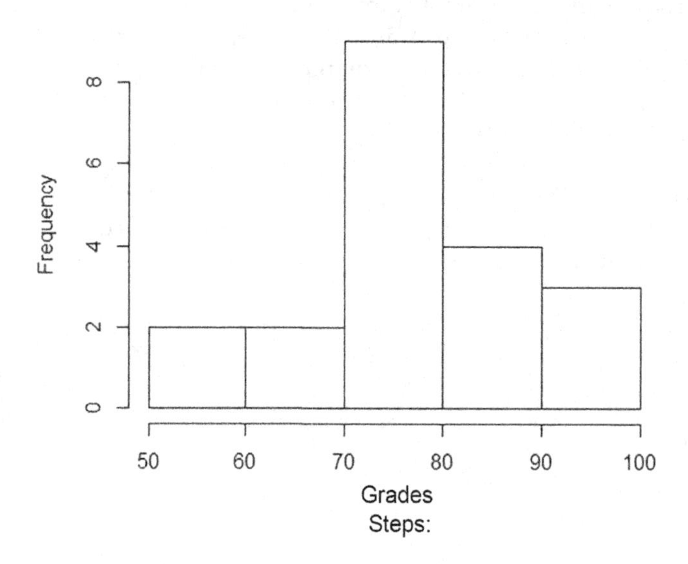

FIGURE 2.6
Histogram of grades.

(1) Obtain descriptive statistics for the data or order the data from smaller to larger.
(2) Determine the interval [smallest, largest].
(3) Calculate the class intervals (largest – smallest)/n, where n is the number of intervals desired. The value of n must be between 5 and 20. Start with 5 and go up until a good view of the histogram is obtained.
(4) List the end points as Bin values.
(5) Go to Data Analysis, Histogram, and bring up the dialog box. Put data in data input and end points in bins.
(6) The output is a table.
(7) Highlight the frequencies of the table, and go to Insert Bar Chart
(8) Right-click in the bar chart (on a bar), and close GAP size to 0.
(9) Comment on shape in regard to symmetry and skewness.

The selection of the "bins" is important, and a poor choice can lead to a poor data display. Bins should be of equal width. R has a built-in algorithm to determine the number of bins and then the corresponding widths. In our example, the bins chosen are the same as the "stems" for the stem and leaf plot. Thus, the histogram and stem and leaf plot are very

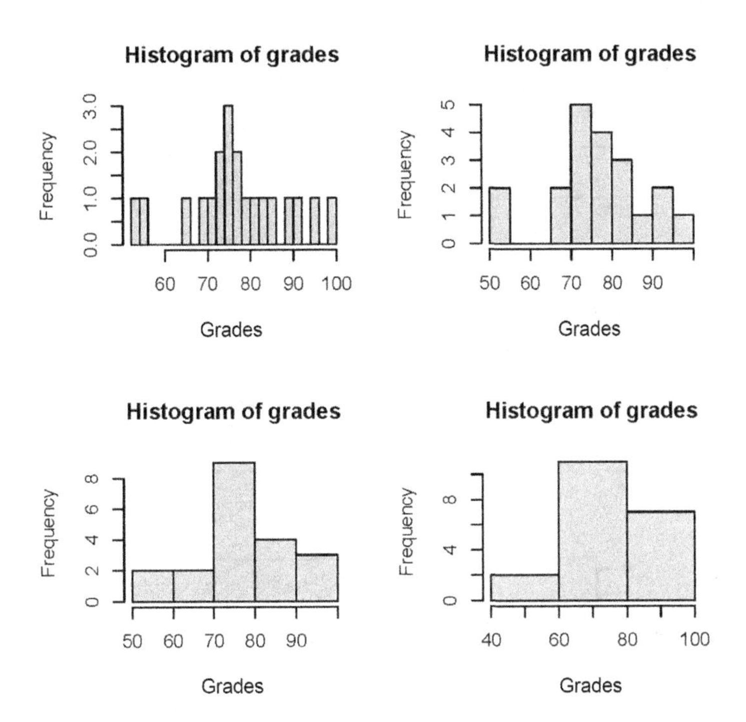

FIGURE 2.7
Histogram in R.

similar: one with bars vertically and the other numbers ("leaves") horizontally. We therefore come to a similar conclusion about the shape of the data; it is somewhat symmetric.

The number of bins chosen can be modified easily in R using several different options. One, shown below, is the "breaks" option. We can give the number of breaks (as below) or provide a vector of values to make the bins.

```
> hist(grades, breaks = 10) # modifying the number of bins
```

The impact of changing the number of bins for our example is shown in Figure 2.8. The default, five bins, is shown in the lower left. How does your assessment of the shape and symmetry change with the number of bins? In this example, there might not be a change; all plots show some evidence of symmetry. In other examples, the shape can change a great deal, so the choice should be made with care.

In general, too many bins lead to a "flat" distribution, as there are very few values in each bin. Too few bins can make it difficult to see any sort of shape in the data. Imagine a single bin!

The command for a histogram in R is

> hist("data")

So

> hist(grades)

If done by hand, here are the steps.

Interval	Tally	Decimal
51–60	2	2/20 = 0.10
61–70	2	2/20 = 0.10
71–80	9	9/20 = 0.45
81–90	4	4/20 = 0.2
91–100	3	3/20 = 0.15
Total	20	20/20 = 1.00

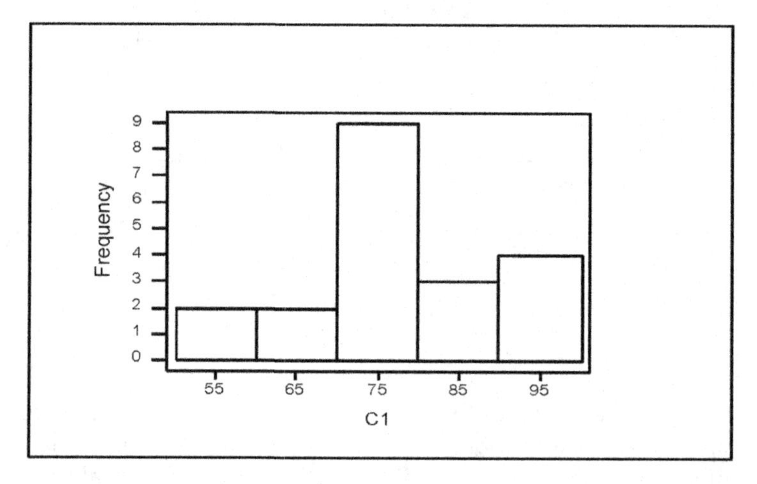

FIGURE 2.8
Histogram of grades for different numbers of bins.

Step 1. Determine, and select the classes, 5–15 classes. Find the range (lowest to highest value). Classes should be evenly spaced if possible.
Step 2. Tally the data in the classes.
Step 3. Find the numerical (relative) frequencies from the tallies.
Step 4. Find the cumulative frequencies.

Histogram: Connects class interval as a base and tallies (or relative frequencies) as the height of the rectangle. The rectangle is centered at the midpoint of class interval.

53, 55, 66, 69, 71, 73, 74, 75, 75, 76, 77, 78,79, 82, 83, 85, 90, 92, 95, 99

Possible class intervals:

(a) Classes 51–60, 61–70, 71–80, 81–90, 91–100
(5 class intervals)

(b) Classes 50–59, 60–69, 70–79, 80–89, 90–99, 100–109
(6 class intervals)

(c) Classes 51–55, 56–60, 61–65, 66–70, 71–75, 76–80, 81–85, 86–90, 91–95, 96–100
(10 class intervals)

Let's use selection (a).

There are other useful options in R, both in the basic command or using packages such as ggplot2 (we will see examples from this package in later chapters). One such option produces the values used in creating the plot but setting the option to plot as FALSE:

```
> hist(grades, plot = FALSE)
$breaks
[1]  50  60  70  80  90 100

$counts
[1] 2 2 9 4 3
```

We can see the breaks chosen and how many observations are in each bin. For larger examples, this data could be useful.

2.10 Chapter 2 Exercises

2.11 Make a stem and leaf plot using R of the following data sets, and comment about the shape.

a. 100, 105, 111, 115, 121, 129, 131, 131, 133, 135, 137, 145, 146, 150, 160, 180

b. 0.10, 0.15, 0.22, 0.23, 0.50, 0.62, 0.62, 0.65, 0.66, 0.69, 0.72

c. 63, 65, 72, 81, 83, 85, 92, 93, 94, 105, 106, 121, 135

Make a histogram by hand and with R using at least five class intervals for each data set in the previous question, and comment about the shape. Use a graphing calculator as well as do these by hand.

2.12 100, 105, 111, 115, 121, 129, 131, 131, 133, 135, 137, 145, 146, 150, 160, 180

2.13 0.10, 0.15, 0.22, 0.23, 0.50, 0.62, 0.62, 0.65, 0.66, 0.69, 0.72

2.14 63, 65, 72, 81, 83, 85, 92, 93, 94, 105, 106, 121, 135

2.15 We found this data, but what does it represent?

1	2	3	4	5	6	7	8	9	10	11	12	13	14	15	16
120	105	112	108	102	117	100	108	103	107	115	143	98	126	103	114

(a) What type of graph would you construct? Produce your plot in R.

(b) What type of analysis needs to be done with this data?

(c) What additional information would you like to have?

(d) Could this be the standardized reading scores of 16 high school students? If not, why not? If so, what does your display tell us about the scores?

2.11 Displaying Quantitative Data Using Boxplots and for Comparisons

A method for presenting quantitative data that is not impacted by the issue of bin size that we saw with histograms is the boxplot. We will present the information on how to construct and use a boxplot. We will use R.

The information in a boxplot comes from the "five number summary": Min, Q1, Median, Q3, Max. We have seen this summary before. Below is the summary command for height data, which contains the "five number summary" as well as the mean (average) of the data. A rectangle is drawn from Q1 to Q3, and the median is drawn with a vertical line in the rectangle. Lines are extended from Q1 to the Min and from Q3 to the Max.

```
> Height<-c(60,62,69,71,71,70,70.5,72,78,73,77,72,72,77,74,70)
> summary(Height)
   Min. 1st Qu.  Median    Mean 3rd Qu.    Max.
  60.00   70.00   71.50   71.16   73.25   78.00
```

The Min and Max are simply the smallest and largest observed values. The median is the value that divides the data in half; 50% of the data is below the median, and 50% above. It is also Q2 (the second quartile) or the 50th percentile. Q1, the first quartile, is the value with 25% of the data below, and Q3, the third quartile, is the value with 75% of the data below.

The boxplot is essentially a rectangle; a box is drawn from Q1 to Q3, so the box includes values from 25% to 75% of the data, known as the "inter quartile range" (IQR). The median is drawn with a line in the rectangle. Lines, sometimes called "whiskers," are extended from Q1 (the bottom of the box) to the Min and from Q3 (the top of the box) to the Max.

In R, a single boxplot is produced with the command boxplot. The example for height shown in Figure 2.9 was produced with the command:

```
> boxplot(Height, range = 0)
```

The option "range = 0" ensures the "whiskers" extend to the Min and Max. We will discuss the default option in R, which produces a modified boxplot.

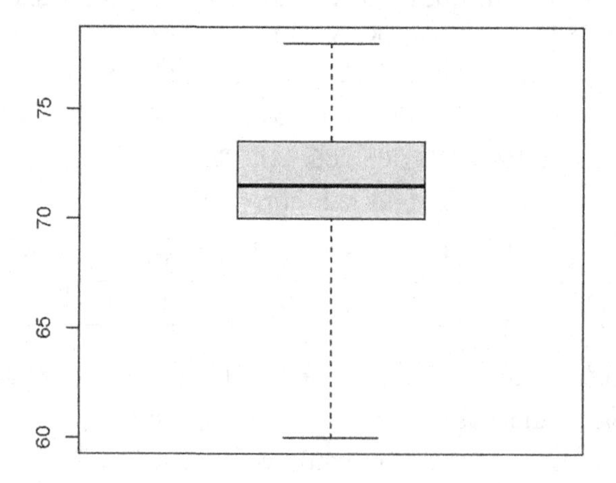

FIGURE 2.9
Boxplot of height data.

We can identify the "five number summary" values for height on the plot in Figure 2.9. The Min was 60, which is the horizontal line showing the bottom of the lower whisker in the plot. The upper whisker line is at the Max of 78. Q1 and Q3 were 70 and 73.25, and these values form the bottom and top of the box. Finally, the horizontal line in the box is the median of 71.5.

As with histograms, the boxplot is useful in determining the shape of the distribution of the data. We look at the relative distance from the median to the Min and Max, as well as from Q1 and Q2, to determine symmetry or skewness. In other words, is the median reasonably in the center of the box, and are the whiskers of approximately the same length? The height data appear to be slightly skewed to the left toward smaller numbers since the bottom whisker is longer than the top.

A useful adaptation of the boxplot is the modified boxplot. Instead of extending the whiskers to the Min and Max, they are only extending to values at most 1.5 times that of the IQR (width of the box). Values beyond these are then shown as dots or circles on the plot. The default command in R—in other words, not using the range option—produces the modified boxplot shown in Figure 2.10:

```
> boxplot(Height, range = 0)
```

Again, we look at the relative distance from the median to Min and the median to the Max to determine symmetry or skewness.

For example, we take some data for students' heights.

```
> Height<-c(60, 62, 69, 71, 71, 70, 70.5, 72, 78, 73, 77, 72, 72, 77, 74, 70)
> Height
[1] 60.0 62.0 69.0 71.0 71.0 70.0 70.5 72.0 78.0 73.0 77.0 72.0 72.0 77.0 74.0
[16]  70.0
```

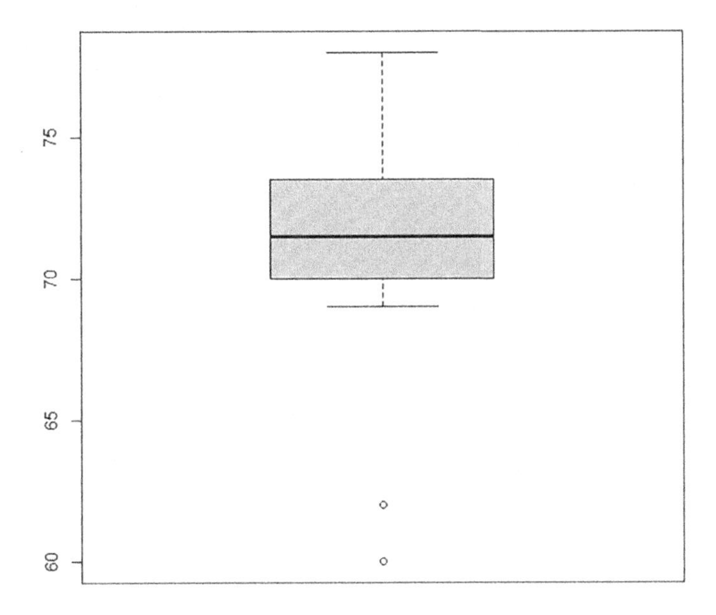

FIGURE 2.10
Modified boxplot of height data.

The width of the box, the IQR or Q3 – Q1, is 73.25 – 70 = 3.25. Thus, the whiskers extend to the last point that is within the distance of 1.5 * 3.25 = 4.875 of the box. For the upper whisker, this is values less than 73.25 + 4.875 = 78.125. The maximum height of 78 is less than this value, so the whisker extends to that point as in the original boxplot.

For the lower whisker, the extension is to values no less–than 70 – 4.875 = 65.125. We see that there are two values lower than this number, 60 and 62, which are shown in Figure 2.10 as circles. Notice that the lower whisker does not extend to 65.125. Instead, it extends to the lowest value not below 65.125, which is 69 in this data set.

The modified boxplot is useful for several reasons. One is that we observe that the appearance of slight left skew in the boxplot may actually be the result of two values. Second, the plot can identify unusual points, or outliers. We have two possible outliers in this data, which may be causing skew.

The data appear to be skewed to the left toward smaller numbers.

Boxplots are a good way to compare data sets from multiple sources. While it is possible to do such comparisons by producing multiple histograms, sometimes it is visually hard to compare. Further, if there are many groups, the graphs quickly become unwieldy.

Suppose we have data of weights from two groups of people: sometimes the data have same length, and sometimes they do not. That requires different commands in R.

Equal lengths

Multiple Boxplot in R

```
(1) Enter the data into R, and check to make sure you did not mistype
something.
> x<-c(111,112,112,113,115,120,121,124,125,125,130,135,140,145)
> y<-c(121,125,138,143,152,168,170,170,170,175,182,185,190)
x
[1] 111 112 112 113 115 120 121 124 125 125 130 135 140
> y
[1] 121 125 138 143 152 168 170 170 170 175 182 185 190
(2) Put the data into a data frame using the command below, and type data
to see that it worked.
> data<-data.frame(x, y)
> data
x y
1 111 121
2 112 125
3 112 138
4 113 143
5 115 152
6 120 168
7 121 170
8 124 170
9 125 170
10 125 175
11 130 182
12 135 185
13 140 190
(3) Obtain a boxplot of the data columns
> boxplot(data)
```

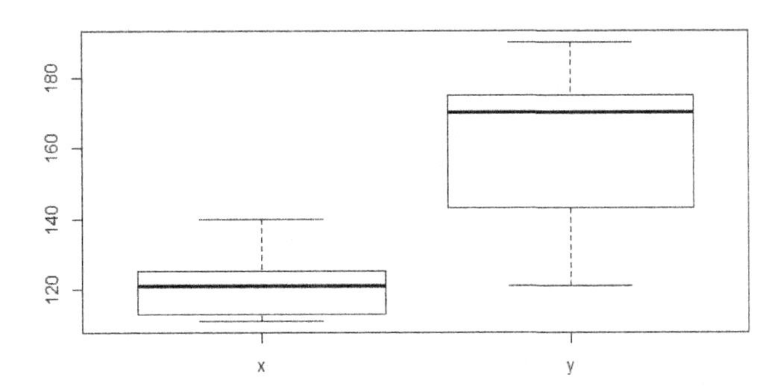

Note that we have groups of different sizes with one more in the x group. This is fine in R. We cannot combine these vectors into a single data frame but will be able to use the boxplot command. We have added some options to label the groups on the x-axis and provide a label for the y-axis:

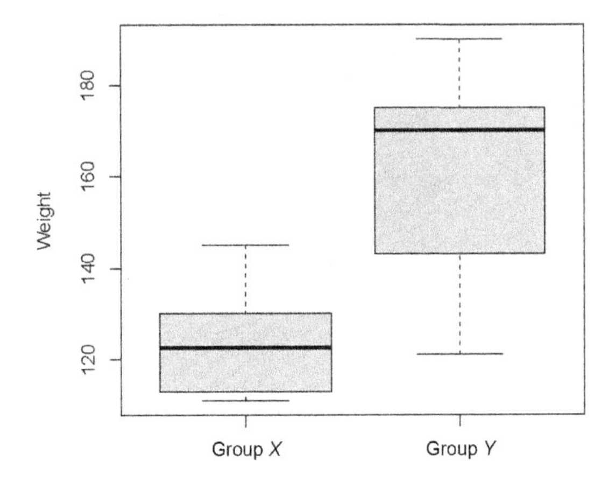

FIGURE 2.11
Boxplot of weight data for two groups.

In this case, we used the default modified boxplot but do not see any outliers. We can readily examine the distributions of the weights in each group and also compare them. We see group X is lighter with slightly right skewed values (longer upper whisker), while the heavier group Y may have left skew (median is toward the top of the box).

(4) Comment on what you see.

2.11.1 Unequal Lengths

For boxplots of data sets that are unequal in length,

In R

(1) Go to Tools, Install packages, Install from CRAN
ggplot2

(2) After ggplot2 is installed, then use the following commands
a<-data.frame(group = *"a"*, value = c(1, 2, 3, 4, 5, 6, 7, 8, 9, 10))

```
> a
group value
1 a 1
2 a 2
3 a 3
4 a 4
5 a 5
6 a 6
7 a 7
8 a 8
9 a 9
10 a 10
> b<-data.frame(group = "b", value = c(1, 3, 5, 7, 9, 11, 13, 15))
> b
group value
1 b 1
2 b 3
3 b 5
4 b 7
5 b 9
6 b 11
7 b 13
8 b 15
> library(ggplot2)
> plot.data<-rbind(a, b)
> ggplot(plot.data, aes(x = group, y = value, fill = group)) + geom_boxplot()
```

The output is shown in Figure 2.12.

As mentioned, an advantage of boxplots is the ability to look at a larger number of groups. For example, let's look at violence, in terms of number of incidents, in 10 regions in Afghanistan as seen in Figure 2.13. Putting the 10 boxplots together allows us to compare many aspects such as medians, ranges, and dispersions.

Boxplot

Step 1. Draw a horizontal measurement scale that includes all data within the domain of the data.
Step 2. Construct a rectangle (the box) whose left edge is the lower quartile value and whose right edge is the upper quartile value.
Step 3. Draw a vertical line segment in the box for the median value.
Step 4. Extend line segments from the rectangle to the smallest and largest data values (these are called whiskers).

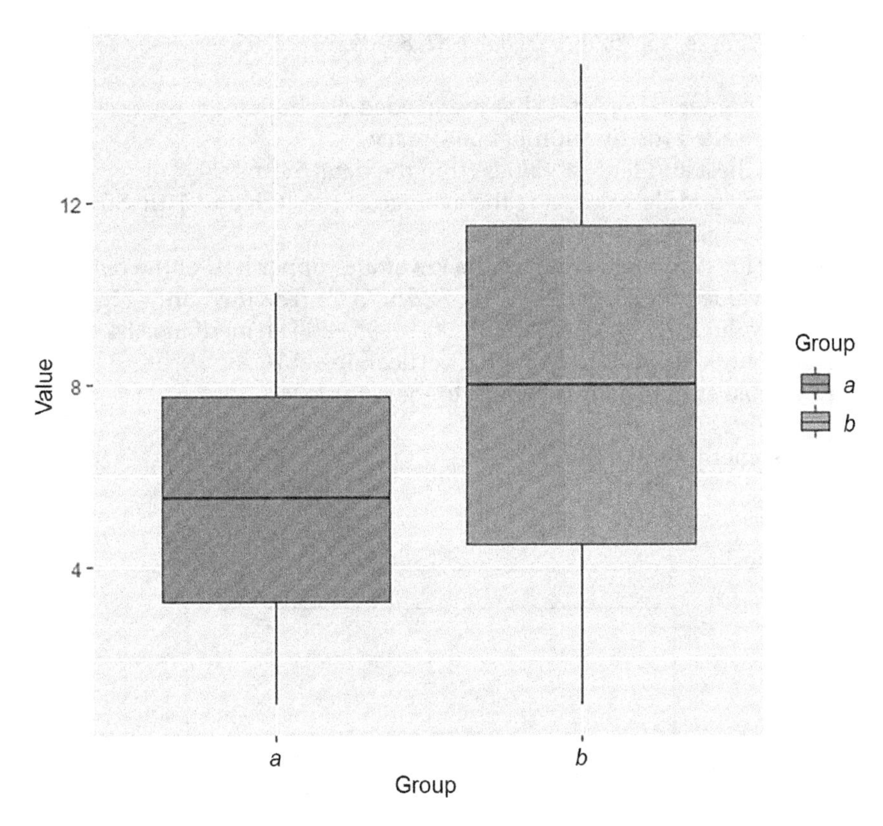

FIGURE 2.12
Boxplots of different length size.

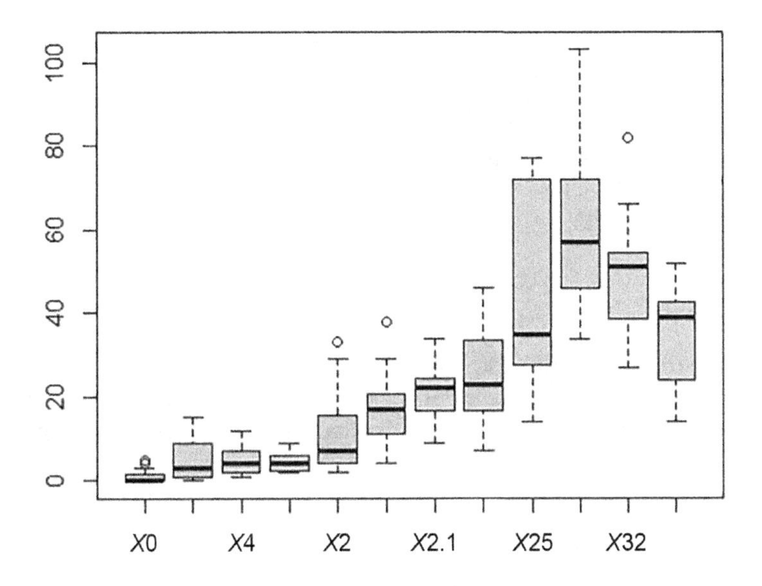

FIGURE 2.13
Boxplot of weight data for two groups.

53, 55, 66, 69, 71, 73, 74, 75, 75, 76, 77, 78, 79, 82, 83, 85, 90, 92, 95, 99

The values are in numerical order. What are needed are the range, the quartiles, and the median. These are called the five number summary.

Range is the smallest and largest values from the data: 53 and 99.

The median is the middle value. It is the average of the 10th and 11th values as we will see later: (76 + 77)/2 = 76.5

The quartiles values are the median of the lower and upper half of the data.

Lower quartile values: 53, 55, 66, 69, 71, 73, 74, 75, 75, 76. The median is 72.

Upper quartile values: 77, 78, 79, 82, 83, 85, 90, 92, 95, 99. The median is 84.

You draw a rectangle from 72 to 84 with a vertical line at 76.5.

Then draw a whisker to the left to 53 and to the right to 99.

It would look something like this:

Using R we use

> boxplot (grades)

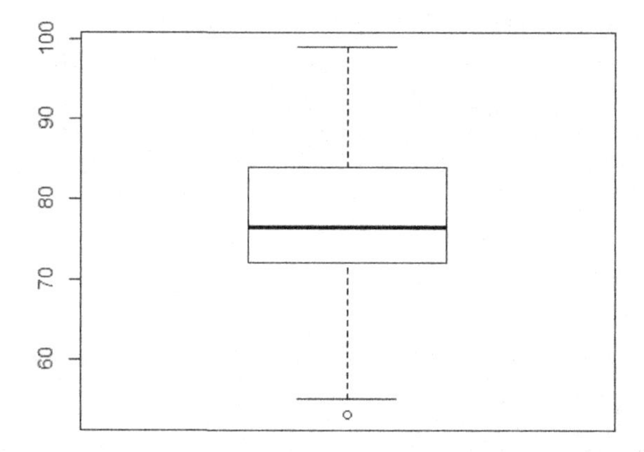

2.11.2 Comparisons

Consider our data for casualties in Afghanistan through the years 2000–2011. This is presented to you as a politician. What information can you interpret from this boxplot?

We had data on casualties that we placed in an Excel, csv file, and used the commands below to produce Figure 2.13.

```
> elements<-read.csv("casualties.csv")
> boxplot(elements)
```

We obtained the following boxplot. What do you conclude from the plot? Are there differences in the number of violent incidents across regions? Notice how easily much information is conveyed in this plot. Imagine if, instead, you were given histograms for each region to compare.

2.11.3 R (Retrieving Data) for Obtaining Displays

Recall from Section 1.4, we discuss manipulating data in R. Now, we take this data and obtain displays.

```
> stem(Hght)
The decimal point is at the |
58 | 00000
60 | 00000
62 | 0000000
64 | 000
66 | 00000
68 | 000000000
70 | 0000000
72 | 0000000
74 | 00
> hist(Hght)
```

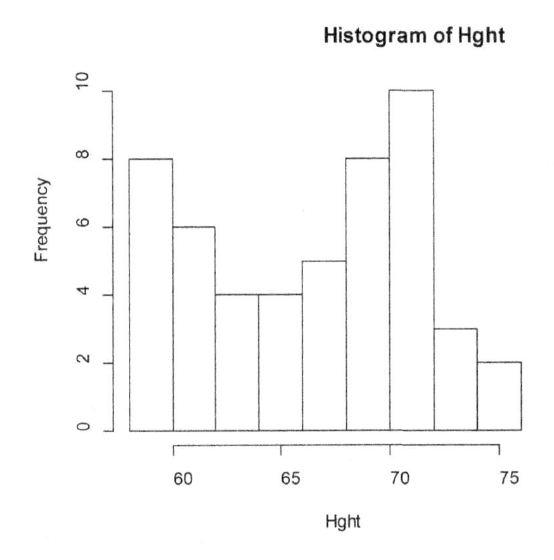


```
> boxplot(Hght)
```

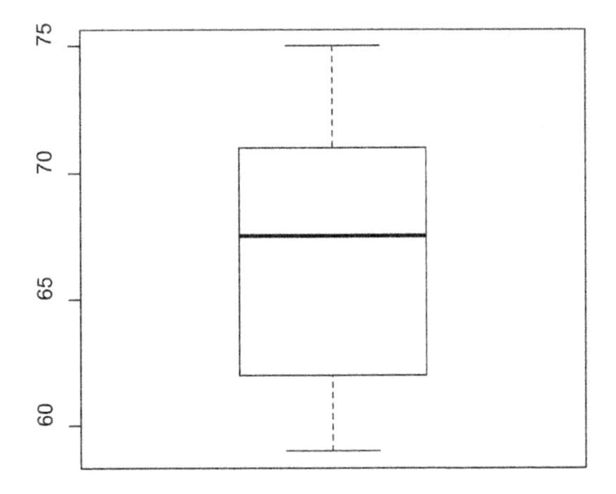

All graph views indicate skewness (skewed left toward smaller numbers)

Statistics

```
> hist(Hght)
> boxplot(Hght)
> summary(Hght)
  Min. 1st Qu.  Median    Mean 3rd Qu.    Max.
 59.00   62.00   67.50   66.66   71.00   75.00
> mean(Hght)
[1] 66.66
> median(Hght)
[1] 67.5
> var(Hght)
[1] 23.90245
> sd(Hght)
[1] 4.889013
# We obtain the coefficient of skewness using the short-cut formula
> sk<-3*(mean(Hght)-median(Hght))/sd(Hght)
> sk
[1] -0.5154415
> # Interpret as skewed left since sk <0
```

We get everything but the mode, which we can see from real data
Put data in numerical order first.

71 65 67 61 69 60 62 67 68 68 73 64 69 71 59 63 63 66 61 70 69 59 72 66 69
[26] 65 69 71 59 73 75 63 59 71 60 62 72 73 72 62 75 62 67 71 59 71 69 72 60 69

We see 59 (five times) 60 (three times), 61 (two times), . . .
In order is

59
59
59
59
59
60
60
60

61
61
62
62
62
62
63
63
63
64
65
65
66
66
67
67
67
68
68
69
69
69
69
69
69
69
70
71
71
71
71
71
71
72
72
72
72
73
73
73
75
75

We count the number of each value.

Hght	Count
59	5
60	3
61	2
62	4
63	3
64	1
65	2
66	2
67	3
68	2
69	7
70	1
71	6
72	4
73	3
74	0
75	2

We see 69 is repeated seven times, so it is the mode.

2.12 Chapter 2 Exercises

Make a boxplot of each of the following data sets. Produce the five number summary as well, and identify these values on each boxplot. Use the boxplots, and comment on the shape of the distribution. Finally, also create modified boxplots. Are there outliers in any of the data sets?

2.16 100, 105, 111, 115, 121, 129, 131, 131, 133, 135, 137, 145, 146, 150, 160, 180

2.17 0.10, 0.15, 0.22, 0.23, 0.50, 0.62, 0.62, 0.65, 0.66, 0.69, 0.72

2.18 63, 65, 72, 81, 83, 85, 92, 93, 94, 105, 106, 121, 135

2.13 Summary: Displays of Data

2.13.1 Categorical Data Displays

Pie Chart:

The circle (pie) represents the whole or 100%. The wedges or pieces of the pie represent the proportion or part of the total for that category.

Bar Chart:

Bars can be horizontal or vertical, but they should be uniformly spaced and of the same width. The lengths of the bars represent the values of the categories we wish to compare.

2.13.2 Quantitative Displays

Stem and leaf plot:

Step 1. Order the data.

Step 2. Separate according to the one or more leading digits. List stems in a vertical column.

Step 3. Leading digit is the stem, and trailing digit is the leaf.

Step 4. Indicate the units for stems and leaves in the display.

Histogram:

Step 1. Determine and select the classes, 5–15 classes. Find the range (lowest to highest value). Classes should be evenly spaced if possible.

Step 2. Tally the data in the classes.

Step 3. Find the numerical (relative) frequencies from the tallies.

Step 4. Find the cumulative frequencies. Classes or bins of the data are created of equal width.

Histogram: Connects class intervals as a base, and tallies of the data (or relative frequencies) in each bin are plotted as the height of a rectangle. The rectangles are centered at the midpoint of each class interval.

Boxplot:

Step 1. Draw a horizontal measurement scale that includes all data within the domain of the data.

Step 2. Construct a box that is a rectangle whose left lower edge is the lower quartile value and whose right edge is the upper quartile value.

Step 3. Draw a vertical line segment in the box for the median value.

Step 4. Extend line segments from the rectangle to the smallest and largest data values (these are called whiskers). In a modified boxplot, the whiskers extend only to the last observation within 1.5 times the box width. Points (circles) then represent points outside this range, which may be outliers.

2.14 Chapter 2 Exercises

2.19 Make a pie chart for the following data. What information is best displayed by a pie chart? Does a pie chart work well for this data?

Never Married	43.9 million
Married	116.7 million
Widowed	13.4 million
Divorced	17.6 million

2.20 Make a bar chart for this data: female doctorates as a percentage of graduates in that field who were females. Can you make a pie chart? What do you have to do first?

Computer Science	15.4%
Education	60.8%
Engineering	11.1%
Life Sciences	40.7%
Physical Sciences	21.7%
Psychology	62.2%
Mathematics	10%

2.21 Display the following data. In 1995, there were 90,402 deaths from accidents in the United States. Among these, 43,363 were from motor vehicles; 10,483 from falls; 9,072 from poisoning; 4,350 from drowning; and 4,235 from fires. How many deaths were due to other unknown causes?

2.22 In a Math I class last semester, the final averages were 88, 63, 82, 98, 89, 72, 86, and 70. Display the data as a stem and leaf plot. Are they symmetric? Are they skewed?

2.23 In a Math II class last semester, the final averages were 66, 61, 78, 54, 75, 40, 78, 91, 84, 82, 76, and 65. Display the data as a histogram. Are they symmetric? Are they skewed?

2.24 Using the grades in Math I (Exercise 2.22) and Math II (Exercise 2.23) from the previous two questions, display the data as two boxplots side by side. Is each display symmetric? Is each display skewed? Can you compare the two data sets? Which class had the higher grades? Which class has grades that are more spread out? Do the symmetry and skewness of the data sets tell us anything about the grades?

2.25 Describe one difference between a frequency histogram and a relative frequency histogram.

2.26 Describe one advantage of a stem and leaf diagram over a frequency histogram. What is an advantage of a histogram?

2.27 Construct a stem and leaf diagram, a frequency histogram, and a relative frequency histogram for each of the following data sets. Compare the plots. For the histograms, use classes 1–10, 11–20, . . . , 51 – 60, 61 – 70, and so on. Which do your prefer?

9	92	68	77	80
70	85	88	85	96
93	75	76	82	100
53	70	70	82	85

2.28 Construct a stem and leaf diagram, a frequency histogram, and a relative frequency histogram for the following data set. For the histograms, use classes 0–0.9,1–1.9, . . . , 6.0 – 6.9, 7.0 – 7.9, and so on.

0.5	8.2	7.0	7.0	4.9
6.5	8.2	7.6	1.5	9.3
9.6	8.5	8.8	8.5	8.7
8.0	7.7	2.9	9.2	6.9

2.29 A data set contains $n = 10$ observations. The values for the random variable x and their frequencies f are summarized in the following data frequency table. Construct a frequency and relative frequency histogram, and comment on its shape.

x | –1 0 1 2

f | 3 4 2 1

2.30 A data set contains 15 observations. The values for the random variable x and their frequencies f are summarized in the following data frequency table. Construct a histogram, and comment on its shape.

2.31 A data set contains $n = 15$ observations. The values for the random variable x and their frequencies f are summarized in the following data frequency table.

x | 0 1 2 3

f | 6 4 3 2

Construct a frequency and relative frequency histogram Comment on the shape.

2.32 A data set contains 20 observations. The values for the random variable x and their frequencies f are summarized in the following data frequency table. Construct a histogram, and comment on its shape.

2.33 A data set contains $n = 20$ observations. The values x and their frequencies f are summarized in the following data frequency table.

x | 0 1 2 3 4

f | 3 4 6 5 2

2.34 The IQ scores of 10 students randomly selected from an elementary school are given. Construct a stem and leaf plot, and comment on the shape.

108, 100, 99, 125, 87, 105, 107, 105, 119, 118

2.35 The IQ scores of 10 students randomly selected from a gifted elementary school are given. Construct a stem and leaf plot of the data and comment on the symmetry or skewness. Compare the scores to those in the previous problem. How easy is it to use these plots to compare? Produce a visualization that more easily allows you to compare the two groups.

133, 140, 152, 142, 137, 145, 160, 138, 139, 138

2.36 Blood donation center. The blood types of these 300 donors are summarized in the table. Construct a frequency and relative frequency histogram. Comment on the shape. Create an appropriate visualization for this data.

Blood Type	O	A	B	AB
Frequency	136	120	33	11

2.37 The weekly sales for a new steamed rice cooker on Amazon for the last 20 weeks are provided. Construct a stem and leaf plot. Comment on the shape. Create two visualizations for the data. Which do you prefer and why?

0, 15, 15, 14, 14, 18, 15, 17, 16, 16, 18, 19, 12, 13, 9, 19, 15, 16, 15, 15

2.38 Random samples, each of size $n = 10$, were taken of the lengths in centimeters of three kinds of commercial fish, with the following results. Grouping the measures by their common digits, construct a stem and leaf diagram, a frequency histogram, and a relative frequency histogram for each of the samples. Compare the histograms, and describe any patterns they exhibit.

Sample 1	16.5	12.4	11.8	19.5	21.5	13.2	10.5	19.7	21	15.5
Sample 2	12.5	13.45	19.6	15.5	12.5	14.5	21.5	14.3	18,7	21.7
Sample 3	17.5	16.5	10.5	13.4	17.5	15.6	18.5	13.5	21.2	13.5

3

Statistical Measures

Objectives

1. **Learn how to use R to obtain and interpret descriptive statistics for the center (location), spread, and shape of the distribution of a data set.**
2. **Learn how to use R to obtain displays and be able to interpret them.**
3. **Know what the measures are and which ones are used for quantitative data and which ones are better for qualitative data.**

3.1 Measures of Central Tendency or Location

3.1.1 Describing the Data

In addition to plots and tables, numerical descriptors are often used to summarize data. Three numerical descriptors, the *mean*, the *median*, and the *mode*, offer different ways to describe and compare data sets. These are generally known as the *measures of location*.

3.1.2 The Mean

The mean is the arithmetic average, with which you are probably very familiar. For example, your academic average in a course is generally the arithmetic average of your graded work. The mean of a data set is found by summing all the data and dividing this sum by the number of data elements.

The following data represent 10 scores earned by a student in a college mathematics course: 55, 75, 92, 83, 99, 62, 77, 89, 91, 72. We find the mean in the same way you would compute the student's average. The mean can be found by summing. We first sum the 10 scores:

$$55 + 75 + 92 + 83 + 99 + 62 + 77 + 89 + 91 + 72 = 795$$

and then divide by the number of data elements (10), $795/10 = 79.5$.

To describe this process in general, we can represent each data element by a letter with a numerical subscript. Thus, for a class of n tests, the scores can be represented by $a_1, a_2, \ldots, a_n$. The mean of these n values of $a_1, a_2, \ldots, a_n$ is found by adding these values and then dividing this sum by n, the number of values. The Greek letter Σ (called sigma) is used

DOI: 10.1201/9781003317906-3

to represent the sum of all the terms in a certain group. Thus, we may see this written as follows:

$$\sum_{i=1}^{n} a_i = a_1 + a_2 + \ldots + a_n$$

$$a_1 + \cdots + a_n = \sum_{i=1}^{n} a_i$$

and the mean then is written as follows:

$$mean = \overline{x} = \frac{\sum_{i=1}^{n} a_i}{n}$$

The symbol for the mean of a sample, as shown in the equation, is $\bar{x}$ or "x bar." Think of the mean as the average. Notice that the mean does not have to equal any specific value of the original data set. The mean value of 79.5 was not a score ever earned by our student.

The mean is a useful and commonly used measure of the center of the data. However, a disadvantage of this measure is that outliers, unusually small or large values, in the data can dramatically impact the value.

3.1.3 The Median

The median locates the true middle of a numerically ordered list. The hint here is that you need to make sure that your data is in numerical order listed from smallest to largest along the x- number line. There are two ways to find the median (or middle value of an ordered list) depending on whether n (the number of data elements) is even or odd.:

(1) If n is odd

If there is an odd number of data elements, then the middle (median) is the exact data element that is the middle value. For example, here are five ordered math grades earned by a student: 55, 63, 76, 84, 88.

The middle value is 76 since there are exactly two scores on each side (lower and higher) of 76. Notice that with an odd number of values that the median is a real data element.

(2) If n is even

If there is an even number of data elements, then there is no true middle value within the data itself. In this case, we need to find the mean of the two middle numbers in the ordered list. This value, probably not a value of the data set, is reported as the median. Let's illustrate with several examples.

Example 1: Here are six math scores for student 1: 56, 62, 75, 77, 82, 85.

The middle two scores are 75 and 77, because there are exactly two scores below 75 and exactly two scores above 77. We average 75 and 77:

$$\frac{75+77}{2} = \frac{152}{2} = 76$$

$$(75+77)\ /\ 2 = 152/2 = 76$$

The median is 76. Note that 76 is not one of the original data values.

Example 2: Here are eight scores for student 2: 72, 80, 81, 84, 84, 87, 88, 89.

The middle two scores are 84 and 84, because there are exactly three scores lower than 84 and three scores higher than 84. The average of these two scores is 84. Note that this median is one of our data elements.

It is also very possible for the mean to be equal to the median. Unlike the mean, the median is not heavily influenced by outliers. Since only the middle one or two observations are used in computing the median, changing the largest (or smallest) value will not impact the result.

3.1.4 The Trimmed Mean

The median addresses the issue of outliers, but does so by only uses one or two point in the data set (while the mean uses all of the data). A "compromise" is the trimmed mean. To compute the trimmed mean, first "trim" or remove a percentage of the smallest and largest values in the data and then calculate the mean of the remaining values. For example, the 10% trimmed mean removes the smallest 10% and the largest 10% of the data.

We return to the scores in the college mathematics course: 55, 75, 92, 83, 99, 62, 77, 89, 91, 72. We found the mean earlier, 79.5. There are 10 scores, so to compute the 10% trimmed mean we remove the largest and smallest 10%, or we remove the largest and smallest observations, which are 55 and 99. This leaves eight observations, and the trimmed mean is then:

$$\frac{75+92+83+62+77+89+91+72}{8} = 80.125$$

The median is actually a trimmed mean where all but the middle values are "trimmed." For this example, we would trim 40% (four values) from both sides, largest and smallest, of the data, leaving the two middle values: 83 and 77. The mean of these two values is 80. You should confirm that this is also the median value.

3.1.5 The Mode

The value that occurs the most often is called the mode. It is one of the numbers in our original data. The mode does not have to be unique. It is possible for there to be more than one mode in a data set. As a matter of fact, if every data element is different from the other data elements then every element is a mode.

For example, consider the following data scores for a mathematics class.

75, 80, 80, 80, 80, 85, 85, 90, 90, 100

The number of occurrences for each value is:

Value	Number of Occurrences
75	1
80	4
85	2
90	2
100	1

Since 80 occurred four times and is the largest value among the number of occurrences, then 80 is the mode.

3.2 Measures of Dispersion

3.2.1 The Variance and Standard Deviation

Measures of variation or *measures of the spread* of the data include the variance and standard deviation. They measure the spread in the data, how far the data are from the mean. The sample variance has notation s^2, and the sample deviation has notation s.

$$s^2 = \frac{\sum_{i=1}^{n}(x_i - \bar{x})^2}{n-1} \text{ where } n \text{ is the number of data elements.}$$

$$s = \sqrt{s^2} = \sqrt{\frac{\sum_{i=1}^{n}(x_i - \bar{x})^2}{n-1}} \text{ where } n \text{ is the number of data elements.}$$

Notice that in the formula for the variance, the difference of each element from the mean is computed and then squared. These squared values, which measure spread, are then summed. Finally, they are "averaged" by dividing by n minus 1. A technical note here is that we divided by $n - 1$ instead of n to get the "average" because the sample mean was estimated in the numerator. The estimate of the variance would be "biased" if the adjustment was not made (we will not explore the mathematical details for this assertion).

The sample standard deviation is then just the square root of the variance. Because the variance involves squared differences, its units are the units of the original data squared. The units of the standard deviation are the same as the original data and, as a result, the standard deviation is usually preferred.

One important consideration for use of the variance and standard deviation as measures of spread is that both are influenced by outliers, in the same fashion as the mean.

Example 3: Consider the following 10 data elements:

50, 54, 59, 63, 65, 68, 69, 72, 90, 90

The mean, $\bar{x}$, is 68. The variance is found by subtracting the mean, 68, from each point, squaring them, adding them up, and dividing by $n - 1$:

$$s^2 = [(50-68)^2 + (54-68)^2 + (59-68)^2 + (63-68)^2 + (65-68)^2 + (68-68)^2 + (69-68)^2 + (72-68)^2 + (90-68)^2 + (90-68)^2]/9 = 180$$

$$s = \sqrt{s^2} = 13.42$$

Example 4: Metabolic rates:

Consider a person's metabolic rate, the rate at which the body consumes energy. Here are 7 metabolic rates for men who took part in a study of dieting. The units are calories in a 24-hour period.

1,792, 1,666, 1,362, 1,614, 1,460, 1,867, 1,439

The researchers wish to report both $\bar{x}$ and s, the mean and standard deviation, for these men.

The mean is computed:

$$\bar{x} = \frac{1792 + 1666 + 1362 + 1614 + 1460 + 1867 + 1439}{7} = 1600$$

To see clearly the nature of the variance, we create a table to compute the deviations and squared deviations of the observations from the mean.

Observations, (x_i)	Deviations $(x_i - \bar{x})$	Squared Deviations
x_i	$x_i - \bar{x}$	$(x_i - \bar{x})^2$
1,792	1792 – 1600 = 192	36,864
1,666	1666 – 1600 = 66	4,356
1,362	1362 – 1600 = –238	56,644
1,614	1614 – 1600 = 14	196
1,460	1460 – 1600 = –140	19,600
1,867	1867 – 1600 = 267	71,289
1,439	1439 – 1600 = –161	25,921
	Sum = 0	Sum = 214,870

The variance is $s^2 = 214{,}870/6 = 35{,}811.67$

The standard deviation is $s = \sqrt{35{,}811.67} = \mathbf{189.24}$

Example 5: Given the following table:

X	1	2	3	4	5
F	3	4	6	3	2

We could expand the data as 1, 1, 1, 2, 2, 2, 2, 3, 3, 3, 3, 3, 3, 4, 4, 4, 5, 5.

To compute the mean, we could add these up and divide by 18, or we could simply find $\Sigma x^* f$ and divide by n as

$\bar{x} = 1(3) + 2(4) + 3(6) + 4(3) + 5(2) / 18 = 51/18 = 2.8333$

The median is 3, and the mode is 6 (repeated 6 times).

In R, we also find: > x<-c(1, 1, 1, 2, 2, 2, 2, 3, 3, 3, 3, 3, 3, 4, 4, 4, 5, 5)

```
> var(x)
[1] 1.558824
> sd(x)
[1] 1.248529
> summary(x)
Min. 1st Qu. Median Mean 3rd Qu. Max.
1.000 2.000 3.000 2.833 3.750 5.000
```

3.2.2 The Range and Interquartile Range (IQR)

In data with a large outlier, we might prefer the median to the mean as a measure of center. There are alternative measures of the spread to the variance and standard deviation as well. The first, the range, is simply a measure that takes the overall spread of the data, from the maximum and to the minimum values of the data. Often, this is provided a single number.

Assume we have the following data: In the data from the previous example of metabolic rates, the maximum value is 1867 and the minimum is 1362. The range is simply the difference of these two values:

$$\text{Range} = 1867 - 1362 = 505$$

What does 505 represent? Sometimes the range is simply given as an interval of the minimum to maximum values: [1362, 1867].

You likely are asking how the range solves the problem of a measure impacted by outliers. You are correct; it too is impacted since it uses both the minimum and maximum values. A commonly used measure when the median is reported is the IQR (interquartile range), which is less impacted by outliers. We discussed the IQR in the last chapter when presenting data using boxplots. It is the length of the "box," or the difference between the 75th and 25th percentiles of the data (the first and third quartiles, Q3 and Q1) or IQR = Q3 – Q1.

There are different methods of computing quantiles, which impact the IQR computation, particularly for small data sets. These details are not critical for our purposes, so we will use R to compute them and give an example in the last section of this chapter. For the metabolic range data, the IQR computed by R is 279.5.

1,792
1,666
1,362
1,614
1,460
1,867
1,439

The maximum value is 1,867, and the minimum value is 1,362. If you take the difference, 1867 – 1362 = 505. What does 505 represent? I suggest you give the range as an interval [1362, 1867].

Some properties of the standard deviation are as follows:

(a) s measures spread about the mean.
(b) $s = 0$ only when there is no spread.
(c) s is strongly influenced by extreme outliers.

3.3 Measures of Symmetry and Skewness

We discussed symmetry and skew when looking at visualizations of the distribution of quantitative variables in the previous chapter.

As a review, we will use the bell-shaped curve to denote symmetry as illustrated in Figure 3.1. The following figures illustrate it.

We saw two types of skew, positive and negative, in the previous chapter. These are illustrated in Figure 3.2.

We can define a measure, the coefficient of skewness, S_k, to capture the amount of skew. Mathematically, we determine this value using a formula. A simple method to estimate skew is Pearson's measure defined as follows:

$$S_k = \frac{3(\bar{x} - \tilde{x})}{s},$$

where $\tilde{x}$ is the median (or, the mode, if there is a single mode).

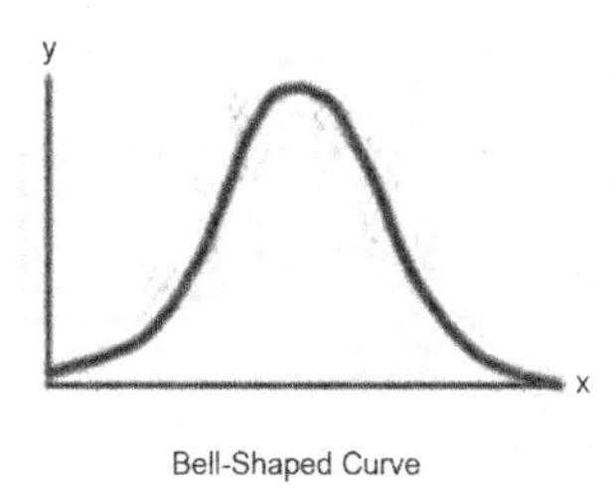

FIGURE 3.1
Bell-shaped curve denotes symmetry.

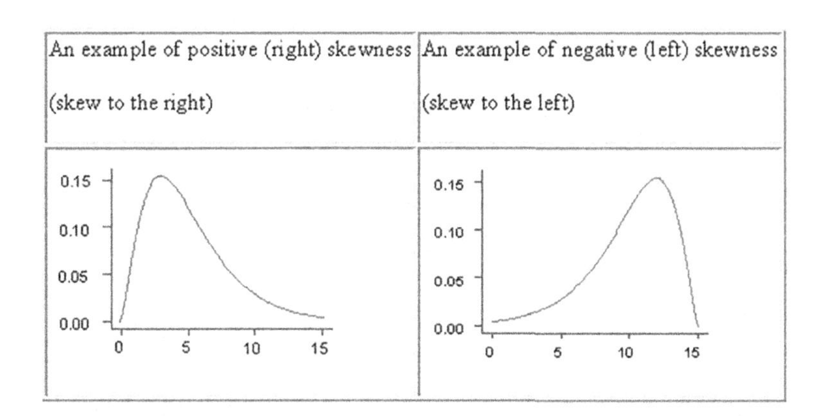

FIGURE 3.2
Shapes of curves that are skewed.

For the metabolic rate data example, we computed $s = 189.24$. You can quickly confirm that the mean of the data is 1,600 and median is 1,614, leading to a Pearson skew estimate of:

$$S_k = \frac{3(1600 - 1614)}{189.24} = -0.22$$

We can develop the following rules for skewness and symmetry based on the measure. The value is zero when the mean and median are equal. This is the case for symmetric data like that of Figure 3.1. If the data is right (positive skewed), as in the left panel of Figure 3.2, the larger values in the right tail of the distribution will "pull" the mean to be larger than the median. Recall that the mean is more impacted by outliers than the median. Thus, the Pearson value will be positive. The reverse occurs for left skew.

Thus, our rules for the measure are as follows:

If $S_k \approx 0$, the data is symmetric.
If $S_k > 0$, the data is positively skewed (skewed right).
If $S_k < 0$, the data is negatively skewed (skewed left).

3.4 Summary with Descriptive Statistics Using R

3.4.1 Measures of Location

We will first look at how to compute measures of location from Section 3.1 in R. First, we input the scores in the math data example from that section into R:

```
> mathscores = c(55, 75, 92, 83, 99, 62, 77, 89, 91, 72)
```

Using the summary command for the data actually provides two of the measures, the mean and the median:

```
> summary(mathscores)
   Min. 1st Qu.  Median    Mean 3rd Qu.    Max.
  55.00   72.75   80.00   79.50   90.50   99.00
```

R has commands to produce each statistic separately as well:

```
> mean(mathscores)
[1] 79.5
> median(mathscores)
[1] 80
```

The mean command includes an option to produce trimmed means. To trim 10% of both the largest and smallest values, use:

```
> mean(mathscores, trim = 0.1)
[1] 80.125
```

R does not have a function for computing the mode in the packages available in the "base" version. However, there are packages in R that we can install to perform the operation. Using packages not loaded by default in R requires two steps: (1) install the package (i.e., download into your R library), and (2) load the package into R for use.

We will use a function from the package "DescTools" to compute the mode. There are several ways to install this package in RStudio. The easiest is to look for the tab "Packages" in the RStudio window where the plots and help appear. Click "Install" in this tab, and start typing the name of the package; RStudio will display a list of packages, and you can pick the one you wish and click install. A second option is using the RStudio menus at the top of the screen. The "install packages" dialog is found on the "Tools" menu. A third approach is to use a command in R:

```
> install.packages("DescTools")
```

Once installed, you need to load the package into your R session. This is required each time that you wish to use the package in a new R session, so we recommend saving this command in your script file. Most R programmers put the list of packages used in a given script at the top of the program; then you simply run all of them at the beginning of using the script. The command is:

```
> library(DescTools)
```

The "Packages" tab in RStudio can also be used to load packages; scroll to find the package, and put a check in the box next to the one you wish to load. This is tedious to do each time you want to use a program, however, so we prefer the "library" command saved to our script files.

With the package loaded, we can use the "Mode" command on our math score data:

```
> Mode(mathscores)
[1] NA
attr(,"freq")
[1] 1
```

What does this result mean? The "NA" is because there is not a single mode or even modes in this data. The "1" for the "freq" tells us why: all 10 values differ. Let's consider the data set used in Section 3.1 to illustrate the mode:

```
> mathscores2 = c(75, 80, 80, 80, 80, 85, 85, 90, 90, 100)
> Mode(mathscores2)
[1] 80
attr(,"freq")
[1] 4
```

Commands and results

mean("data") → gives the mean

median("data") → gives the median

mode—R does not do well. Perhaps do a Stem and Leaf plot and manually examine it for the mode.

sd("data")

var("data")

You can use
Summary("data") that gives results as we will illustrate.

Coefficient of skewness =

SK<-3*(mean("data")-median("data")/sd("data")

For grades

```
mean(grades)
[1] 77.35
> median(grades)
[1] 76.5
> sd(grades)
[1] 11.79775
> SK<-3*(mean(grades)-median(grades))/sd(grades)
> SK
[1] 0.216143

> summary(grades)
```

In this data, the mode is 80, which occurs four times.

3.4.2 Measures of Spread

In Section 3.2, we looked at examples using the metabolic rate data input and summarized in R:

```
> metabolicrate = c(1792,1666,1362,1614,1460,1867,1439)
> summary(metabolicrate)
         Min.   1st Qu.   Median   Mean   3rd Qu.   Max.
  1362    1450      1614      1600   1729      1867
```

We can obtain the variance, standard deviation, and IQR in R for this data:

```
> var(metabolicrate)
[1] 35811.67
> sd(metabolicrate)
[1] 189.2397
> range(metabolicrate)
[1] 1362 1867
> IQR(metabolicrate)
[1] 279.5
```

Note that there are different methods of computing the quantiles, so the IQR is slightly different than the one we computed "by hand." R returns the interval for the range rather than the difference between the two values.

3.4.3 Measures of Symmetry and Skewness

In Section 3.3, continued using the metabolic rate data to consider measures of skew. The Pearson measure is easily computed using the mean, median, and standard deviation functions. We save the values needed and then create the formula:

```
> xbar = mean(metabolicrate)
> med = median(metabolicrate)
> s = sd(metabolicrate)
> 3*(xbar - med)/s
[1] -0.2219407     Min. 1st Qu.  Median    Mean 3rd Qu.    Max.
```

We can also compute the skew using the more complicated formula by installing and loading the "moments" R package:

```
> library(moments)
> skewness(metabolicrate)
[1] 0.1594803
```

In this data, the two measures "disagree" about the direction of the skew. One measure is positive, the other negative. The reason is twofold. First, in both cases, the values are small in absolute value (near 0). We actually have very little skew in this data, near symmetric. You should produce visualizations to confirm this: a histogram or boxplot. Second, the data set is very small. It is difficult to really determine the shape of a distribution from such a small sample, so the skew measure is really not useful. . `53.00    72.50`
`76.50    77.35    83.50    99.00`

```
> var(grades)
[1] 139.1868
Mode is a bit more difficult in R.
```

```
getmode <- function(v) {
uniqv <- unique(v)
uniqv[which.max(tabulate(match(v, uniqv)))]
}

# Create the vector with numbers.
v <- c(2, 1, 2, 3, 1, 2, 3, 4, 1, 5, 5, 3, 2, 3)

# Calculate the mode using the user function.
result <- getmode(v)
print(result)

# Create the vector with Characters.
charv <- c(1, 1, 1, 2, 2, 3, 4, 5, 6, 6, 6, 6, 7)

# Calculate the mode using the user function.
result <- getmode(charv)
print(result)
[1] 6
```

3.5 Summary of Measures

Now let's summarize the interpretations of the statistics.

We can calculate various **measures of location** (center) for a set of data. The **mean** is the simple mathematical average of a set of two or more numbers. The mean for a given set of numbers can be computed in more than one way, including the arithmetic mean method, which uses the sum of the numbers in the series, and the geometric mean method. However, all the primary methods for computing a simple average of a normal number series produce the same approximate result most of the time.

Definition of the **"median"**

The median is the middle number (or average of the middle two numbers for n being even) in a sorted list of numbers. To determine the median value in a sequence of numbers, the numbers must first be arranged in value order from lowest to highest. If there is an odd number of numbers, the median value is the number that is in the middle, with the same number of numbers below and above. If there is an even number of numbers in the list, the middle pair must be determined, added together, and divided by two to find the median value. The median does not use all the data, as does the mean, but may be preferred if there are outliers. For symmetric data, the two measures are the same. The trimmed mean is the average of the "middle values," trimming off some percentage of the very small and very large numbers. It is "between" the median and mean in terms of how much data is used and the impact of outliers.

Mode refers to the most frequently occurring number found in a set of numbers. The mode is generally useful in describing where the graphical "peak" of a data set occurs or whether there is more than one peak. The mode is also the only measure that can be used with qualitative data.

We can also calculate various measures of spread for a set of data. The definition of "variance":

A measure of the dispersion of a set of data points around their mean value. Variance is a mathematical expectation of the average squared deviations from the mean. The square root of the variance is the definition of "standard deviation," which has units that are the same as the data.'.

A measure of the dispersion of a set of data from its mean. The more spread apart the data, the larger these deviations. A simple measure of spread is the **range**, the difference between the two extreme points of the data. When outliers are present in the data, the IQR (inter quartile range) is often used. The IQR is the difference between the 75th (Q3, third quartile) and 25th (Q1, first quartile) percentile of the data. When the median is used as the measure of center, the IQR is usually given as the measure of spread.

Measures that relate to the **shape of a distribution** are estimates of **skewness**. The standard deviation is calculated as the square root of variance.

3.5.1 Definition of "Skewness"

We generally describe asymmetry from the normal distribution in a set of statistical data. Skewness can come in the form of "negative skewness" or "positive skewness," depending on whether data points are skewed to the left (negative skew) or to the right (positive skew) of the data average.

3.5.2 Definition of "Mean"

The mean is the average value of a data set.

Definition of the "mode"

A statistical term that refers to the most frequently occurring number found in a set of numbers. The mode is found by collecting and organizing the data in order to count the frequency of each result. The result with the highest occurrences is the mode of the set.

3.5.3 Statistics and Measures Summary

Type of Data		Best Measures of Location	Best Measures of Dispersion	Measures of Shape
Quantitative	No outliers/skew	Mean, median, mode	Range, variance, standard. deviation, skewness	Skewness
	Outliers/skew	Median, trimmed mean	IQR	
Qualitative		Median, mode		

3.5.4 R (Retrieved Data and Descriptive Statistics)

All graph views indicate skewness (skewed left toward smaller numbers)

Statistics

```
> summary(Hght)
Min. 1st Qu. Median Mean 3rd Qu. Max.
59.00 62.00 67.50 66.66 71.00 75.00
> mean(Hght)
[1] 66.66
> median(Hght)
[1] 67.5
> var(Hght)
[1] 23.90245
> sd(Hght)
[1] 4.889013
# We obtain the coefficient of skewness using the shortcut formula
> sk<-3*(mean(Hght)-median(Hght))/sd(Hght)
> sk
[1] −0.5154415
> # Interpret as skewed left since sk <0
```

We get everything, but the mode, which we can see from real data
Put data in numerical order first.

71 65 67 61 69 60 62 67 68 68 73 64 69 71 59 63 63 66 61 70 69 59 72 66 69
[26] 65 69 71 59 73 75 63 59 71 60 62 72 73 72 62 75 62 67 71 59 71 69 72 60 69

We see 59 (5 times) 60 (3 times), 61 (2 times), . . .
In order is

59
59
59
59
59
60
60
60
61
61
62
62
62
62
63
63
63
64
65
65
66
66
67
67
67
68
68
69
69
69
69
69
69
69
70
71
71

71
71
71
71
72
72
72
72
73
73
73
75
75

We count the number of each value.

Hght	Count
59	5
60	3
61	2
62	4
63	3
64	1
65	2
66	2
67	3
68	2
69	7
70	1
71	6
72	4
73	3
74	0
75	2

We see 69 is repeated 7 times, so it is the mode.

3.6 Chapter 3 Exercises

3.1 For the sample data set {1, 2, 6}, find

a. $\sum x$
b. $\sum x^2$

c. $\sum(x-3)$

d. $\sum(x-3)^2$

3.2 For the sample data set {–1, 0, 1, 4}, find

a. $\sum x$

b. $\sum x^2$

c. $\sum(x-1)$

d. $\sum(x-1)^2$

3.3 Find the mean, median, and mode by hand for the samples below, and use R to confirm your results. Data: 1, 2, 3, 4.

3.4 Find the mean, median, and mode for the sample data: 3, 3, 4, 4.

3.5 Find the mean, median, and mode for the sample data: 2, 1, 2, 7.

3.6 Find the mean, median, and mode for the sample data: –1, 0, 1, 4, 1, 1.

3.7 Find the mean, median, and mode for the sample data represented by the table.

x	1	2	7
f	1	2	1

3.8 Batting average is the total number of hits divided by the total number of official at bats. Is batting average a mean? Explain.

3.9 Find the mean, median, and mode for the sample data represented by the table. The "rep" command in R may be useful (rep(3,4) will produce the vector 3,3,3,3).

x	–1	0	3	4
frequency	1	1	5	6

3.10 Create a sample data set of size $n = 3$ for which the mean $\bar{x}$ is greater than the median $\tilde{x}$.

3.11 Create a sample data set of size $n = 3$ for which the mean $\bar{x}$ is less than the median $\tilde{x}$.

3.12 Create a sample data set of size $n = 4$ for which the mean $\bar{x}$, the median $\tilde{x}$, and the mode are all identical.

3.13 Create a sample data set of size $n = 4$ for which the median $\tilde{x}$ and the mode are identical but the mean $\bar{x}$ is different.

Find the mean, median, and mode for the following applied data to discuss the differences in LDL cholesterol between those on a diet and those not on a diet in the sample:

3.14 LDL cholesterol levels in 10 patients (no diet): 132, 162, 133, 145, 148, 139, 147, 162, 150, 153.

3.15 LDL cholesterol levels in 10 patients on a diet: 1 27, 132, 138, 100, 132, 110, 131, 140, 135, 130

For problems 3.16–3.18, find the mean, median mode.

3.16 The frequency of the number of vehicles owned in a survey of 60 households is provided below. Compute the mean, median, and mode of this data. What do these values suggest about the shape of the distribution?

x	0	1	2	3	4	5	6
frequency	2	14	18	13	8	5	1

The "rep" command in R may be useful (rep(3,4) will produce the vector 3,3,3,3).

3.17 The number of passengers in 120 randomly selected vehicles during morning rush hours is provided below. Compute the mean, median, and mode of this data. What do these values suggest about the shape of the distribution?

x	1	2	3	4	5
frequency	80	33	3	3	1

The "rep" command in R may be useful (rep(3,4) will produce the vector 3,3,3,3).

3.18 Thirty boxes of 1 lb. penny nails were randomly selected by quality control, and the number of nails is counted with the following results. What do these values suggest about the shape of the distribution?

x	45	46	47	48	49	50	51
frequency	1	2	3	18	4	1	1

The "rep" command in R may be useful (rep(3,4) will produce the vector 3,3,3,3).

3.19 Five laboratory mice with thymus leukemia are observed for a predetermined period of 500 days. After 500 days, four mice have died, but the fifth one survives. The recorded survival times for the five mice are as follows:

493, 421, 222, 378, 500*,

where 500* indicates that the fifth mouse survived for at least 500 days, but the survival time (i.e., the exact value of the observation) is unknown.

a. Can you find the sample mean for the data set? If so, find it. If not, why not?

b. Can you find the sample median for the data set? If so, find it. If not, why not?

3.20 Five laboratory mice with thymus leukemia are observed for a predetermined period of 500 days. After 450 days, three mice have died, and one of the remaining mice is sacrificed for analysis. By the end of the observational period, the last remaining mouse still survives. The recorded survival times for the five mice are 222, 421, 378, 450*, 500* (see previous problem).

a. Can you find the sample mean for the data set? If so, find it. If not, why not?

b. Can you find the sample median for the data set? If so, find it. If not, why not?

3.21 Given the following stem and leaf from R, compute the mean, median, and mode.

```
The decimal point is 1 digit(s) to the right of the |
3 | 00
```

```
4  | 25688
5  | 02334467789
6  | 0122223445777788
7  | 0001124456667777889
8  | 011223457889
9  | 111123
10 | 00
```

3.22 A man tosses a coin repeatedly until it lands heads and records the number of tosses required. (For example, if it lands heads on the first toss, he records a 1; if it lands tails on the first two tosses and heads on the third, he records a 3.) The data are shown. Find the mean and median and compare.

x	1	2	3	4	5	6	7	8	9	10
frequency	384	208	98	56	28	12	8	2	3	1

a. Find the mean of the data.

b. Find the median of the data.

3.23 Construct a data set consisting of 10 numbers, all but one of which is above average, where the average is the mean. Is it possible to construct a data set as in part (a) when the average is the median? Explain.

3.24 Show that no matter what kind of average is used (mean, median, or mode), it is impossible for all members of a data set to be above average.

3.25 Twenty sacks of grain weigh a total of 1,003 lbs. What is the mean weight per sack?

Can the median weight per sack be calculated based on the information given? If not, explain why.

3.26 Find the range, variance, and standard deviation by hand for the following samples, and use R to confirm.

1, 2, 3, 4

3.27 Find the range, variance, and standard deviation for the following sample.

2, –3, 6, 0, 3, 1,

3.28 Find the range, variance, and standard deviation for the following sample.

2, 1, 2, 7

3.29 Find the range, variance, and standard deviation for the following sample.

–10, 14, 11, 20,

3.30 Find the range, variance, and standard deviation for the sample represented by the data frequency table.

x	1	2	3	5
f	1	2	1	1

3.31 Find the range, variance, and standard deviation for the sample represented by the data frequency table.

x	–1	0	1	4
f	1	1	3	1

3.32 Find the mean, median, mode, range, variance, and standard deviation, and use them to compare for the samples of 10 IQ scores randomly selected from two different schools for academically gifted students.

School 1: 132, 139, 162, 147, 133,160, 145,150, 148,153

School 2: 142, 139, 152, 147, 138, 155, 145, 150, 148, 153

3.33 Find the range, variance, and standard deviation for the sample of 10 IQ scores randomly selected from a school for academically gifted students.

142, 139, 152, 147, 138, 155, 145, 150, 148, 153

3.34 Use the data sets represented by the tables below to compute the sample mean and standard deviations:

x	0	2	4	6
f	8	4	5	3

3.35 Find the sample standard deviation for the data

x	10	12	14	16
f	8	4	5	3

3.36 A random sample of 24 invoices for meals at a local restaurant. The data are arrayed in the stem and leaf diagram shown. (Stems are tens of dollars, leaves are ones, so that, for example, the largest observation is 50).

```
The decimal point is 1 digit(s) to the right of the |

  1 | 000234
  2 | 011112367
  3 | 2345
  4 | 2355
  5 | 0
```

a. Compute the mean, median, and mode.

b. Compute the range.

c. Compute the sample standard deviation.

3.37 What must be true of a data set if its standard deviation is 0?

3.38 A data set consisting of 25 measurements has standard deviation 0. One of the measurements has value 17. What are the other 24 measurements?

3.39 Create a sample data set of size $n = 3$, for which the range is 0 and the sample mean is 2.

3.40 Create a sample data set of size $n = 3$ for which the sample variance is 0 and the sample mean is 1.

3.41 The sample {−1, 0, 1} has mean $\bar{x} = 0$ and standard deviation $s = 1.0$. Create a sample data set of size $n = 5$ for which the mean and standard at 0 and 1, respectively.

3.42 Create a sample data set of size $n = 3$ for which the mean $\bar{x} = 0$ and the standard deviation s is greater than 1.

3.43 The sample {−1, 0, 1} has mean $\bar{x} = 0$ and standard deviation $s = 0$. Create a sample data set of size $n = 3$ for which the mean is $\bar{x} = 0$ and the standard deviation s is less than 1.

3.44 The 1994 live birth rates per thousand population in the mountain states of Idaho, Montana, Wyoming, Colorado, New Mexico, Arizona, Utah, and Nevada were 12.9, 15.5, 13.5, 14.8, 16.7, 17.4, 20.1, and 16.4, respectively. What are the mean, variance, and standard deviation?

3.45 In five attempts, it took a person 11, 15, 12, 8, and 14 minutes to change a tire on a car. Find the mean, variance, and standard deviation.

3.46 A soldier is sent to the range to test a new bullet that the manufacturer says is very accurate. You send your best shooter with his weapon. He fires 10 shots using the standard ammunition and then 10 shots with the new ammunition. We measure the distance from the bull's eye to each shot location. Which appears to the better ammunition? Explain.

Standard ammunition: −3, −3, −1, 0, 0, 0, 1, 1, 1, 2

New ammunition: −2, −1, 0, 0, 0, 0, 1, 1, 1, 2

3.47 AGCT Scores: AGCT stands for Army General Classification Test. These scores have a mean of 100, with a standard deviation of 20.0. Here are the AGCT scores for a unit:

79, 100, 99, 83, 92, 110, 149, 109, 95, 126, 101, 101, 91, 71, 93, 103, 134, 141, 76, 108, 122, 111, 97, 94, 90, 112, 106, 113, 114, 117

Find the mean, median, mode, standard deviation, variance, and coefficient of skewness for the data. Provide a brief summary to your S-1 about this data in relation to the known values.

4

Classical Probability

Objectives

1. **Know the addition rules and axioms for probability.**
2. **Know unions and intersections.**
3. **Know independence.**
4. **Know conditional probability.**
5. **Know Bayes' Theorem and the Law of Total Probability.**

A disaster happens on a cruise ship in the Mediterranean Sea. The results are shown in Table 4.1.

TABLE 4.1

Disaster at Sea

	Men	Women	Boys	Girls	Total
Survived	332	318	29	27	706
Died	1,360	104	35	18	1,517
Total	1,692	422	64	45	2,223

One rule of a disaster at sea is rescue women and children first. Was this rule followed?

Some basic calculations, that we will show the formulas later for, reveal that only 19.6% of the men (332 out of 1,692 survived), and 70.4% of the woman and children survived (374 out of 531). Such simple calculations are powerful tools in analyzing information and providing insights into results.

4.1 Introduction to Classical Probability

Probability is a measure of the likelihood of a random phenomenon or chance behavior. Probability describes the long-term proportion with which a certain **outcome** will occur

DOI: 10.1201/9781003317906-4

in situations with short-term uncertainty. Probability deals with **experiments** that yield random short-term results or outcomes yet reveal long-term predictability.

The long-term proportion with which a certain outcome is observed is the probability of that outcome.

4.1.1 The Law of Large Numbers

> As the number of repetitions of a probability experiment increases, the proportion with which a certain outcome is observed gets closer to the probability of the outcome.

In probability, an **experiment** is any process that can be repeated in which the results are uncertain. A **simple event** is any single outcome from a probability experiment. Each simple event is denoted *ei.*

The **sample space,** S **,** of a probability experiment is the collection of all possible simple events. In other words, the sample space is a list of all possible outcomes of a probability experiment. An **event** is any collection of outcomes from a probability experiment. An event may consist of one or more simple events. Events are denoted using capital letters such as E.

Example 1: Consider the probability experiment of flipping a fair coin twice.

(a) Identify the simple events of the probability experiment.
(b) Determine the sample space.
(c) Define the event E = "have only one head."

Solution:

(a) Events for two flips: on each flip, there are two outcomes:
H = head
T = tail

We will introduce a tool helpful for probability problems: a tree diagram. The diagram for the two coin flip problem is shown in Figure 4.1. We begin from the left with a split shown by two "branches" representing the first coin flip. One branch is H for a head, the other T for tail, the two possible outcomes of the first flip. For each outcome on the first flip there are two subsequent branches representing the second flip. For the total of four paths through the tree, the sequence of flip outcome is shown at the end of the second branch.

The simple events for the two flip problem are the outcomes for each of the branches, such as HH.

A tree diagram, Figure 4.1, is created to assist for each flip

The sample space is the set of all possible outcomes or S = {HH, HT, TH, TT}.

We see that there are two simple events exactly having one head, so E = {HT, TH}.

Example 2: Flip a fair coin three times.

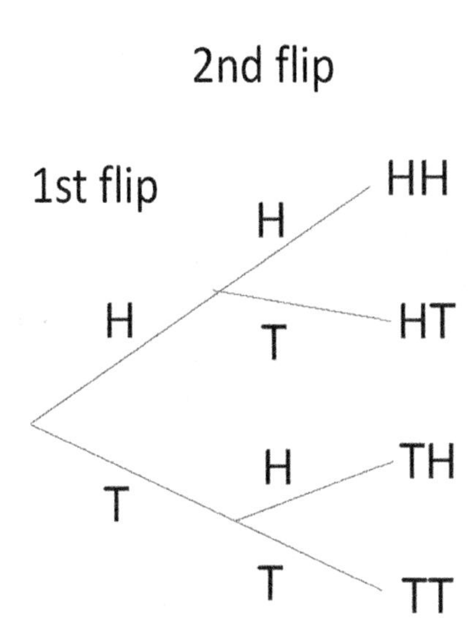

FIGURE 4.1
Tree diagram for two flips of a coin.

There are eight outcomes that you can easily obtain from a tree diagram. You will be asked to draw the tree diagram for this example as an exercise. The sample space then consists of outcomes such as HHH and is defined as:

S = {HHH, HHT, HTH, THH, TTH, THT, HTT, TTT}

Example 3: A fair coin is flipped once, and we also roll a fair die once. List the elements in the sample space.

Once again, a tree diagram can be drawn. Simple events have two values such as H1 for head on the coin and a 1 on the die. With two branches for the coin flip and then six subsequent branches after each for the die roll, the resulting sample space will have $2 \times 6 = 12$ outcomes:

S = {H1, H2, H3, H4, H5, H6, T1, T2, T3, T4, T5, T6}

Now, let's develop definition for the probability of an event. The **probability of an event**, denoted $P(E)$, is the likelihood of that event occurring. We define $P(E)$ as the number of ways that event E can happen divided by the total number of outcomes. Probabilities have properties that we must know.

Properties and Axioms of Probabilities

1. The probability of any event E, $P(E)$, must be between 0 and 1 inclusive. That is, $0 \leq P(E) \leq 1$
2. If an event is **impossible**, the probability of the event is 0.

3. If an event is a **certainty,** the probability of the event is 1.
4. If $S = \{e1, e2, \ldots, en\}$, then

$$P(e1) + P(e2) + \ldots + P(en) = 1$$

where S is the sample space and e_i are the events.

The classical method of computing probabilities requires *equally likely outcomes.* An experiment is said to have **equally likely outcomes** when each simple event has the same probability of occurring. An example of this is a flip of a fair coin where the chance of flipping a head is 1/2 and the chance of flipping a tail is 1/2.

Formally, if an experiment has n equally likely simple events and if the number of ways that an event E can occur is m, then the probability of E, $P(E)$, is

$$P(E) = \frac{\text{Number of way that } E \text{ can occur}}{\text{Number of Possible Outcomes}} = \frac{m}{n}$$

So, if S is the sample space of this experiment, then

$$P(E) = \frac{N(E)}{N(S)}$$

Example 4: Determine $P(E)$ for Example 1 (E = only one head in two flips).

We refer back to Example 1. We defined E, as one head in two flips; in Example 1, it is easily seen that the set consists of two events $E = \{HT, TH\}$. The total number of outcomes was four. So, $P(E)$ is the number of events with only one head divided by the total number of outcomes, or $= 2/4 = 1/2$.

The classical method of computing probabilities requires *equally likely outcomes.*

An experiment is said to have **equally likely outcomes** when each simple event has the same probability of occurring. An example of this is a flip of a fair coin where the chance of flipping a head is 1/2 and the chance of flipping a tail is 1/2.

If an experiment has n equally likely simple events and if the number of ways that an event E can occur is m, then the probability of E, $P(E)$, is

$$P(E) = \frac{\text{Number of way that } E \text{ can occur}}{\text{Number of Possible Outcomes}} = \frac{m}{n}$$

So, if S is the sample space of this experiment, then

$$P(E) = \frac{N(E)}{N(S)}$$

Example 5: Suppose a "fun size" bag of M&Ms contains nine brown candies, six yellow candies, seven red candies, four orange candies, two blue candies, and two green candies. Suppose that a candy is randomly selected.

(a) What is the probability that it is brown?
(b) What is the probability that it is blue?
(c) Comment on the likelihood of the candy being brown versus blue.

Solution:

(a) $P(\text{brown}) = 9/30 = 0.3$
(b) $P(\text{blue}) = 2/30 = 0.066666$
(c) Since there are more brown candies than blue candies, it is more likely to draw a brown candy than a blue candy.

In many of the previous examples, we provided a tree diagram in order to display the elements in a sample space. Very often, it is impractical to list and physically count them. Consider dealing 5 cards out of a deck of 52 cards. Imagine trying to draw a tree diagram to represent this experiment. It is therefore important to provide alternative methods to enumerate the sample space or an event. Approaches to this problem are known as counting methods, which we cover in the next section.

6! is a factor of 8!, so it will cancel, leaving $7 \cdot 8$ in the

numerator. We will have $\dfrac{7 \cdot 8}{1 \cdot 2} = 28$

To assist us with the rules for counting, we provide Figure 4.2.

4.2 Counting

Why counting methods?
Answer: Many times, the sample space is too large to list all the possible outcomes in S and then count them by hand. We need methods to assist us when the tree diagram becomes too large to handle!
Several questions can help in selecting the counting method to use:

(1) Is there replication (replacement)?
  (a) Yes, we use the multiplication rule described below.
  (b) No, then ask if there are distinct outcomes (order is important) or not (order is not important).
(2) Distinct outcomes require a permutation, while if there are no distinct outcomes, then we could use a combination.

To assist us with the rules for counting, we can use the flowchart in Figure 4.2.

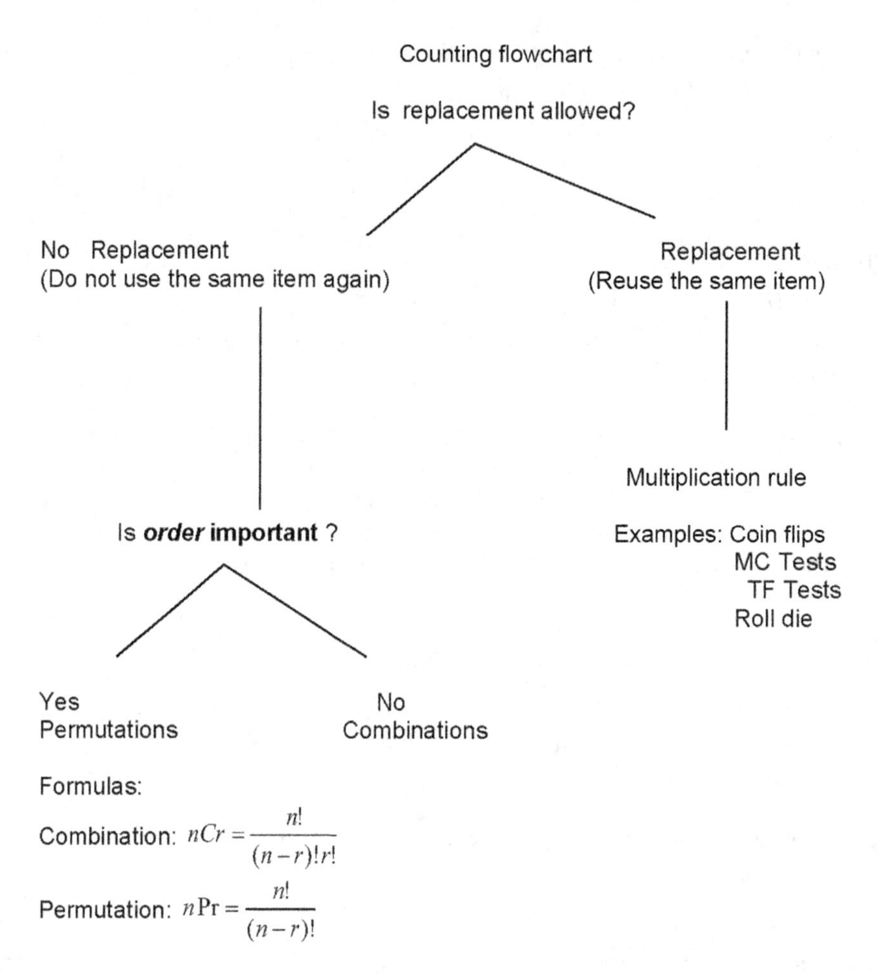

FIGURE 4.2
Counting flowchart.

When considering counting as a technique to get the number of outcomes in the sample space or the number of elements in event *A*, we need to consider a few fundamental questions so that we can apply the correct technique. Consider the following flowchart (Figure 4.2) as a guide:

4.2.1 The Multiplication Rule

Let us begin with the multiplication rule. Basically, an example of this rule occurs if we are determining the number of ways that we could serve three sandwiches, four different salads, and five types of chips. The answer is that we need only to multiply $3 * 4 * 5 = 60$ ways. We note that each sandwich gets paired with each salad and with each type of chips option. We do not discard anything, so there is replacement here. Another way to think of this is that if we pick sandwich A, we could pair it with chip B. For the next sandwich of type B, we could also use chip B. The chip type was "replaced" in the pool of options for use with the next sandwich.

As a matter of fact, the flip of a coin and the roll of a die are also a multiplication roll rule example. We can flip a H and then roll a 1. If we change the flip to a T, we still have the option to roll a 1, so we have replaced that value as a possible die roll. Thus, we apply the multiplication rule as 2 ways to flip a coin and 6 sides to a die, so 2 * 6 = 12 ways. Even the number of outcomes from flipping a fair coin m times is another multiplication rule problem with the answer: 2 * 2 * 2 * . . . *m times or 2^m possible outcomes. We already saw that there are 4 ways to flip a coin twice (2^2). We can use the result to determine the number of ways and of flipping a fair coin 3 times is 8 and flipping a fair coin 10 times is 2^{10} = 1,024.

4.2.2 Permutations and Combinations

The left side of the counting method diagram involves situations with no replacement. Let's consider a new example. We have a student club with leadership positions or committees to fill. Once a student is selected for a position or committee, they cannot be selected again. For example, if there is a committee with four people, Steve cannot be two of the people. Thus, there is no replacement. Once on the committee, Steve is not replaced in the pool for selection.

There is a second split on the left side of the tree based on ordering. Before we discuss the rules for no replacement, let's illustrate what is meant by distinct versus not distinct. First, in returning to our example, practical sense says that if we are considering a student club that will have the positions of president, vice president (VP), and secretary, then identifying those positions is a **distinct** reference. Thus, selection of Steve as president, Brittany as VP, and Jane as secretary is different than picking Jane as president, Brittany as VP, and Steve as secretary. Since the positions are distinct, the order the students are selected in matters. If we denote the students as S (Steve), J (Jane), and B (Brittany), then SJB as president, VP, and secretary is different than JSB, BJS, and so on.

On the other hand, if I am selecting a committee of three people without titles, the positions are not distinct. Each is a committee member. If I only want three students chosen for some reason, it would then make no difference the order of the students, as they are the same three students. Consider the three letters c, a, b. A committee selection of SJB and JSB is the same, as it is the same three students. Order does not matter in this case. If we have a distinct ordering, then we have ABC, ACB, BCA, BAC, CBA, CBA, and if we do not have a distinct ordering, then we have all the same three letters (CAB).

When we have or want distinct results, the order matters, and we use a permutation to compute the number of ways; when we want nondistinct results, order doesn't matter, and we use a combination.

To compute permutations and combinations, we need to first introduce the concept of a factorial.

Factorials

The symbol $n!$ ("n factorial") is defined as the product of consecutive numbers 1 through n.

$$\begin{aligned} n! &= 1 \cdot 2 \cdot 3 \cdots (n-2)(n-1)n \\ &= n(n-1)(n-2) \cdots 3 \cdot 2 \cdot 1 \end{aligned}$$

The order of the factors does not matter, whether backward or forward.

0! is defined as 1.

0! = 1 (The usefulness of this definition will become clear as we continue.)

Example 6: Compute 6!

$$6! = (6) \cdot (5) \cdot (4) \cdot (3) \cdot (2) \cdot (1) = 720$$

Example 7: 6! Is a factor of 10!

$$\begin{aligned} 10! &= (1) \cdot (2) \cdot (3) \cdot (4) \cdot (5) \cdot (6) \cdot (7) \cdot (8) \cdot (9) \cdot (10) \\ &= (6!) \cdot (7) \cdot (8) \cdot (9) \cdot (10) \end{aligned}$$

Example 8: Evaluate $\frac{8!}{5!}$.

$$\frac{8!}{5!} = \frac{1 \cdot 2 \cdot 3 \cdot 4 \cdot 5 \cdot 6 \cdot 7 \cdot 8}{1 \cdot 2 \cdot 3 \cdot 4 \cdot 5} = 6 \cdot 7 \cdot 8 = 336$$

Example 9: Evaluate $\frac{8!}{5!3!}$.

From Example 8, we see that 5!, a factor of 8!, cancels, leaving just the (6) · (7) · (8) in the numerator:

$$\frac{8!}{5!} = \frac{6 \cdot 7 \cdot 8}{1 \cdot 2 \cdot 3} = 7 \cdot 8 = 56$$

Permutations

We are now ready to discuss the counting methods for no replacement. We begin with a brief discussion of permutations.

A *permutation*, is denoted by **nPr**. A permutation answers the question: "From a set of **n** different items, how many ways can you select *and* order (arrange) **r** of these items?" One thing to keep in mind is that order is important or distinct when working with permutations.

Permutations are used for questions like "In how many ways could 10 runners end up on the Olympic medal stand (gold, silver, or bronze)?" Is order important? Yes, so we use:

nPr with **n = 10** and **r = 3**

The formula for a permutation is:

$$nPr = \frac{n!}{(n-r)!}$$

For the Olympic medal stand example, then, we would compute the number of possible medal stand orderings as:

$$\frac{10!}{(10-3)!} = 10 \cdot 9 \cdot 8 = 720$$

4.2.3 Combinations

When order doesn't matter, we use a *combination*, denoted by **nCr**. Combinations answer questions like "From a set of **n** different items, how many ways can you select (independent or order) **r** of these items?" Order is not important with combinations.

An example of a combination question is "A subcommittee made up of 3 people must be selected from a group of 10." Is order important? No; subcommittee of three people from a group of 10, n = 10, r = 3. Thus, we use **nCr** with **n = 10** and **r = 3**

The formula for a combination is:

$$nCr = \frac{n!}{r!(n-r)!}$$

The combination formula is often read "*n* choose *r*."

For the committee example, the number of different committees is then "10 choose 3":

$$\frac{10!}{3!(10-3)!} = \frac{10 \cdot 9 \cdot 8}{3 \cdot 2 \cdot 1} = 10 \cdot 3 \cdot 4 = 120$$

Looking at these two examples, are there more possible permutations or combinations? Clearly, there are more permutations. Why? The combinations ABC and BCA are not considered different; they are the same committee members. However, they are different if order matters, leading to more permutations.

4.2.4 Computing Permutations and Combinations in R

Several packages in R perform more advanced counting operations. We will use base commands to help us with the computation of the formulas.

A factorial in R is computed:

```
> factorial(3)
Combination in R:
```

A combination in R is computed using the "choose" function. For the example the above ${}_{10}C_3$ is:

```
> choose(n = 10, k = 3)
[1] 120
```

Permutation in R

A permutation in R is computed by noting that the permutation formula is the formula for a combination removing the *r*! term in the denominator. Thus, for our example above, ${}_{10}P_3$ is:

```
> choose(10, 3) * factorial(3)
[1] 720
```

We can use permutations and combinations to help us answer more complex probability questions.

Example 10: Four-digit pin.

A four-digit PIN is selected. How many PINs have no repeated digits? What is the probability of a random PIN having no repeated digits?

There are 10 possible values for each digit of the PIN (namely: 0, 1, 2, 3, 4, 5, 6, 7, 8, 9), so there are 10 × 10 × 10 × 10 = 10^4 = 10,000 total possible PINs.

To have no repeated digits, all four digits would have to be different, which is selecting digits from the possible numbers without replacement. Further, the order matters: the pin 8765 is not the same as 5678. Thus, we have 10 × 9 × 8 × 7, or a permutation:

$${}_{10}P_4 = 5{,}040$$

To find the probability of such a PIN, we need to find the total number of possible PINs. Suppose we draw a 6 for the first digit of the PIN. Can we draw a 6 again for the second digit? Yes. Thus, we are selecting values with replacement from the set of 10 possible numbers and we can compute the total possible using the multiplication rule:

$$10 \cdot 10 \cdot 10 \cdot 10 = 10{,}000$$

The probability is then:

$$\frac{N(E)}{N(S)} = \frac{5040}{10000} = 0.504$$

The probability of no repeated digits is the number of four-digit PINs with no repeated digits divided by the total number of four-digit PINs. This probability is

$${}_{10}P_4/10^4 = 5040/10000 = 0.504$$

Example 11: Lottery.

In a certain state's lottery, 48 balls numbered 1–48 are placed in a machine, and six of them are drawn at random. If the six numbers drawn match the numbers that a player had chosen, the player wins $1,000,000. In this lottery, the order the numbers are drawn in doesn't matter. Compute the probability that you win the million-dollar prize if you purchase a single lottery ticket.

In order to compute the probability, we need to count the total number of ways six numbers can be drawn, and the number of ways the six numbers on the player's ticket could match the six numbers drawn from the machine. Since there is no stipulation that the numbers be in any particular order, the number of possible outcomes of the lottery drawing is:

$${}_{48}C_6 = 12{,}271{,}512$$

Of these possible outcomes, only one would match all six numbers on the player's ticket, so the probability of winning the grand prize is:

$${}_6C_6/{}_{48}C_6 = 1/12{,}271{,}512 \approx = 0.0000000815$$

Example 12: Lottery II.

In the state lottery from the previous example, if 5 of the 6 numbers drawn match the numbers that a player has chosen, the player wins a second prize of $1,000. Compute the probability that you win the second prize if you purchase a single lottery ticket.

As above, the number of possible outcomes of the lottery drawing is ${}_{48}C_6 = 12{,}271{,}512$. In order to win the second prize, 5 of the 6 numbers on the ticket must match 5 of the 6 winning numbers; in other words, we must have chosen 5 of the 6 balls containing winning numbers and 1 of the 42 balls with losing numbers. The number of ways to choose 5 out of the 6 winning numbers is given by ${}_6C_5 = 6$ and the number of ways to choose 1 out of the 42 losing numbers is given by ${}_{42}C_1 = 42$. Thus, the number of favorable outcomes is then given by the basic counting multiplication rule: ${}_6C_5 \times {}_{42}C_1 = 6 \times 42 = 252$. So the probability of winning the second prize is $P(\text{winning second prize}) = 6/252/12{,}271{,}512 \approx= 0.00002053809$.

Example 13: Drawing 5 cards.

Compute the probability of randomly drawing 5 cards from a deck and getting exactly one ace.

In many card games (such as poker), the order in which the cards are drawn is not important (since the player may rearrange the cards in his hand any way he chooses); in the problems that follow, we will assume that this is the case unless otherwise stated. Thus, we use combinations to compute the possible number of 5-card hands, ${}_{52}C_5$. This number will go in the denominator of our probability formula, since it is the number of possible outcomes.

For the numerator, we need the number of ways to draw 1 ace and 4 other cards (none of them aces) from the deck. Since there are 4 aces and we want exactly 1 of them, there will be ${}_4C_1$ ways to select 1 ace; since there are 48 non-aces and we want 4 of them, there will be ${}_{48}C_4$ ways to select the 4 non-aces. Now we use the basic counting multiplication rule to calculate that there will be ${}_4C_1 \times {}_{48}C_4$ ways to choose 1 ace and 4 non-aces.

Putting this all together, we have

$$P(1 \text{ ace}) = ({}_4C_1)\ ({}_{48}C_4)\ /\ {}_{52}C_5 = 778{,}320/2{,}598{,}960 \approx 0.299$$

In R:

```
> choose(4,1)*choose(48,4)/choose(52,5)
[1] 0.2994736
```

Example 14: Two aces in a draw of 5 cards.

Compute the probability of randomly drawing 5 cards from a deck and getting exactly 2 aces.

The solution is similar to the previous example, except now we are choosing 2 aces out of 4 and 3 non-aces out of 48; the denominator remains the same:

$$P(2 \text{ aces}) = ({}_4C_2)\ ({}_{48}C_3)\ /\ {}_{52}C_5$$

```
> choose(4,2)*choose(48,3)/choose(52,5)
[1] 0.03992982
```

It is useful to note that these card problems are remarkably similar to the lottery problems discussed earlier.

Example 15: Birthday Problem part one.

Let's pause to consider a birthday problem, which is a famous problem in probability theory.

Suppose you have a room full of 30 people. What is the probability that there is at least one shared birthday?

Take a guess at the answer to the above problem. Was your guess fairly low, like around 10%? That seems to be the intuitive answer (30/365, perhaps?). Let's see if we should listen to our intuition. Let's start with a simpler problem, however.

Example 16: Birthday problem revisited.

Suppose three people are in a room. What is the probability that there is at least one shared birthday among these three people?

There are a lot of ways there could be at least one shared birthday. Fortunately, there is an easier way. We ask ourselves, "What is the alternative to having at least one shared birthday?" In this case, the alternative is that there are no shared birthdays. In other words, the alternative to "at least one" is having none. In other words, since this is a complementary event (we will discuss this formally later),

$$P(\text{at least 1}) = 1 - P(\text{none})$$

We will start, then, by computing the probability that there is no shared birthday. We can get the denominator using the multiplication rule since when picking each person's birthday, all 365 birthdays are available; they could be replicated. The total number of possible sets of three birthdays is then:

$$N(\text{3 birthdays}) = (365)(365)(365)$$

To obtain the numerator, let's imagine that you are one of these three people. Your birthday can be anything without conflict, so there are 365 choices out of 365 for your birthday. What is the probability that the second person does not share your birthday? There are 365 days in the year (let's ignore leap years), and removing your birthday from contention, there are 364 choices that will guarantee that you do not share a birthday with this person, so the probability number of possible birthdays that the second person does not share your birthday is 364/365. Now we move to the third person. How many birthdays are there that ensure the third person does not have the same birthday as either you or the second person? There are 363 days that will not duplicate your birthday or the second person's, so the probability that the third person does not share a birthday with the first two is 363/365.

We want the second person not to share a birthday with you *and* the third person not to share a birthday with the first two people, so we use the multiplication rule to get the numerator as:

$$N(\text{3 different birthdays}) = (365)(364)(363)$$

We can now compute the probability of no shared birthday as:

$$\frac{365(364)(363)}{365(365)(365)} \approx 0.9918$$

$P(\text{no shared birthday}) = 365/365 \cdot 364/365 \cdot 363/365 \approx 0.9918$

And if we then subtract from 1 we to get $P(\text{shared birthday}) = 1 - P(\text{no shared birthday}) = 1 - 0.9918 = 0.0082$.

This is a pretty small number, so maybe it makes sense that the answer to our original problem will be small. Let's make our group a bit bigger.

Example 17: Birthday problem part 2.

Suppose five people are in a room. What is the probability that there is at least one shared birthday among these five people?

Continuing the pattern of the previous example, the answer should be

$$1 - \frac{365(364)(363)(362)(361)}{365(365)(365)(365)(365)} \approx 0.0271$$

P(shared birthday) = 1 – 365/365·364/365·363/365·362/365·361/365 ≈ 0.0271.

Note that the numerator is just a permutation, so we could rewrite this more compactly as

$$P(\text{shared birthday}) = 1 - ({}_{365}P_5 \;/\; 365^5) \approx 0.0271,$$

which makes it a bit easier to type into a calculator or computer and which suggests a nice formula as we continue to expand the population of our group.

Full birthday problem solution:

We return to the original problem and suppose 30 people are in a room. What is the probability that there is at least one shared birthday among these 30 people?

Here we can calculate

$$P(\text{shared birthday}) = 1 - ({}_{365}P_{30} \;/\; 365^{30}) \approx 0.706$$

which gives us the surprising result that when you are in a room with 30 people, there is a 70% chance that there will be at least one shared birthday!

If you like to bet, and if you can convince 30 people to reveal their birthdays, you might be able to win some money by betting a friend that there will be at least two people with the same birthday in the room anytime you are in a room of 30 or more people. (Of course, you would need to make sure your friend hasn't studied probability!) You wouldn't be guaranteed to win, but you should win more than half the time.

This is one of many results in probability theory that is counterintuitive; that is, it goes against our gut instincts. If you still don't believe the math, you can carry out a simulation. Just so you won't have to go around rounding up groups of 30 people, someone has kindly developed a Java applet so that you can conduct a computer simulation. Go to this web page: *http://www-stat.stanford.edu/~susan/surprise/Birthday.html*, and once the applet has loaded, select 30 birthdays and then keep clicking Start and Reset. If you keep track of the number of times that there is a repeated birthday, you should get a repeated birthday about 7 out of every 10.

4.3 Chapter 4 Exercises

4.1 Compute the following.

a. 0!

b. 1!

c. 11!

d. 4 52!

e. 11! – 7!

f. 20!

g. ${}_{10}P_2$

h. ${}_{20}P_5$

i. ${}_{10}C_2$

j. ${}_{20}C_5$

4.2 A multiple-choice question on an economics quiz contains 10 questions with five possible answers each. Compute the probability of randomly guessing the answers and getting exactly 9 questions correct.

4.3 Compute the probability of randomly drawing five cards from a deck of cards and getting three aces and two kings.

4.4 Suppose 10 people are in a room. What is the probability that there is at least one shared birthday among these 10 people?

4.5 Draw the tree diagram for Example 2.

4.6 Draw the tree diagram for Example 3.

4.7 If Sam has four shirts, five pairs of trousers, and six ties, assuming everything matches, how many ways can he wear a shirt, trousers, and a tie?

4.8 A student club has 20 members; 8 are males, and 12 are females. We want to elect a president and vice president. Find the probability that the president is a girl and the vice president is a male.

4.9 Compute the following:

(a) 10! / 7!

(b) 12! / (10! 2!)

4.10 How many ways can we build a computer?

In putting together a computer if we have the following choices for components:

monitors (13″, 15″, 17″, 19″)

processors (500 MHz, 600 MHz, 800 MHz)

RAM (128, 256, 512)

Memory (5 GB, 10 GB, 20 GB, 40 GB)

How many ways can we build a computer?

4.11 Joan is gambler. Her game is flipping a coin and matching the coin (heads to heads or tails to tails). If they do not match, then she loses. She plays this game three times. List the outcomes. How many outcomes are there?

4.12 Psychology gives a 20-question true-or-false exam. How many ways can students provide answers assuming that they answer every question.

4.13 There are 11 members in the club, we want to select 4 of the members.

a) How many ways can we select a president, vice president, secretary, and treasurer for the club at your college?

b) How many ways can we select the 4 to go to a conference in Charlotte?

c) If 5 members are male and 6 members are female, how many ways can 2 of each be selected to attend the conference?

d) What is the probability of 2 of each attending the conference?

4.14 Determine the following:

a) 7!

b) 25! / 20!

c) ${}_{12}P_4$

d) ${}_7P_7$

e) 0!

4.15 There are 25 members in the math club; we want to select 4 of the members.

a) How many ways can we select a president, vice president, secretary, and treasurer for the math club at your college?

b) How many ways can we select the 4 to go to a conference in Charlotte?

c) How many ways can we select 6 to go to a conference?

d) If 15 members are female and 10 members are male, how many ways can 3 females and 3 males be selected to attend the conference?

e) What is the probability of 3 females and 3 males being selected to attend the conference?

4.16 Calculate the following values from their expression:

a. ${}_{10}P_4$

b. ${}_{10}C_4$

c. $\binom{8}{4}$

d. ${}_{5}P_5$

e. 5!

f. 11! / (11 − 6)!

0!

4.17 State whether each of the following is true (T) or false (F):

a. A tree diagram is used to obtain a list all the possible outcomes or to count the number of possible outcomes.

b. Counting is used when the number of outcomes is too large to draw a tree diagram.

c. The number of four-letter permutations from the letters in the word *TALK* is 24.

d. The number of four-letter permutations of the letters in the word *SEEN* is 24.

4.18 Consider households in the Pee Dee that own two vehicles. Assume that cars are purchased on an equally likely basis. We want to know about these cars being American, Japanese, German, or Korean engineering.

Make a tree diagram to list all the elements in the sample space.

Let's define the following events and then determine the probability of the event.

a. Event A = exactly one car is an America car

b. Event B = at least one car is a Japanese car

c. Event C = no cars are American cars

d. Event D = one car is Korean

e. Event E = both cars are German

Determine the elements in each event.

Find the probabilities of

P(Event A)

P(Event B)

P(Event C)

P(Event D)

P(Event E)

4.19 A probability experiment is conducted. Which of the following CANNOT be considered a probability of an outcome?

a) 1/3, b) −0.59, c) 1, d) −1/2, e) 0, f) 1.45, g) 122, h) 0.8123, i) 23.75%, j) 133%, k) 24/20, l) 5.0E−02

4.20 In each situation given below, first clearly circle the correct response. Determine if there is replacement (R) or no replacement (NR). Next tell the method you will use to solve the problem: general multiplication rules (M), permutation (P), or Combination (C). Finally, calculate the number of ways.

a. Twelve people want to play in a bingo game. There are only four seats available. How many ways can 4 of these 12 people be seated?

b. You flip a fair coin 10 times. How many ways can this be done?

c. You roll a die five consecutive times. How many ways can this be done?

d. You flip a coin twice and then roll a die. How many ways can this be done?

e. Jack has five dress shirts, four sport coats, four ties, and six pairs of trousers. How many ways can he wear a shirt, a coat, trousers, and a tie if they all can match?

f. In a board of directors composed of eight people, how many ways can we select and name the CEO, director, and treasurer if the same person cannot be selected to more than one position?

g. How many four-letter permutations can be made from the letters in the word *hexagons*?

h. How many permutations can be formed from all the letters in the word *Mississippi?*

i. How many distinct student ID cards can be made if there are nine digits (1, 2, 3, 4, 5, 6, 7, 8, 9) and no digit can be used more than once?

j. How many different student ID cards can be made if the digits (1, 2, 3, 4, 5, 6, 7, 8, 9) can be repeated?

k. A police chief has 21 cases and 15 detectives. How many ways can the cases be assigned if only one case can be assigned to each detective and the order of cases does not matter?

l. How many ways can four people be selected out of eight people to serve on a committee?

m. You give out nine essay questions and a student can do any five of them. How many ways can this be done if order does not matter?

n. In a train yard there are seven tank cars, 11 boxcars, and five flatcars. How many trains can be made up consisting of three tank cars, 4 boxcars, and two flatcars?

o. How many ways can a 15-question true-or-false exam be answered?

p. We are playing the SC lottery, Pick 4. Numbers from 0 to 9 are used {0, 1, 2, 3, 4, 5, 6, 7, 8, 9}. It is possible to have all the same number such as 1, 1, 1, 1 as a winning number. How many possible outcomes are there?

Read the following questions carefully before answering.

4.21 You have 15 marbles in a bag of which 6 are green and 9 are blue. You want to select 2 marbles from the bag one at a time without replacement. (Assume order does not matter).

(a) How many ways can we select 2 marbles from the bag of marbles?

(b) How many ways can both marbles drawn be blue?

(c) What is the probability that both marbles were blue?

(d) How many ways can you draw 1 marble of each color? What is the probability of drawing 1 marble of each color?

4.22 You have nine marbles in a bag of which five are green and four a blue. You want to select two marbles one at a time from the bag without replacement, recording the color as they are drawn (so now order matters).

(a) How many ways can this be done?

(b) How many ways can the first marble drawn be green?

(c) How many ways can the second marble be green (think)?

(d) What is the probability that both marbles drawn were green?

4.23 Consider the following problem. We have 35 students in class, and we want to select 5 to represent the need for statistics to the psychology department. Within the class, there are 14 males and 21 females. Find the probability that exactly 3 males are chosen.

4.24 You are playing cards with friends. You want to tell them how hard it is to get 3 aces and 2 8s in a card hand dealt by the dealer. What is the probability of being dealt this hand if you use a standard 52-card deck and deal 5 cards?

4.25 A woman has three skirts, five blouses, and four scarves. How many different outfits can she wear, assuming that they are all color-coordinated?

4.26 How many five-digit ZIP codes are possible if digits can be repeated?

4.27 How many five-digit ZIP codes are possible if digits cannot be repeated?

4.28 How many ways can six speakers be seated in a row on a speaker's platform?

4.29 How many ways can a baseball manager arrange a batting lineup with nine players?

4.30 Three dice are rolled. How many outcomes can be there?

4.31 There are eight different statistics books, six different probability books, and three different probability and statistics books. A student must select one of each. How many different ways can this be done?

4.32 A coin is flipped two times. List the sample space. How many ways were there?

4.33 Ten students are finalists for the positions of president, vice president, and secretary for a class. How many ways can this be done?

4.34 At Burger King, you can get a whopper your way. The choices of topping are ketchup, mayonnaise, onion, tomato, lettuce, and cheese. How many ways can you order the whopper, assuming you get the burger and the bun included?

4.35 A coin is tossed eight times. How many ways can this be done?

4.36 The call letters of a radio station must begin with a *K* or a *W* and have four letters. How many different call letters can be made if replications of letters are allowed? If replications of letters are not allowed?

4.37 Make a tree diagram to determine the possible genders of a family with three children.

4.38 How many ways can a person select one or more coins if he has two nickels, one dime, and one quarter?

4.39 In a barnyard there is an assortment of chickens and cows. Counting heads, one gets 15; counting legs, one gets 46. How many of each are there?

4.4 Probability from Data

The probability of an event E is approximately the number of times event E is observed divided by the number of repetitions of the experiment:

$$P(E) \approx \textbf{relative frequency of } E = \frac{\textbf{frequency of } E}{\textbf{number of trials of experiment}}$$

Now, lets' return to our disaster at sea on the cruise ship. We can use this method to compute the probabilities. Recall Table 4.1:

	Men	Women	Boys	Girls	Total
Survived	332	318	29	27	706
Died	1,360	104	35	18	1,517
Total	1,692	422	64	45	2,223

$P(\text{Survived the attack}) = 706/2{,}223 = 0.3176$

$P(\text{Died}) = 1{,}517/2{,}223 = 0.6824$

$P(\text{Woman and children survived}) = (318+29+27)\ /\ (422+64+45) = 374/531 = 0.704$

$P(\text{Men survived}) = 332/1{,}692 = 0.196$

We see that women and children do appear to have been rescued first. We will return to this table of data later.

We now consider some additional important aspects of computing probabilities.

4.4.1 Intersections and Unions

Now, let suppose E and F are two events.

- ***E* and *F*** is the event consisting of simple events that belong to both E and F. The notation is $\cap$ (intersection), $E \cap F$.
- ***E* or *F*** is the event consisting of simple events that belong to either E or F or both.
- The notation is $\cup$ (union), $E \cup F$.

Example 18: Dice rolls.

Suppose that a pair of die is thrown. Let the two events be:

E = "the first die is a two"

F = "the sum of the dice is less than or equal to five"

Find $P(E \cap F)$ and $P(E \cup F)$ directly by counting the number of ways E or F could occur and dividing this result by the number of possible outcomes.

We will solve this problem by explicitly writing the entire sample space for the roll of two dice, as shown in Figure 4.3.

The sets of rolls that define the two events are as follows:

Event $E = \{(2,1), (2,2), (2,3), (2,4), (2,5), (2,6)\}$

Event $F = \{(1,1), (1,2), (1,3), (1,4), (2,1), (2,2), (2,3), (3,1), (3,2), (4,1)\}$

There are 36 total outcomes above, so we can compute the probabilities of the events:

$P(E) = 6/36 = 1/6$

$P(F) = 10/36 = 5/18$

The intersection can be found by identifying the rolls that are in E and in F. There are three:

$$(E \cap F) = \{(2,1), (2,2), (2,3)\}$$

For the union, we look at the rows in the table with sums five or less (event F), 10 rolls:

2	(1,1)			
3	(1,2)	(2,1)		
4	(1,3)	(2,2)	(3,1)	
5	(1,4)	(2,3)	(3,2)	(4,1)

Roll total							Total ways	Relative frequency
2	(1,1)						1	1/36
3	(1, 2)	(2,1)					2	2/36
4	(1,3)	(2,2)	(3,1)				3	3/26
5	(1,4)	(2,3)	(3,2)	(4,1)			4	4/36
6	(1,5)	(2,4)	(3,3)	(5,1)	(4,2)		5	5/36
7	(1,6)	(2,5)	(3,4)	(4,3)	(5,2)	(6,1)	6	6/36
8	(2,6)	(3,5)	(4,4)	(5,3)	(6,2)		5	5/36
9	(3,6)	(4,5)	(5,4)	(6,3)			4	4/39
10	(4,6)	(5,5)	(6,4)				3	3/36
11	(5,6)	(6,5)					2	2/36
12	(6,6)						1	1/36
Total								1

FIGURE 4.3
Outcomes for a roll of a pair of die.

We then add the rolls from other rows that have a "2" for die one. There are three additional rolls: (2,4), (2,5), and (2,6).

The union of the events is then:

$$(E \cup F) = \{(1,1), (1,2), (1,3), (1,4), (2,1), (2,2), (2,3), (3,1), (3,2), (4,1), (2,4), (2,5), (2,6)\}$$

The probabilities are then:

$$P(E \cap F) = 3/36 = 1/12$$
$$P(E \cup F) = 13/36$$

4.4.2 The Addition Rule

In the previous example, we counted 13 elements in the union of E and F. However, there were 6 outcomes in event E and 10 in event F for 16 total events. Thus, if we had just added the outcomes for the two events, we would overestimate the union. Why? The reason is that there are outcomes that are in both E and F, namely the rolls: (2,1), (2,2), and (2,3). If we added, we would double count these three events! The addition rule provides a formula for the union that avoids double counting.

For any two events E and F,

$$P(E \text{ or } F) = P(E) + P(F) - P(E \text{ and } F)$$
$$P(E \cup F) = P(E) + P(F) - P(E \cap F)$$

Checking this formula with our example,

$$P(E \cup F) = 6/36 + 10/36 - 3/36 = 13/36$$

Example 19: Newspapers to introduce Venn diagrams.

Let's consider the following example. Let event A be the event that a student in the dorms takes the local newspaper, and let event B be the event that a student in the dorms takes the *USA Today*. There are 1,000 students living in the dorms, and we know 750 take the local paper, and 500 take *USA Today*. We are told 450 take both papers. We are interested in the probability a student takes a paper.

The event that a student takes a paper is the union, they take the local newspaper or the USA today. We can compute probabilities from the given information:

$$P(A \cap B) = 450/1000 = 0.45$$
$$P(A) = 0.75$$
$$P(B) = 0.50$$

We then can find the union using the addition formula, $P(A \cup B) = -P(A) + P(B) - P(A \cap B)$.

$$P(A \cup B) = 0.75 + 0.50 - 0.45 = 0.8$$

Thus, 80% of the students take at least one of the two newspapers.

Computations such as those in this example are sometimes made easier with a visualization of the situation. A commonly used tool is a Venn diagram.

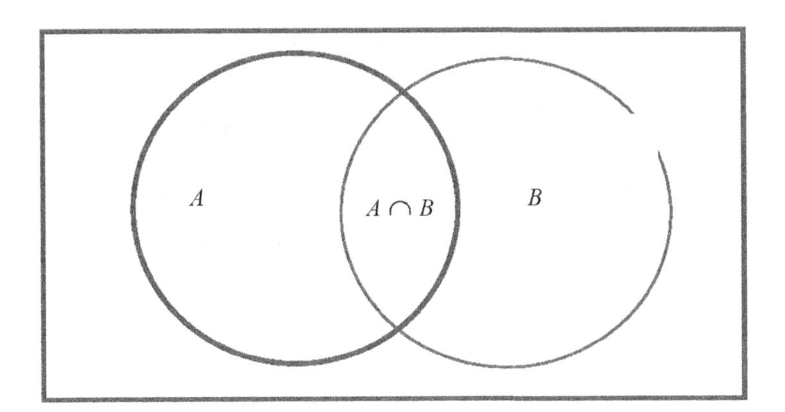

FIGURE 4.4
Venn diagram for events A and B.

Venn diagrams represent events as circles enclosed in a rectangle. The rectangle represents the sample space, and each circle represents an event. The overlap in circles is then the intersection of the events. A typical Venn diagram of two events is shown in Figure 4.4.

4.4.3 Rule for Mutually Exclusive Sets

If events E and F have no simple events in common or cannot occur simultaneously, they are said to be **disjoint** or **mutually exclusive**, so their intersection does not exist:

$$E \cap F = \varnothing \text{ (the null set)}$$

4.4.4 Addition Rule for Mutually Exclusive Events

In the case when events E and F are mutually exclusive events, then we can simply add the probabilities of the events since there are no outcomes in common to double count. The addition rule becomes:

$$P(E \text{ or } F) = P(E) + P(F)$$

In general, if $E, F, G, \ldots$ are mutually exclusive events, then

$$P(E \text{ or } F \text{ or } G \text{ or } \ldots) = P(E) + P(F) + P(G) + \ldots$$

The addition rule for mutually exclusive events is made very clear from the Venn diagrams for such events illustrated in Figure 4.5.

4.4.5 Complement Rule

Let S denote the sample space of an experiment, and let E denote an event., The complement of E or "not E," denoted as E', is all the simple events in the sample space S that are not simple events in the event E.

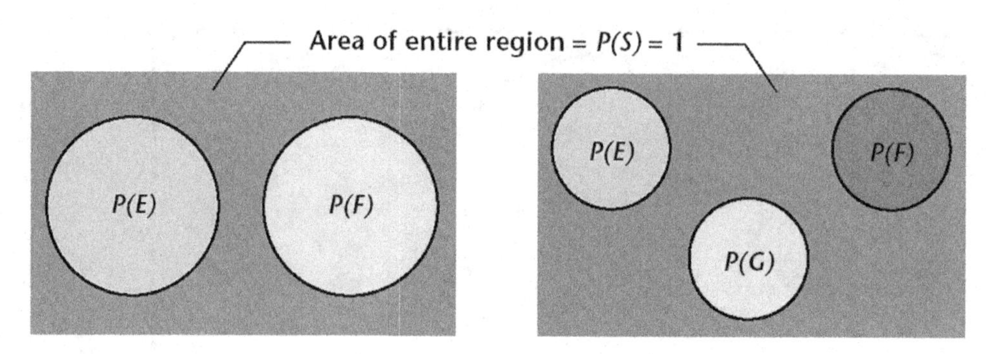

FIGURE 4.5
Venn diagram for mutually exclusive events.

The complement rule for an event E is an event not E, written as E', is that

$$P(E) + P(E'E') = 1$$
or
$$P(E'E') = 1 - P(E)$$

Example 20: Consider a roll of a single die.
The sample space, let S = {1, 2, 3, 4, 5, 6}.
Let event A be the roll in an even number.

$$A = \{2, 4, 6\}$$

The event $A'A'$ would be {1, 3, 5}.

The following is a useful method for determining probabilities with Venn diagrams and involves shading the regions of interest. The diagrams in Figure 4.6 illustrate all the above various set operations by shading the Venn diagrams: complement, intersection, and union. Venn diagrams are also very useful to find probabilities. Here, sets are represented by simple plane areas and U, the universal set, by the area in the entire rectangle.

Example 21: Newspaper II.
We wish to create a Venn diagram for the newspaper example. The way to build the diagram, after the circles are drawn, is from the "inside out." We start with the intersection, which has probability 0.45. For event A, the probability is 0.75. Thus, the probabilities in that circle must sum to 0.75. With 0.45 already in circle A for the intersection, this means we have 0.3 in the rest of circle A. Similar logic leads to placing 0.05 in the portion of circle B outside the intersection.

Finally, the probability outside the circles is calculated by first summing the total probability in the two circles: 0.3 + 0.45 + 0.05 = 0.8. The probability outside the circles must be 0.2, so the total probability in the sample space is 1. The Venn diagram for the problem would look like in Figure 4.7.

The following probabilities can be used or found from the Venn diagram. We always start filling in probabilities from inside the intersection of the events and move our way out. The sum total of all probabilities within the Venn diagram rectangle, S, the sample set is 1.0.

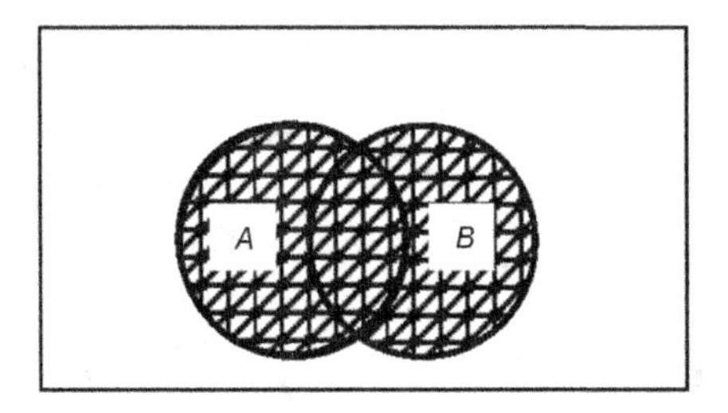

(a) $A \cup B$ is shaded.

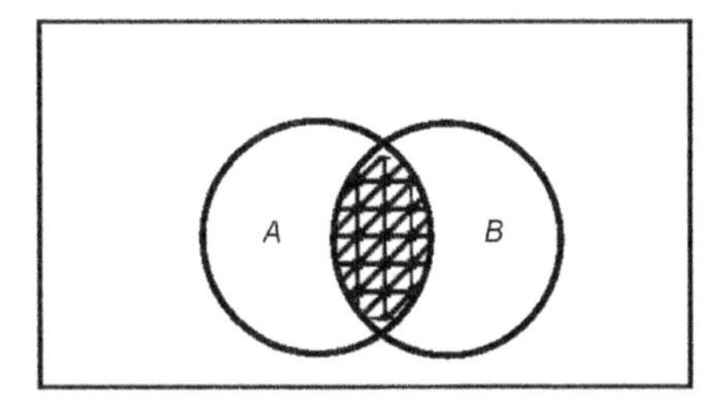

(b) $A \cap B$ is shaded.

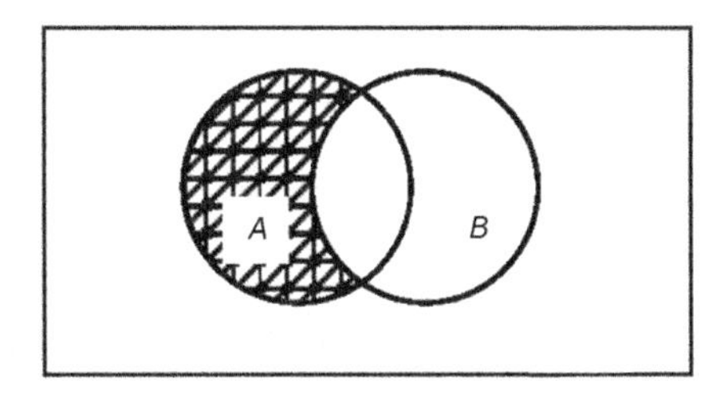

(c) $A - B$ is shaded. This is only A

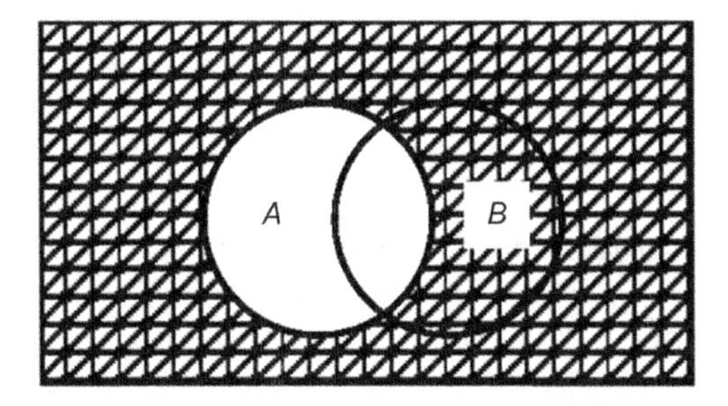

(d) A' A' is shaded. Everything that is NOT A is shaded.

FIGURE 4.6
Venn diagrams for (a) Union, (b) Intersection, (c) *A* only, (d) not *A*.

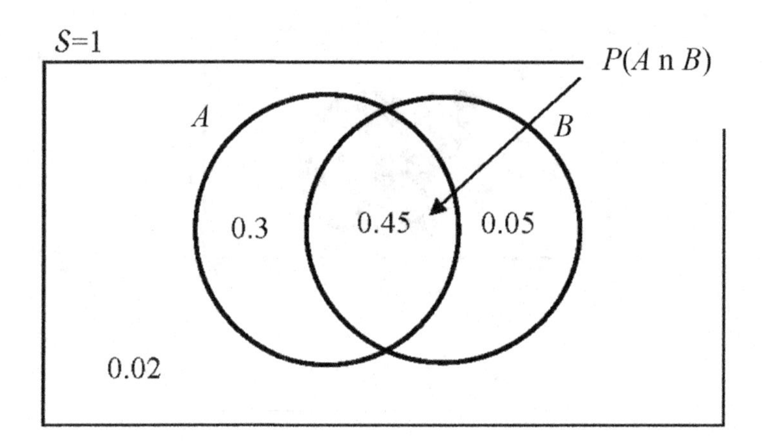

FIGURE 4.7
Venn diagram for newspaper II example.

Example 22: Newspaper II.

Let's revisit the newspaper example. We can use the diagram to easily compute various probabilities. For example, the union is the sum of values in the circles as shown in the shaded area of Figure 4.7. What are the probabilities y that a student only takes one paper and the probability that a student does not take a paper at all? What is the probability of the complement of A, and what does it represent?

We already computed the probability of getting only paper A and only paper B in creating the diagram:

Solution:

$$P(A) = 0.75$$
$$P(B) = 0.5$$
$$P(A \cap B) = 0.45$$
$$P(A \cup B) = P(A) + P(B) - P(A \cap B) = 0.8$$
$$P(\text{only } A) = 0.3$$
$$P(\text{only } B) = 0.05$$

Thus, the answer to the first question is:

$$P(\text{only take 1 paper}) = P(\text{only } A) + P(\text{only } B) = 0.3 + 0.05 = 0.35$$

The second question is the area outside the circles:

$$P(\text{a student does not take either paper}) = 0.2$$

The probability of the complement of A is:

$$P(A'A') = 1 - 0.75 = 0.25$$

From the Venn diagram, we see this is the event that a student only takes paper B or doesn't take a paper at all.

Example 23: Three-variable Venn diagram.

Given that we know that we have three operating systems (OSs): Linux, macOS, and Windows. A survey of 50 random college students is taken, and 23 know and use Linux, 18 know and use macOS, 30 know and use Windows as an OS, and 4 do not know to use any of the three. As a matter of fact, 9 said they knew and used all three, 12 know and use both Linux and macOS, 11 know and use both Linux and Windows, and 11 know and use both macOS and Windows.

$$\text{Let } L = \text{Linux}, M = \text{macOS, and } W = \text{Windows}$$

Fill out the Venn diagram, and use it to help answer the following probability questions:

a) $P(L \text{ È } M \cup W) = 46/50$
b) $P(\text{only } L) = 9/50$
c) $P(\text{only } W) = 4/50$
d) $P(\text{know and use only 1 of the 3}) = 9/50 + 4/50 + 17/50 = 30/50$
e) $P(L \text{ È } M \cup W)' = 4/50$

To complete the Venn diagram, we first we need to compute some useful probabilities:

We start in the middle with $P(L \cap M \cap W)$ and work out.

$$\text{Inside: } P(L \cap M \cap W) = 9/50$$

We will next look at the intersections for pairs of events. For example:

$$P(L \cap M) = 12/50$$

Thus, the portion of the $L \cap M$ area not in the three-event intersection must be 3/50. A similar logic places 2/50 in the corresponding regions of the other two event intersections.

Next, we move out to the circles for each event. For event L, the probability in the circle must be:

$$P(L \cap M) = 11/50, P(W \cap M) = 11/50$$
$$P(L) = 23/50, P(M) = 18/50, P(W) = 30/50$$

We already have $9/50 + 3/50 + 2/50 = 14/50$ in this circle, which means 9/50 is the amount in L but not in any of the intersections. For M, the corresponding value is 4/50; and for W, it is 17/50.

Finally, the probabilities in the circles sum to 46/50, meaning the amount outside the circles is 4/50. Note that we were given this value, but it is nice to confirm that our numbers make sense.

The final Venn diagram, Figure 4.8, is:

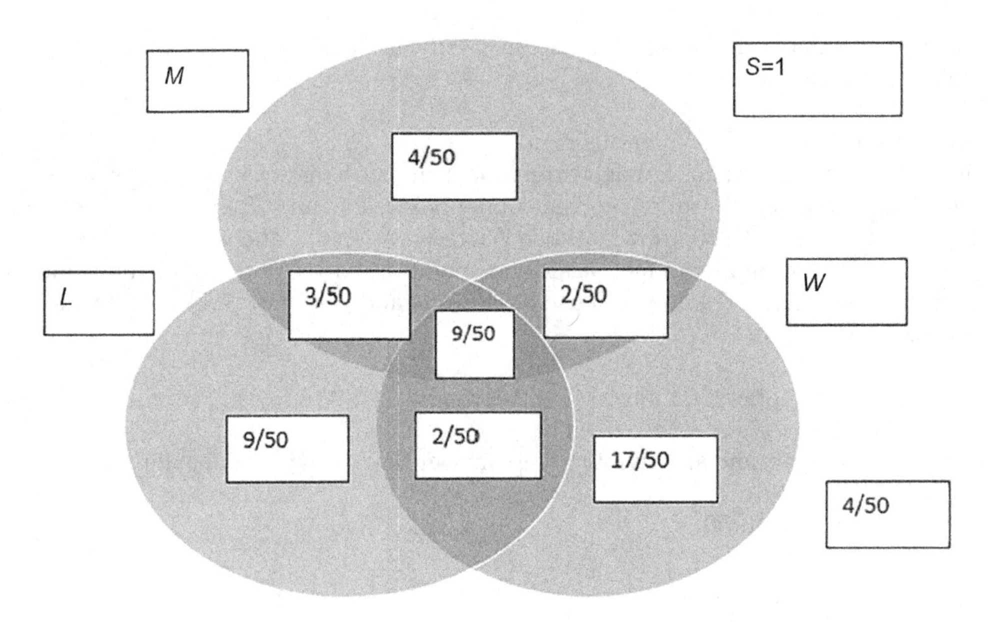

FIGURE 4.8
Venn diagram for Example 23.

We now address the probability questions:

a) $P(L \cup M \cup W) = 46/50$. This is the area inside all three circles.
b) $P(\text{only } L) = 9/50$
c) $P(\text{only } W) = 4/50$
d) $P(\text{use only 1 of the 3}) = 9/50 + 4/50 + 17/50 = 30/50$
e) $P(L \cup M \cup W)' = 4/50$

4.4.6 Conditional Probability

The notation $P(F \mid E)$ is read "the probability of event F given event E has happened". It is the probability of an event F given the occurrence of the event E. The idea in a Venn diagram here is again useful in computing and understanding conditional probabilities. If an event has happened, then we only consider that circle of the Venn diagram, and we look for the **portion of that circle that is intersected** by another event's circle.

The formula for computing conditional probabilities for events A and B are:

$$P(A \mid B) = \frac{P(A \cap B)}{P(B)}$$

$$P(B \mid A) = \frac{P(A \cap B)}{P(A)}$$

In most cases, these conditional probabilities lead to different probabilities, as answers so the order of the conditioning is important.

Example 24: Revisit newspaper example.

Let's return to our newspaper example. Find $P(A \mid B)$ and $P(B \mid A)$.

We replicate the Venn diagram, Figure 4.9, here for convenience.

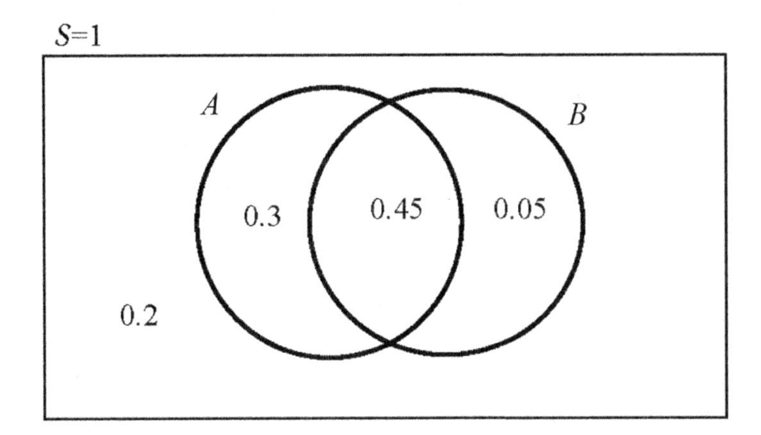

FIGURE 4.9
Revisit newspaper example.

$P(A \mid B)$ is the probability of A occurring, given we know B occurred. Since B has occurred, we are "inside" the B circle, which has total probability 0.5. We see that of that 0.5, 0.45 represents event A occurring. That is, 0.45/0.5 = 0.9 is the percent of the circle that is A. This is the conditional probability. A similar argument is used to find the other conditional probability. Using the formulas:

$$P(A \cap B) = 0.45$$
$$P(A) = 0.75$$
$$P(B) = 0.5$$

$$P(A \mid B) = \frac{P(A \cap B)}{P(B)} = \frac{.45}{.50} = .9$$

$$P(B \mid A) = \frac{P(A \cap B)}{P(A)} = \frac{.45}{.75} = .60$$

Recall that the unconditional probabilities are $P(A) = 0.75$ and $P(B) = 0.5$. In both cases, notice that the conditional probabilities increased as we obtained more additional information about the other events after we knew information about the events that did occur. The probabilities do not always increase; they could decrease or remain the same. They do not have to be affected the same way.

Example 25: Revisit disaster at sea.

We replicate our data in the table below. Given that the person is a man, what is the probability that person survived? Died? Given the person died, what is the probability that person was a woman?

We have our data in Table 4.2. Given that the person is a men what is the probability that person survived? Died?

TABLE 4.2

Disaster at Sea

	Men	Women	Boys	Girls	Total
Survived	332	318	29	27	706
Died	1,360	104	35	18	1,517
Total	1,692	422	64	45	2,223

We can define the events M = men, W = women, B = boy, G = girl, S = survive, and D = died. For the first question,

We want to find the probability, $P(S \mid M)$. Using the formula:

$$P(S \mid M) = P(S \cap M) \,/\, P(M)$$

We need two probabilities. The probability a passenger is a man is 1,692/2,223. The number in the intersection of being a man and surviving (i.e., both a man and a survivor) is 332, so the probability of the intersection is 332/2,223. The desired probability is computed:

$$P(S \mid M) = (332/2{,}223) \,/\, (1{,}692/2{,}223) = 332/1{,}692 = 0.19622$$

The probability the passenger died given they were a man is then 1 – 0.19622 = 0.80378.

The second question is the probability the person was a woman given they died, or:

$$P(M' | S') = \frac{P(M' \cap S')}{P(S')}$$

$$P(M' \# S') = \frac{P(M' \cap S')}{P(S')}$$

The event in the numerator is the intersection of being a woman and dead. The probability of this is 104/2,223. The denominator is the probability of dying, which is 1,517/2,223). Thus, the Example 2. Given the person died, what is the probability that person was a woman?

Died?

Let M = men, W = women, B = boy, G = girl, S = survive, and D = died

The probability we want to find is:

$$P(M' | S')\; P(M' | S'),\; P(W \mid D) = P(W \text{Ç} D) \,/\, P(D) = (104/2{,}223) \,/\, (1{,}517/2{,}223)$$
$$= 104/1{,}516 = 0.068556$$

4.4.7 Independence

Two events E and F are **independent** if the occurrence of event E in a probability experiment does not affect the probability of event F. Two events are **dependent** if the occurrence of event E in a probability experiment affects the probability of event F.

4.4.8 Definition of Independent Events

Two events E and F are independent if and only if

$$P(F \mid E) = P(F) \text{ or } P(E \mid F) = P(E)$$

Recall the definition of conditional probability:

$$P(E|F) = \frac{P(E \cap F)}{P(F)}$$

If we solve this equation for the intersection, we obtain the expression:

$$P(E \cap F) = P(F)P(E|F)$$

However, if the events are independent, $P(E \mid F) = P(E)$, which implies that another way to see if the events are independent is as follows:

If $P(A \cap B) = P(A) \cdot P(B)$, then the events A and B are independent.
If $P(A \cap B) \neq P(A) \cdot P(B)$, then the events are dependent.

This result is known as the

Multiplication Rule for Independent Events. It is a helpful result for computing the intersection of two events. However, be careful to ensure the events are, in fact, independent. If they are not, to compute the intersection requires using the conditional probability expression, $(E \cap F) = P(F)P(E|F)\ \sqrt{P(E \cap F) = P(E|F)P(F)}$

If E and F are independent events, the probability that E and F both occur is

$$P(E \cap F) = P(E) * P(F)$$

Example 26: Newspaper example revisited.

Are the events of getting the local newspaper and *USA Today* independent events?

Solution:

We have probabilities of the events from the original problem:

$$P(A \cap B) = 450/1{,}000 = 0.45$$
$$P(A) = 0.75$$
$$P(B) = 0.50$$

In our last example, we computed conditional probabilities:

$$P(A \mid B) = 0.9$$
$$P(B \mid A) = 0.6$$

Since $P(A)$ is not equal to $P(A \mid B)$, the events are not independent (same for event B).

We can also check this using the multiplication rule: $P(A) = 0.75$, $P(B) = 0.5$.

$$P(A)\ P(B) = 0.75(0.5) = 0.375$$
$$P(A) * P(B) = (0.75)*(0.5) = 0.375$$

We are given that $P(A \cap B) = 0.45$

Since $P(A \cap B) \neq P(A) \cdot P(B)$, these events are not independent.

Example 27: Given the following information:

$$P(E) = 0.2,\ P(F) = 0.6,\ P(E \cup F) = 0.68$$

are E and F independent events?

Solution:

We can multiply the probabilities of the two events:

$$P(E) * P(F) = 0.2(0.6) = 0.12$$

We do not assume independence and use the product rule to obtain the intersection.

$P(E \cap F)$ is not given and must be found first. We do not assume independence and use the product rule. We can use the addition rule:

where

$$P(E \cup F) = P(A) + P\text{–}(B) - P(E \cap F)$$

and solve for $P(E \cap F)$. The calculation is:

$$0.68 = 0.2 + 0.6 - P(E \cap F)$$
$$P(E \cap F) = 0.12$$

Since $P(E$ Ç $F) = 0.12$ and $P(A) * P(B) = 0.12$, then events E and F are independent.

Example 28: Suppose we have a box full of 500 golf balls. In the box, there are 50 titlist golf balls.

Suppose a golf ball is selected at random and then replaced. A second golf ball is then selected. What is the probability they are both titlists? Note: When sampling with replacement, the events are independent.

Solution:

When selecting two golf balls, the following can occur: both are titlists, both are other, one of each.

We assume independence, so

$$P(\text{both titlists}) = P(T \cap T) = P(T) * P(T) = 0.1 * 0.1 = 0.01$$

Note: Mutually exclusive and independent are not synonymous.

4.4.9 Mutually Exclusive and Independence

Independence is not the same as mutually exclusive. In fact, mutually exclusive events are essentially as far from independence as possible. We illustrate with the Venn diagram, Figure 4.10, below for two events A and B.

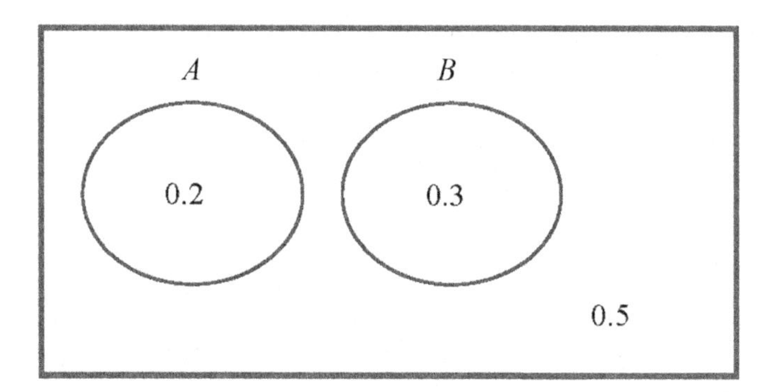

FIGURE 4.10
Two mutually exclusive events.

The two events have no intersection and, therefore, are mutually exclusive. However, clearly they are not independent. If we know B occurs, then A cannot occur. In other words:

$$P(A \mid B) = 0$$

The same is true in reverse. Clearly, knowing an event happens changes the probability we assign to the other event!

We can confirm with the multiplication rule: 0.2 * 0.3 = 0.06, not 0, so the events are not independent.

4.4.10 Review/Summary

Elementary probability theory is required for understanding this chapter in discrete *stochastic models.* We will provide a quick review of some important concepts in probability.

An **event** is any collection of results or outcomes of a procedure or experiment.

A **simple event** is an outcome that cannot be broken down into simpler components.

The **sample space** for a procedure or an experiment consists of all possible outcomes (as simple events).

Examples

Experiment	Example of an event	Sample Space
Flip of a coin	Head	{Head, Tail}
Roll of a die	5 (simple event)	{1, 2, 3, 4, 5, 6}
Roll of two die	7 (not a simple event)	{1–1, 1–2, . . . , 6–5, 6–6}

Explanation: If we have only one die, we can roll a die to get a 5 only one way. If we have two die and roll a 7, it could be done as 1–6, 2–5, 3–4, 4–3, 5–2, 6–1, so it is not a simple event.

We define the **probability** that **event** A occurs, $P(A)$, as the number of times A occurs out of the total number of possible outcomes in the **sample space:**

$$P(A) = \frac{\textit{number of times Aoccurred}}{\textit{number of events in the sample space}}$$

4.4.11 Sampling and Experiments

An event is any collection of results or outcomes of an experiment. A sample space is a listing of all possible outcomes from an **experiment**. An experiment is any process that allows researchers to obtain observations (data). In a flip of a fair coin two times, the sample space are all the possible outcomes of two flips of a fair coin. If we call a head, H, and a tail, T, then the possible outcomes are:

HH HT TH TT

This set constitutes the entire sample space. Let's call event A the event that exactly one head appeared in the two flips. That occurred in flip HT and flip TH, or two times. Since there were four possible outcomes, the probability of event A, $P(A)$, is 2/4 = 0.5.

Let's consider a tennis match between player A and player B where the winner is the first to win three sets. The sample space for the winner is:

{AAA, ABAA, ABBAA, AABA, AABBA, ABABA, BBB, BBAB, BAABB, BABAB, BAABA, BBAAB, BABB, BAAA, ABBB, ABABB, BABAA, BBAAA, AABBB, ABBAB}. If A and B were equally likely to win a set, then we can compute the probability of each event in the sample space. Thus, the probability that A wins is 0.5 or 10/20.

4.4.12 Tree Diagrams

Tree diagrams are a useful way to delineate the outcomes of a sample space. For example, consider wanting to find out what happened after only three sets of the match. Each branch of the tree signifies the winner of that set. For three sets, we could have the following tree diagram, Figure 4.11 (note there are only two outcomes that show a winner at that time).

There are eight simple events in three sets. Of the three sets, only two results in a identifying a winner. The events {AAA} and {BBB} are winners. All other outcomes must continue to a fourth set or more.

Thus, the probability that A wins in three sets is (1/8).

Show that the probability that A wins in four sets is 3(1/16) and in five sets is 6(1/32).

Now let's assume that A is a higher-ranked player with odds to win a set as 3:1. We can recompute the probabilities that A wins from the given sample space in three, four, or five sets. Odds are a way of weighting the outcomes so that they are no longer equally likely.

$$P(A \text{ wins in 3 sets}) = (0.75^3) = 0.421875$$
$$P(A \text{ wins in 4 sets}) = 3(0.75^{3)}\ (0.25) = 0.3164065$$
$$P(A \text{ wins in 5 sets}) = 6(0.75^3)\ (0.25^2) = 0.158203125$$

The probability that A wins this match, $P(A$ wins the match), is the sum of the probability that A wins in three sets + probability that A wins in four sets + probability that A wins in five sets = 0.42187" + 0.3164065 + 0"158203125 = 0.89648625.

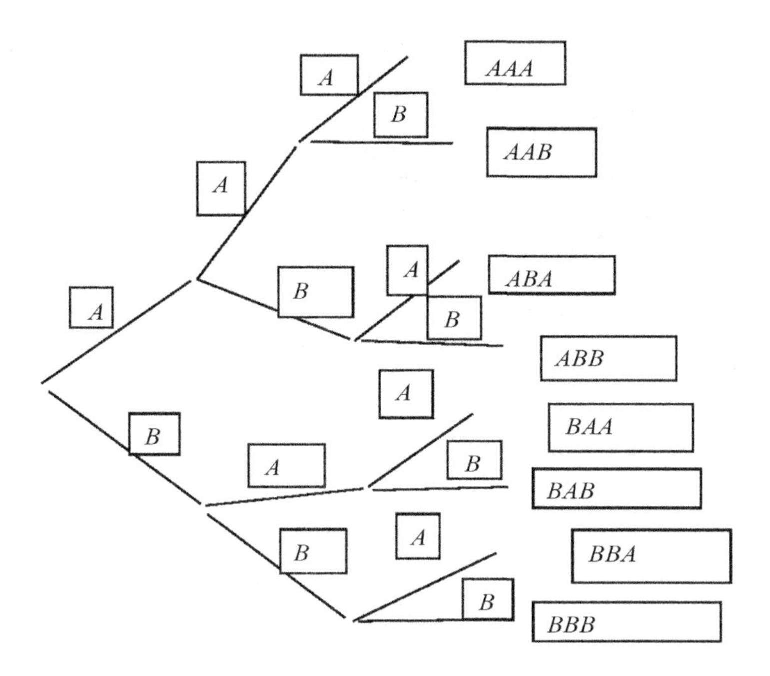

FIGURE 4.11
Tree diagram.

4.4.13 Review of Probability Laws

The are some important rules for probability. These rules that we will use, include the following:

1. The Law of Large Numbers states that if an experiment is repeated again and again, the relative frequency probability of an event approaches its probability. We will see this in our chapter on simulations.
2. Addition Rule: $P(A \text{ or } B) = P(A) + P(B) - P(A \text{ and } B)$
3. For any event A in the sample space, S,
   a. $P(A) \geq 0$
   b. $P(A) \leq 1$
   c. $P(S) = 1$
   d. $P(\text{not } A) = 1 - P(A)$
4. For any events A and B in the sample space, S,
   a. If mutually exclusive, $P(A \text{ and } B) = 0$
   b. If independent, $P(A \text{ and } B) = P(A) * P(B)$
   c. Otherwise, $P(A \text{ and } B) = P(A) + P(B) - P(A \text{ or } B) = P(A \mid B)P(B)$
5. Conditional probability: $P(A$ given B has occurred already) $P(A \mid B) = P(A \text{ and } B) / P(B)$

What's the point of probability (and statistics) anyway? **One of its most important applications is decision-making under uncertainty**. When you decide on an action (assuming you are a rational human being), **you are betting** that completing the action will leave you better off than had you not done it. **But bets are inherently uncertain**, so how do you decide whether to go ahead with it or not?

Implicitly or explicitly, you estimate a probability of success—and if the probability is higher than some threshold, you forge ahead.

So being able to accurately estimate this success probability is critical to making good decisions. While chance will always play a role in the outcome, if you can consistently stack the odds in your favor, then you should do very well over time.

That's where Bayes' Theorem comes in—it gives us a quantitative framework for updating our beliefs as the facts around us change, which in turn allows us to improve our decision-making over time.

4.5 Chapter 4 Exercises

4.40 Let A be McDonald's and B be the movies with $P(A) = 0.655$, $P(B) = 0.738$, $P(A \cap B) = 0.55$.

Draw the Venn diagram. Find the following probabilities.

a) $P(A \cup B)$

b) $P(\text{only } A)$

c) $P(\text{only } B)$

d) $P(A \cup B)''$

e) $P(A \cap B)''$

f) $P(A \cap B')$

4.41 For a visit to Disneyland, the probabilities that a person will go on various rides are:

The Jungle Cruise is 0.74.

The Monorail is 0.70.

The Matterhorn is 0.62.

The Jungle Cruise and the Monorail is 0.52.

The Jungle Cruise and the Matterhorn is 0.46.

The Matterhorn and the Monorail is 0.44.

The Jungle Cruise, Matterhorn, and the Monorail is 0.34.

Let event A be the Jungle Cruise, event B be the Monorail, and event C be the Matterhorn.

Draw a Venn diagram, and then find the following probabilities?

a) $P(A \cup B \cup C)$

b) Only A

c) Only B

d) Only C

e) A and B, not C

f) B and C, not A

g) A'

h) B'

i) C'

4.42 A person visits the dentist. They have the following probabilities of work to be done on them:

Teeth cleaned is 0.47.

Cavity filled is 0.29.

Tooth pulled is 0.09.

Teeth cleaned and cavity filled is 0.13.

Teeth cleaned and tooth pulled is 0.03.

Cavity filled and tooth pulled is 0.02.

Teeth cleaned, tooth pulled, and cavity filled is 0.01.

Draw a Venn diagram, and find the following probabilities.

a) At least one procedure is done.

b) Only teeth cleaning is done.

c) Only a cavity is filled.

d) Teeth cleaned and cavity filled, no tooth pulled.

e) No cleaning, filling cavities, or pulling teeth is done at the dentist.

4.43 During the past hailstorm in Florence, South Carolina, State Farm Insurance found that 40% of their Florence clients suffered window damage, 70% suffered roof damage, and 28% suffered both roof and window damage.

Let event A = Window damage

and event B = Roof damage

a. Complete a Venn diagram, and fill in all probabilities.

b. Using either the Venn diagram or the rules of probability, compute the following:

(i) $P(A \text{ only})$

(ii)

(iii) $P(B \text{ only})$

(iv)

(v) $P(A \cap B)'$ (Not the intersection of A and B)

(vi)

(vii) $P(A \cup B)$

(viii)

(ix)

(x)

c. Are these events independent? Show work.

4.44 Use the following information concerning Diet Coke and Pepsi One drinkers that were surveyed from 300 people to find the probabilities

Container

Drink	Cans (12 oz.)	Bottles (20 oz.)
Diet Coke	125	65
Pepsi One	75	35

a) *P*(drink Diet Coke)

b) *P*(drink Diet Coke from a can)

c) *P*(drink Pepsi One from a can)

d) *P*(drink a 20 oz. bottle | drink Pepsi One)

e) *P*(drink Diet Coke | drink a 20 oz. bottle)

f) Are the events 12 oz. cans and 20 oz. bottles independent? Show work.

4.45 Considering the purchase of gasoline (*unleaded, unleaded plus,* and *supreme*) by Florence gasoline stations. The following probabilities are found:

Purchase *unleaded* gasoline is 0.62.

Purchase *unleaded plus* gasoline is 0.59.

Purchase *supreme* gasoline is 0.53.

Purchase *unleaded* and *supreme* is 0.25.

Purchase *unleaded* and *unleaded plus* is 0.33.

Purchase *unleaded plus* and *supreme* is 0.30.

Purchase *unleaded, unleaded plus,* and *supreme* is 0.19.

Complete the Venn diagram, and then answer the probability questions:

a. *P*(only unleaded purchased)

b. *P*(only unleaded plus purchased)

c. *P*(only supreme purchased)

d. *P*(either of the three types of gasoline is purchased)

e. *P*(none of three types of gasoline is purchased)

4.6 Bayes' Theorem

One of the most important applications of probability and statistics is decision-making under uncertainty. When you decide on an action (assuming you are a rational human being), you are betting that completing the action will leave you better off than had you not done it. But bets are inherently uncertain, so how do you decide whether to go ahead with it or not?

Implicitly or explicitly, you estimate a probability of success—and if the probability is higher than some threshold, you move forward.

So being able to accurately estimate this success probability is critical to making good decisions. While chance will always play a role in the outcome, if you can consistently stack the odds in your favor, then you should do very well over time.

That's where Bayes' Theorem comes in—it gives us a quantitative framework for updating our beliefs as the facts around us change, which in turn allows us to improve our decision-making over time.

In this section we really want to make sure we understand and will be able to demonstrate how to use the Law of Total Probability and Bayes' Theorem to solve problems. We start with the Theorem Law of Total probability.

Let's start with a graphical view using a Venn diagram in Figure 4.12. Let's say we know the probabilities of A_1 and A_2. We also know some information about event E. Event E intersects only both A_1 and A_2, and not with any other events, as shown in the diagram. Further, suppose we know the conditional probabilities of E, given each of those events. This is shown in Figure 4.12. Is this enough information to find the probability of event E? Yes, we can use the Law of Total Probability.

LAW OF TOTAL PROBABILITY

Let E be an event that is a subset of a sample space S. Let $A_1, A_2, \ldots, A_n$ be a partition of the sample space, S. Then, $P(E) = P(A_1) \cdot P(E \mid A_1) + P(A_2) \cdot P(E \mid A_2) + \ldots + P(A_n) \cdot P(E \mid A_n)$.

To illustrate where the formula comes from, using our example in Figure 4.12, if we define E to be any event in the sample space S, then we can write event E as the union of the intersections of event E with A_1 and event E with A_2:

$$E = (E \cap A_1) \cup (E \cap A_2)$$

If we have more events, we just expand the union of the number of events that E intersects. The rules of probability we learned previously allow us to rewrite the intersections as a product:

$$P(E \cap A_1) = P(A_1) P(E \mid A_1)$$

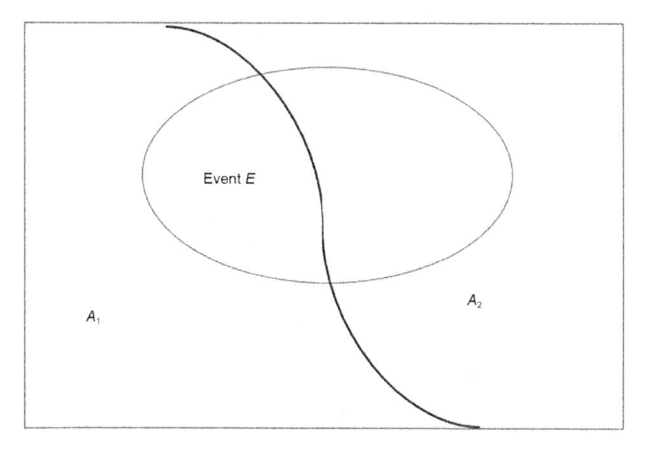

FIGURE 4.12
Venn diagram partition of E by A_1 and A_2.

Thus, we can write the probability of E as:

$$
\begin{aligned}
P(E) &= P(E \cap A_1) + P(E \cap A_2) \\
&= P(A_1)P(E|A_1) + P(A_2)P(E|A_2) \\
P(E) &= P(A_1 \cap E) + P(A_2 \cap E) + P(A_3 \cap E)\ P(E) = P(A_1 \cap E) + P(A_2 \cap E) + P(A_3 \cap E) \\
&= P(E \cap A_1) + P(E \cap A_2) + P(E \cap A_3) \\
&= P(A_1) \cdot P(E \mid A_1) + P(A_2) \cdot P(E \mid A_2) + P(A_3) \cdot P(E \mid A_3)
\end{aligned}
$$

This idea might become clearer in a tree diagram of the situation, as shown in Figure 4.13.

We can see that there are only two paths through the tree that involve event E. So, to get the probability of E, we sum the probabilities of these two paths. The result is the expression given by the Law of Total Probability.

Once we find events E and its complement E′, we are ready to proceed to the Law of Total Probability, which is an important part of what is known as **Bayes' Theorem**. In our example, we know the conditional probability of event E given A_1 and A_2. But, what if what

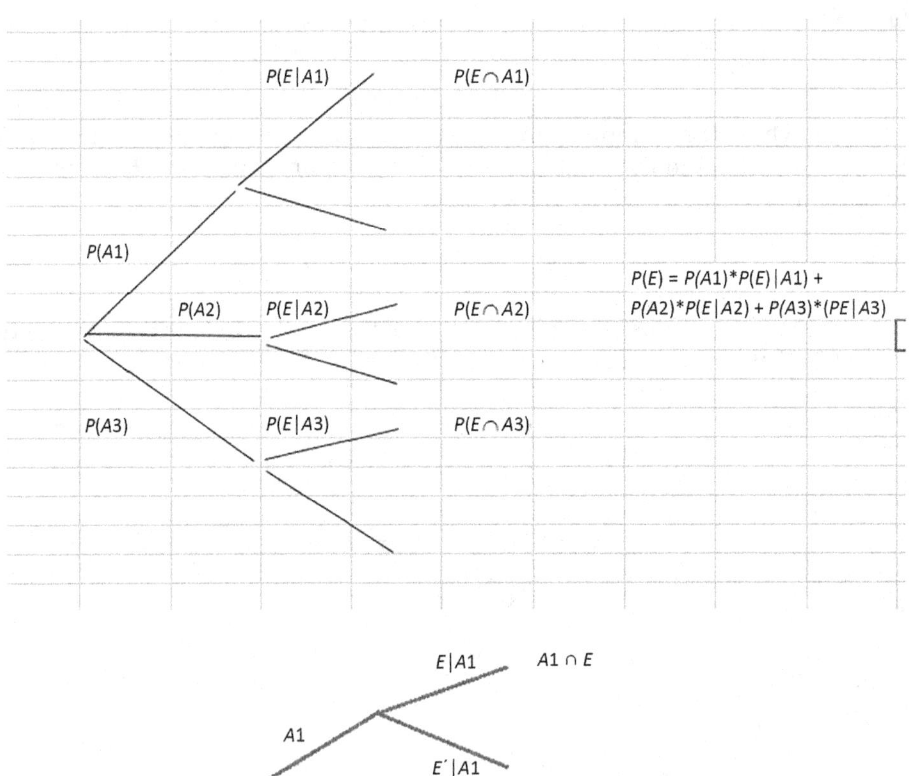

FIGURE 4.13
Tree diagram of E by A_1 and A_2.

we really care about is the conditional probability of A_1 given E or of A_2 given E. Bayes' Theorem allows us to obtain these conditional probabilities; essentially, we can "flip" the order of the conditional probabilities using the following formula.

Formally, Bayes' Theorem is defined below. Let's look at this formula using what we know about probability. We are interested in the conditional probability of event A_i given E. The definition of conditional probability allows us to write this as:

$$P(A_i \mid E) = \frac{P(A_i \cap E)}{P(E)}$$

Since we know the "reverse" conditional probabilities (i.e., E given A_1, etc.), we can calculate the denominator using the Law of Total Probability. We know how to compute the probability of an intersection, and note that we can change the order of the events listed in the intersection, so we obtain the numerator as:

$$P(A_i \cap E) = P(E \cap A_i) = P(A_i)P(E \mid A_i)$$

Tree diagrams can help to make this formula easy to obtain, as we will illustrate in examples. If obtaining more information improves our ability to make a better decision under uncertainty, then we want to use the information we found.

4.6.1 Bayes' Theorem

Let $A_1, A_2, \ldots, A_n$ be a partition of a sample space S. Then for any event E that is a subset of S for which $P(E) > 0$, the probability of event A_i for $i = 1, 2, \ldots, n$, given the event E, is

$$P(A_i \mid E) = \frac{P(A_i) \cdot P(E \mid A_i)}{P(E)}$$

$$= \frac{P(A_i) \cdot P(E \mid A_i)}{P(A_1) \cdot P(E \mid A_1) + P(A_2) \cdot P(E \mid A_2) + \ldots + P(A_n) \cdot P(E \mid A_n)}$$

Example 29: Government service.

According to the 2019 report from the Bureau of Labor and Statistics, employment percentages by the federal government were 43.54% women and 56.46% men. Most positions were either in cabinet positions or other federal positions. The analysis shows that 9.88% of women were in other positions, and 6.56% of men were in other positions. Suppose that a randomly selected cabinet member is selected, what is the probability that the cabinet member is female?

Solution:

The most difficult part of solving Bayes' Rule problems is defining the events and determining which probability we want to know and which we do know in terms of those events. Here, we have two events of interest: one is the gender (F or M) and the other the type of position (C = cabinet or O = other).

What do we want to know? The probability a selected person is female, but notice this is conditional. Female, given they are a cabinet member or:

$P(F \mid C)$ is what we want to know.

What probabilities do we know? The first two values do not have a condition attached, so they are:

$$P(F) = 0.4354$$
$$P(M) = 0.5646$$

The other two values given are conditional, in the reverse order of the condition we want to know (this is always the case in Bayes' problems). The "of women" and "of men" indicates conditioning here. Thus, we are given:

$$P(O \mid F) = 0.0988$$
$$P(O \mid M) = 0.0656$$

Note that while not given, we also know by the complements of these probabilities, the conditional probabilities for being on the cabinet. Since we want to know the probability of F given cabinet, these are the conditional probabilities that are important to us:

$$P(C \mid F) = 1 - 0.0988 = 0.9012$$
$$P(C \mid M) = 1 - 0.0656 = 0.9344$$

We now have translated

The question put into symbols, and we can use Bayes' Rule to write the expression for what we want to know, *is* $P(F \mid C)$. Before we do so, we will draw a tree diagram to illustrate the formula. This is the diagram of Figure 4.14 with the events of our problem and probabilities added. Since we want to know a probability conditional on being on the cabinet, the paths in the tree that matter are the two involving C; we compute the probabilities for those two paths, see Figure 4.15.

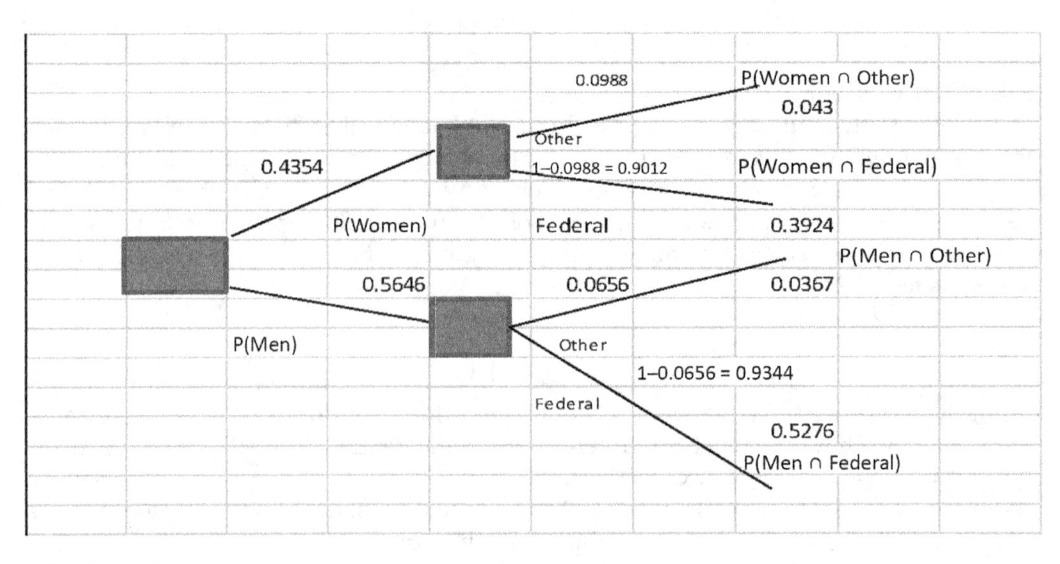

FIGURE 4.14
Tree diagram for Example 29.

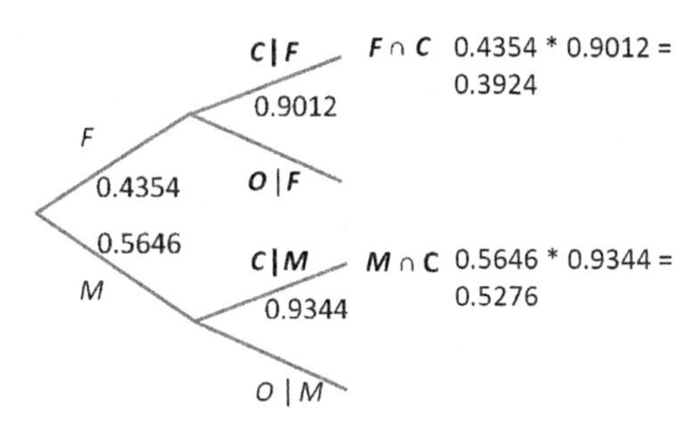

FIGURE 4.15
Simplified view of tree diagram.

Let's look at the definition of Bayes' Theorem for finding $P(F \mid C)$:

$$P(F|C) = \frac{P(F \cap C)}{P(C)} = \frac{P(F)P(C|F)}{P(F)P(C|F) + P(M)P(C|M)}$$

Notice, again, how this is equal to $P(F \cap C)/P(C)$. We need to find $P(C)$. Using the Law of Total Probability, we find the denominator. Using the known values:

$$P(F|C) = \frac{0.4354 \cdot 0.9012}{0.4354 \cdot 0.9012 + 0.5646 \cdot 0.9344} = \frac{0.3924}{0.3924 + 0.5276}$$

Let's pause here and look at this expression and back at our tree diagram. What do you notice? The denominator is the two paths (branches) of the tree involving C. In other words, this sum is the probability of being on the cabinet. The numerator is the cabinet path that is F. Thus, our final formula is the proportion of all on the cabinet who are female, which is:

$$P(C) = P(F \cap C) + P(M \cap C) = P(F)\,(P(C \mid F) + P(M)\,P(C \mid M) =$$
$$(0.4354)(0.9012) + (0.5646)(0.9344) = 0.90256$$
$$P(F \mid C) = P(F \cap C) \,/\, P(C) = 0.3924/0.90256 = 0.4347$$

We will not show tree diagrams for subsequent examples but suggest drawing them as you read and also using them when you solve problems.

Example 30: Coronavirus (COVID-19) by race.

In state x, the population breakdown is approximately 24.3% African American, 29.1% Hispanic, and 46.6% other. In the latest statistics on deaths due to COVID-19, 21% were African American, 17% were Hispanic, and 62% other. If a random patient died of COVID-19, what was the probability that they were African American, Hispanic, or other? (Data from U.S. Census Bureau, 2019.)

Solution:

We again need to define our events and identify what we know and want to know. Here, one event is the race, and it has three instead of two possibilities: We need to find all the intersection of COVID-19 deaths with race. We do this using the Law of Total Probability.

Let AA = African American, H = Hispanics, and O = Other. The other event is death (or not). We can let C = death by COVID-19 and then C' would be no death if needed.

What do we want to know? Conditionally on death what race, or: $P(AA \mid C)$, $P(H \mid C)$, and $P(O \mid C)$.

What values do we know? We know some unconditional probabilities for race in the state:

$$P(AA) = 0.243$$
$$P(H) = 0.291$$
$$P(O) = 0.466$$

We also have conditional probabilities of death given the race (again, note that this is suggested by what we want to know; the given values are in the opposite order of conditioning).

$$P(C \mid AA) = 0.21$$
$$P(C \mid H) = 0.17$$
$$P(C \mid O) = 0.62$$

If you draw a tree diagram, Figure 4.16, for this problem, notice it will have more branches. How many? Three initial splits (race) followed by two splits for C or not. The three paths involving C are the ones of interest here.

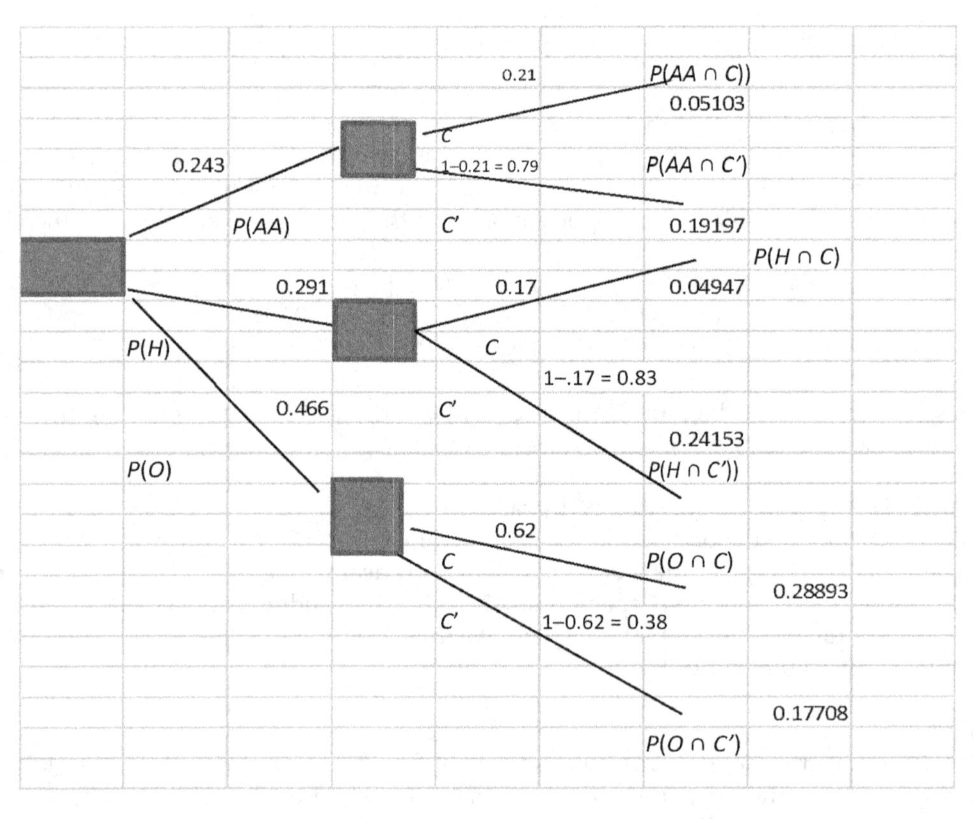

FIGURE 4.16
Tree diagram for Example 30.

We need to find all the intersection of COVID-19 deaths with race. We do this using the Law of Total Probability.

We can first get the denominator for each of the three Bayes' Theorem computations since it will be the same (it is the sum of the probabilities on the three paths of interest):

$$P(C) = P(AA)P(C \mid AA) + P(H)P(C \mid H) + P(O)P(C \mid O)$$
$$= 0.243 * 0.21 + 0.291 * 0.17 + 0.466 * 0.62 = 0.3485$$

Now we use the theorem to compute the three probabilities of interest. For $P(AA \mid C)$:

$$P(AA \mid C) = \frac{P(AA \cap C)}{P(C)} = \frac{P(AA)P(C \mid AA)}{P(C)} = \frac{0.243 \cdot 0.21}{0.3485} = 0.1464$$

Similarly, the other two values are:

$$P(AA \mid C) = (0.243 * 0.21) / 0.3485 = 0.1464$$
$$P(H \mid C) = (0.291 * 0.17) / 0.3485 = 0.1419$$
$$P(O \mid C) = (0.466 * 0.62) / 0.3485 = 0.829$$

We can see the importance of drawing the tree diagram to assist us.

Example 31: COVID-19 by age.

In state x, the population breakdown is approximately 25% are over 60, 27.4% are under 18, and 58.5% are between 18 and 60 years old. In the latest statistics on deaths due to COVID-19, 85% were 60 and older, 0.2% were 18 and younger, and 14.8% were between 18 and 60 years old. If a random patient died of COVID-19, what was the probability that they were older than 60 years?

Solution:

This example is similar to the previous one, so we give a summary. For the age groups, let $A = > 60$, $B = [18, 60]$, and $C = < 18$. Let D = death by COVID-19.

We need to find all the intersection of COVID-19 deaths with age. We again do this using the Law of Total Probability.

Let $A = > 60$, $B = [18, 60]$, and $C = < 18$. Let D = death by COVID-19.

$$P(D) = P(A)P(D \mid A) + P(B)P(D \mid B) + P(C)P(D \mid C)$$
$$= 0.25 * 0.85 + 0.274 * 0.02 + 0.585 * 0.148$$
$$= 0.30456$$

The Bayes' Theorem calculation is then:

$$P(A \mid D) = (0.25 * 0.85) / 0.30456 = 0.6977$$

The result here points out a very important use of Bayes' Theorem. If we knew nothing about COVID-19 and only had the distribution of ages, our best estimate of the probability of death in the A group (over 60) would just be $P(\text{A}) = 0.25$. Remember, we know nothing about COVID-19, so since 25% are over 60, that would be our best guess. We say that all the probabilities $P(A_i)$ are a priori probabilities. These are probabilities of events prior to any knowledge regarding the event, $P(A)$ in this example.

However, the probabilities $P(A_i \mid E)$ are a posteriori probabilities because they are probabilities computed after some knowledge regarding the event. The calculation we did uses data we have obtained about COVID-19 deaths in age groups. Thus, we "update" our belief about the probability the person is over 60 to the "posterior" estimate: $P(A \mid D) = 0.70$. We now estimate a 70% chance the person is over 60! Why? We have incorporated the information about COVID-19 and age, which is that it is much, much more dangerous for older than for younger people.

Example 32: Unemployed women.

According to the data collected in 2018 by the U.S. Census Bureau (www.census.gov/newsroom/stories/2019/unmarried-single-americans-week.html), 30% of American adult women are single, 51% of American adult women are married, and 19% of American adult women are widowed or divorced (other). Of the single women, 3.8% are unemployed; of the married women, 2.7% are unemployed; of the "other" women, 4.2% are unemployed. Suppose that a randomly selected American adult woman is determined to be unemployed. What is the probability that she is single?

Solution:

Again, we first define the following events.
Let U = unemployed.
For the marital status,
let

S = single
M = married
O = other

We are given the following probabilities:

$$P(S) = 0.300,\ P(M) = 0.51,\ P(O) = 0.19$$
$$P(U \mid S) = 0.038,\ P(U \mid M) = 0.027,\ P(U \mid O) = 0.042$$

and from the Law of Total probability, we know the probability of being unemployed:

$$P(U) = P(S) * P(U \mid S) + P(M) * P(U \mid M) + P(O)\, P(U \mid O) = 0.03315$$

We wish to determine the probability that a woman is single given the knowledge that she is unemployed. That is, we wish to determine $P(S \mid U)$. We will use Bayes' Theorem as follows:

$$P(S \mid U) = \frac{P(S \cap U)}{P(U)} = \frac{P(S) \cdot P(U \mid S)}{P(U)}$$

$P(S \mid U) = (0.30) * (0.038) / 0.03315 = 0.3439$

There is a 34.39% probability that a randomly selected unemployed woman is single.

We say that all the probabilities $P(A_i)$ are a priori probabilities. These are probabilities of events prior to any knowledge regarding the event. However, the probabilities $P(A_i \mid E)$ are a posteriori probabilities because they are probabilities computed after some knowledge regarding the event. In our example, the a priori probability of a randomly selected woman being single is 0.30. The a posteriori probability of a woman being single knowing that she is unemployed is 0.3439. Notice the information that Bayes' Theorem gives us. Without any knowledge of the employment status of the woman, there is a 30% probability that she is single. But, with the knowledge that the woman is unemployed, the likelihood of her being single increases to 34.9%. In this case we increased our chances by 4.9%.

Example 33: Work disability.

A person is classified as work disabled if they have a health problem that prevents them from working in the type of work they can do. Table 4.3 contains the proportion of Americans that are 16 years of age or older that are work disabled by age.

TABLE 4.3

Work Disability Proportions

Age	Event	Proportion Work Disabled
16–24	A_1	0.058
25–34	A_2	0.143
35–44	A_3	0.205
45–54	A_4	0.289
55 and older	A_5	0.305

Source: Modified from U.S. Census Bureau data.

If we let M represent the event that a randomly selected American who is 16 years of age or older is male, then we can also obtain the following probabilities:

$$P(\text{male} \mid 16\text{–}24) = P(M \mid A_1) = 0.471$$
$$P(\text{male} \mid 25\text{–}34) = P(M \mid A_2) = 0.496$$
$$P(\text{male} \mid 35\text{–}44) = P(M \mid A_3) = 0.485$$
$$P(\text{male} \mid 45\text{–}54) = P(M \mid A_4) = 0.497$$
$$P(\text{male} \mid 55 \text{ and older}) = P(M \mid A_5) = 0.460$$

(a) If a work disabled American aged 16 years of age or older is randomly selected, what is the probability that the American is male?

(b) If the work disabled American who is randomly selected is male, what is the probability that he is 25–34 years of age?

Solution:

(a) We will use the Law of Total Probability to compute $P(M)$ as follows:

$$P(M) = P(A_1) \cdot P(M \mid A_1) + P(A_2) \cdot P(M \mid A_2) + P(A_3) \cdot P(M \mid A_3)$$
$$+ P(A_4) \cdot P(M \mid A_4) + P(A_5) \cdot P(M \mid A_5)$$

(b) We use Bayes' Theorem to compute P(25–34 | male) as follows:

$$P(A_2 \mid M) = \frac{P(A_2) \cdot P(M \mid A_2)}{P(M)}, \text{ where } P(M) \text{ is found from part (a).}$$

The calculations are:

0.058
0.143
0.205
0.289
0.305

(a) $P(M) = P(A_1) \cdot P(M \mid A_1) + P(A_2) \cdot P(M \mid A_2) + P(A_3) \cdot P(M \mid A_3)$
$+ P(A_4) \cdot P(M \mid A_4) + P(A_5) \cdot P(M \mid A_5)$
= (0.058) (0.471) + (0.143) (0.496) + (0.205) (0.485) + (0.289) (0.497) + (0.305) (0.460)
= 0.4816

There is a 48.16% probability that a randomly selected work disabled American is male.

(b) $P(A_2 \mid M)$ = (0.1430) (0.496) / (0.4816) = 0.14723

There is a 14.723% probability that a randomly selected work disabled American who is male is 25–34 years of age.

Notice that the a priori probability (0.143) and the a posteriori probability (0.14723) do not differ much. This means that the knowledge that the individual is male does not yield much information regarding the age of the work disabled individual.

4.7 Chapter 4 Exercises

In problems 4.46–4.67, find the indicated probabilities by referring to the tree diagram given below and by using Bayes' Theorem.

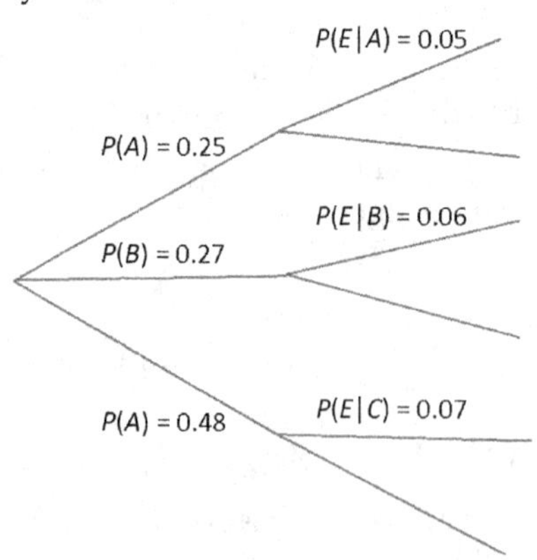

4.46 $P(E \mid A)$

4.47 $P(E \mid B)$

4.48 $P(E \mid \text{A})$

4.49 $P(E \mid \text{B})$

4.50 $P(E \mid \text{C})$

4.51 $P(E \mid \text{C})$

4.52 $P(E)$

4.53 $P(E)$

4.54 $P(A \mid E)$

4.55 $P(A \mid E)$

4.56 $P(C \mid E)$

4.57 $P(B \mid E)$

4.58 $P(B \mid E)$

4.59 $P(C \mid E)$

4.60 Suppose that events A_1 and A_2 form a partition of the sample space S with $P(A_1) = 0.48$ and $P(A_2) = 0.52$. If E is an event that is a subset of S and $P(E \mid A_1) = 0.05$ and $P(E \mid A_2) = 0.10$, find $P(E)$.

4.61 Suppose that events A_1 and A_2 form a partition of the sample space S with $P(A_1) = 0.356$ and $P(A_2) = 0.644$. If E is an event that is a subset of S and $P(E \mid A_1) = 0.13$ and $P(E \mid A_2) = 0.20$, find $P(E)$.

4.62 Suppose that events A_1, A_2, and A_3 form a partition of the sample space S with $P(A_1) = 0.33$, $P(A_2) = 0.47$, and $P(A_3) = 0.2$. If E is an event that is a subset of S and $P(E \mid A_1) = 0.10$, $P(E \mid A_2) = 0.15$, and $P(E \mid A_3) = 0.18$), find $P(E)$ and $P(E')$.

4.63 Suppose that events A_1, A_2, and A_3 form a partition of the sample space S with $P(A_1) = 0.3$, $P(A_2) = 0.55$, and $P(A_3) = 0.15$. If E is an event that is a subset of S and $P(E \mid A_1) = 0.05$, $P(E \mid A_2) = 0.20$, and $P(E \mid A_3) = 0.15$, find $P(E)$ and $P(E')$.

4.64 **Color-blindness**. The most common form of color-blindness is so-called red-green color-blindness. People with this type of color-blindness cannot distinguish between green and red. Approximately 7% of all males have red-green color-blindness, while only about 0.4% of women have red-green color-blindness. In 2019, 51.9% of all Americans were male, and 48.1% were female according to the U.S. Census Bureau.

(a) What is the probability that a randomly selected American is color-blind?

(b) What is the probability that a randomly selected American who is color-blind is female?

4.65 **The coronavirus test**. The standard test for the coronavirus is a test that tests for the presence of COVID-19 antibodies. If an individual does not have the coronavirus, the test will come back negative for the presence of COVID-19 antibodies 99.8% of the time and will come back positive for the presence of COVID-19 antibodies 0.2% of the time (a false positive). If an individual has the coronavirus, the test will come back positive 99.8% of the time and will come back negative 0.2% of the time (a false negative). The latest reports available indicate that approximately 0.7% of the world population has the coronavirus.

(a) What is the probability that a randomly selected individual has a test that comes back positive?

(b) What is the probability that a randomly selected individual has the COVID virus if the test comes back positive?

4.66 **Educational attainment.** The data in the following table represent the proportion of Americans 25 years of age or older at the various levels of educational attainment in 2000.

Level	Event	Proportion
Not a high school graduate	A_1	0.165
High school graduate	A_2	0.348
Some college, no degree	A_3	0.169
Associate's degree	A_4	0.090
Bachelor's degree	A_5	0.151
Advanced degree	A_6	0.077

If we let M represent the event that a randomly selected American who is 25 years of age or older is male, then we can also obtain the following probabilities:

$$P(M \mid A_1) = 0.477 \; P(M \mid A_2) = 0.460 \; P(M \mid A_3) = 0.472$$

$$P(M \mid A_4) = 0.434 \; P(M \mid A_5) = 0.500 \; P(M \mid A_6) = 0.555$$

(a) What is the probability that a randomly selected American 25 years of age or older is male?

(b) What is the probability that an American male 25 years of age or older has an advanced degree?

(c) What is the probability that an American male 25 years of age or older is not a high school graduate?

4.67 **Educational attainment.** Refer to Problem 4.66. If we let E represent the event that a randomly selected American who is 25 years of age or older is employed, then we can also obtain the following probabilities:

$$P(E \mid A_1) = 0.402 \; P(E \mid A_2) = 0.628 \; P(E \mid A_3) = 0.699$$

$$P(E \mid A_4) = 0.755 \; P(E \mid A_5) = 0.785 \; P(E \mid A_6) = 0.791$$

(a) What is the probability that a randomly selected American 25 years of age or older is employed?

(b) What is the probability that an employed American 25 years of age or older has a bachelor's degree?

(c) What is the probability that an employed American 25 years of age or older is not a high school graduate?

4.68 **Voting pattern.** The following data represent the proportion of Americans who are voting age at the various levels of educational attainment in 2019.

Level	Event	Proportion
Grade school	A_1	0.163
High school graduate	A_2	0.600
College graduate	A_3	0.237

Source: Statistical Abstract, 2002.

If we let D represent the event that a randomly selected American who is voting age voted Democratic in the 2000 presidential election, then we can also obtain the following probabilities:

$$P(D \mid A_1) = 0.74\ P(D \mid A_2) = 0.540\ P(D \mid A_3) = 0.500$$

(a) What is the probability that a randomly selected American who is voting age voted Democratic in the 2000 presidential election?

(b) What is the probability that an American who is voting age and voted Democratic has graduated from college?

(c) What is the probability that an American who is voting age and voted Democratic has a grade school education?

4.69 **Murder victims.** The following data represent the proportion of murder victims at the various age levels.

Level	Event	Proportion
Less than 17 years	A_1	0.082
17–29	A_2	0.424
30–44	A_3	0.305
45–59	A_4	0.125
At least 60 years	A_5	0.064

Source: Federal Bureau of Investigation.

If we let M represent the event that a randomly selected murder victim was male, then we can also obtain the following probabilities:

$$P(M \mid A_1) = 0.622\ P \quad (M \mid A_2) = 0.843\ P \quad (M \mid A_3) = 0.733$$
$$P(M \mid A_4) = 0.730\ P \quad (M \mid A_5) = 0.577$$

(a) What is the probability that a randomly selected murder victim was male?

(b) What is the probability that a randomly selected male murder victim was 17–29 years of age?

(c) What is the probability that a randomly selected male murder victim was less than 17 years of age?

4.70 **Espionage.** Suppose that the CIA suspects that one of its operatives is a double agent. Past experience indicates that 95% of all operatives suspected of espionage are, in fact, guilty. The CIA decides to administer a polygraph to the suspected spy. It is known that the polygraph returns results that indicate a person is guilty 90% of the time if they are guilty. The polygraph returns results that indicate a person is innocent 99% of the time if they are innocent. What is the probability that this particular suspect is innocent given that the polygraph indicates that he is guilty?

5

Discrete Distributions

Objectives

1. **Know the basic rules for a distribution and a cumulative distribution function.**
2. **Know how to compute the mean and variance using expected values for general discrete distribution.**
3. **Know the binomial distribution and how to use it.**
4. **Know the Poisson distribution and how to use it.**
5. **Be familiar with hypergeometric, geometric, negative binomials, to be used in applications.**
6. **Be able to apply distributions in applications.**
7. **Apply R as necessary.**

5.1 Introduction to Discrete Random Variables and Distributions

A **random variable** is a rule that assigns a number to every outcome of a sample space. Then, a **probability distribution** gives the probability for each value of the random variable. A **discrete random variable** takes on counting numbers 0, 1, 2, 3, ... These are either finite or countable. We will also discuss several probability distributions that are used as models for discrete random variables. Then, a probability distribution gives the probability for each value of the random variable.

Let's consider a return to our coin-flipping example earlier. Let the random variable F be the number of heads of the two flips of the coin. The possible values of the random variable F are 0, 1, and 2.

We can count the number of outcomes that fall into each category of F as shown in the probability mass function (PMF) table below.

Random Variable	0	1	2
Occurrences	1	2	1
Corresponding to events	TT	TH, HT	HH
$P(F)$	1/4	1/2	1/4

DOI: 10.1201/9781003317906-5

Note that the $\Sigma P(F) = 1/4 + 2/4 + 1/4 = 1$. This is a rule for any probability distribution. Specifically, there are two rules for valid probability distributions:

1. $P(\text{each event}) \geq 0$
2. $\Sigma P(\text{events}) = 1$

Thus, the coin-flip experiment is a valid probability distribution.

Most probability distributions have means, μ, and variances, σ^2. We can find the mean and the variance for a random variable X using the following formulas, what is known as the expected value function, $E[X]$s.

We define the mean (or expected value) for discrete random variables as follows:

$$\mu = E[X] = \sum x P(X = x).$$

The expected value is a weighted average of the possible values of X. The weights, $P(X = x)$, are the probabilities for each value.

The variance is defined as the expected squared difference of each x value from the mean:

$$\sigma^2 = E\left[(X - \mu)^2\right] = \sum (x - \mu)^2 P(X = x).$$

The variance can also be computed with the "shortcut" formula: $\boldsymbol{m = E[X] = \Sigma x P(X = x)}$

$$\sigma^2 = E\left[X^2\right] - \left(E[X]\right)^2$$

For our two-coin example, we compute the mean and variance as follows:

$$\mu = E[X] = S\, x\, P(X = x) = 0(1/4) + 1(2/4) + 2(1/4) = 1$$

We use the shortcut formula by first finding the expected value of the squared values of X. To compute the $E[X^2]$, replace each X with X^2 in the expected value formula, as follows:

$$E[X^2] = 0^2\,(1/4) + 1^2\,(2/4) + 2^2\,(1/4) = 1.5$$
$$\sigma^2 = E[X^2] - (E[X])^2 = 0(1/4) + 1(2/4) + 4(1/4)1.5 - 1^2 = 0.5$$

We can also find the standard deviation, σ, as the square root of the variance as follows:

$$\sigma = \sqrt{\sigma^2}$$

Thus, we find the variance first and then take its square root. For our example,

$$\sigma = \sqrt{.5}$$

Example 1: Going to the store.

Compute the mean and standard deviation of the following probability distribution, which represents the number of trips to a store in a given day.

x	0	1	2	3	4	5
$P(X = x)$	0.06	0.58	0.22	0.10	0.03	0.01

We check to make sure that this is a valid probability distribution. Each $P(X = x)$ is greater than or equal to 0 and less than or equal to one. The sum of all $P(X = x) = 1$, so this is a valid probability distribution.

The expected value or mean of the distribution is:

$$E[X] = (0 * 0.06) + (1 * 0.58) + (2 * 0.22) + (3 * 0.10) + (4 * 0.03) + (5 * 0.01) = 1.49$$

The variance, using the shortcut formula, is:

$$E[X^2] = (0^2 * 0.06) + (1^2 * 0.58) + (2^2 * 0.22) + (3^2 * 0.10) + (4^2 * 0.03) + (5^2 * 0.01) = 3.09$$

Let $V[X]$ be the variance.

$$V[X] = E[X^2] - (E[X])^2 = 3.09 - 1.49^2 = 0.8699$$

Finally, the standard deviation is:

$$\sigma = \sqrt{0.8699} = 0.9327$$

Another important function that describes a distribution is $V[X] = E[X^2] - (E[X])^2$

In our previous example, we found $E[X] = 1.49$, so $(E[X])^2 = 1.49^2 = 2.2201$

$$E[X^2] = (0^2 * 0.06)+(1^2 * 0.58)+(2^2 * 0.22)+(3^2 * 0.10)+(4^2 * 0.03)+(5^2 * 0.01) = 3.09, \text{ so } V[X] = E[X^2] - (E[X])^2 = 3.09 - 2.2201 = 0.8699$$

We can compute the cumulative distribution function (CDF), defined as:

$$P(X \le x) = \sum_{i=-\infty}^{x} P(X = i).$$

$$P(X \le x) = \sum_{i=-\infty}^{x} P(X = i)$$

Essentially, the CDF gives the probability of less than or equal to a given x value.

We put this into the table for the store example by summing the probabilities from the PMF at or below each x, or simply adding the probability for a given x to the CDF values up to that point (the "accumulation" of probability), as:

X	0	1	2	3	4	5
$P(X = x)$ (PMF)	0.06	0.58	0.22	0.10	0.03	0.01
$P(X \le x)$ (CDF)	0.06	0.64	0.86	0.96	0.99	1.0

Notice the key features of the CDF. The probability starts at 0. In our example, we began with the first x value but for values below $X = 0$, the probability is zero. For example, $P(X \le -1) = 0$. The values of the CDF never decrease and eventually reach a value of 1, but never higher. For example, $P(X \le 100) = 1$.

The CDF is sometimes easier to use than the PMF in computation as we illustrate in next example.

Example 2: Going to the store example revisited. Compute the following probabilities.

a. $P(X < 2)$ using both the PMF and CDF.

Using the PMF: $P(X < 2) = P(X = 0) + P(X = 1) = 0.06 + 0.58 = 0.64$

Using the CDF, $P(X < 2) = P(X \leq 1) = 0.64$

Note that using the PMF here actually repeats a calculation done for the CDF. However, in using the CDF, we must be careful about the equal sign in the probability expression!

b. $P(X \leq 2)$.

The CDF is clearly the best choice; we read the value from the table $P(X \leq 2)$ $P(X \leq 2) = 0.86$.

c. $P(X > 2)$.

This calculation can be done with the PMF by adding the probabilities for X above 2. The CDF is perhaps simpler if we rewrite the expression appropriately:

$$P(X > 2) = 1 - P(X \leq 2) = 1 - 0.86 = 0.14$$

Again, be careful with the equal sign. Since we want greater than two, we use one minus the less than or equal to 2, which is then a value given by the CDF found in part (b).

d. $P(1 \leq X \leq 4)$ using the CDF.

To write in terms of the CDF, we start with $P(X \leq 4)$. The problem is this probability includes $X = 0$. We need to subtract all values of 0 (or below) giving us:

$$P(1 \leq X \leq 4) = P(X \leq 4) - P(X\ x \leq 0) = 0.99 - 0.06 = 0.93$$

e. $P(1 < X < 4)$ using the CDF.

Can you see why the expression below is correct?

$$P(1 < X < 4) = P(X \leq 3) - P(X \leq 1) = 0.96 - 0.54 = 0.42$$

One way is to write the numbers for X (0, 1, 2, 3, 4, 5) and circle those given by the first probability:

0 1 2 3

These are correctly less than 4, but include 0 and 1. We can cross those two values off using the second probability expression. Use this technique to confirm the next two results are correct.

f. $P(1 \leq X < 4) = P(X \leq 3) - P(X \leq 0) = 0.96 - 0.10 = 0.86$

g. $P(1 < X \leq 4) = P(X \leq 4) - P(X \leq 1) = 0.99 - 0.64 = 0.35$

5.2 Bernoulli Distribution

The examples in the introduction section involved "unnamed" discrete distributions. There are some distributions that will arise commonly in our modeling and are "named" discrete distributions such as: Bernoulli, binomial and Poisson. Each arises from a specific set of assumptions for an experiment.

The Bernoulli distribution is the discrete probability distribution of a random variable, which takes a binary, Boolean output: 1 with probability p, and 0 with probability $(1 - p)$. The idea is that, whenever you are running an experiment that might lead either to a success or to a failure, you can associate with your success (labeled with 1) a probability p, while your failure (labeled with 0) will have probability $(1 - p)$. Note the assumption here: the probability of success is constant for each run of the experiment.

Formally, the probability function associated with a Bernoulli variable is the following:

$$f(x) = \begin{cases} p & \text{if } x = 1 \\ 1-p & \text{if } x = 0 \end{cases}$$

The Bernoulli distribution is simple to define since there are only two possible x values. For other distributions, it is useful to write the definition as a single function. For the Bernoulli distribution, this function is:

$$f(x) = \begin{cases} p^x (1-p)^{1-x} & \text{if } x = 0,1 \\ 0 & \text{otherwise} \end{cases}$$

We can see that the function produces the correct values by substituting the two possible x values (remember raising a term to the power 0 yields one):

$$f(0) = p^0 (1-p)^{1-0} = 1-p$$

$$f(1) = p^1 (1-p)^{1-1} = p$$

Probabilities are equally simple using the Bernoulli distribution. For example, imagine the experiment consists of flipping a coin, and you will win if the output is a tail. Furthermore, since the coin is fair, you know that the probability of having tail is $p = 1/2$. Hence, if we define the random variable as $X = 1$ for a tail (success in this case) and $X = 0$ for a head, once set tail = 1 and head = 0, you can compute the probability of success; winning (obtaining a tail) is simply as follows:

$$P(X = 11) = f(11) = p = 1/2$$

As another example, imagine you are playing a game. Suppose we toss a dice, and you bet your money on the number 1: hence, number 1 will be your success (labeled with 1), while any other number will be a failure (labeled with 0). The probability of success is then 1/6. If you want to compute the probability of failure, we find:

$$P(X = 0) = f(0) = 1 - \text{p} = 1 - 1/6 = 5/6$$

The probability of success p is the parameter of the Bernoulli distribution, and if a discrete random variable X follows this distribution, and we can compute probabilities in R using special packages. However, for the Bernoulli distribution, this is not needed for most instances; there may be occasions where computing a Bernoulli probability is part of a larger computation. Thus, we will show how to compute Bernoulli probabilities in the next section.

```
dbern(x, prob, log = FALSE)
pbern(q, prob, lower.tail = TRUE, log.p = FALSE)
qbern(p, prob, lower.tail = TRUE, log.p = FALSE)
rbern(n, prob)
```

Arguments.

x, q

Vector of quantiles.

p

Vector of probabilities.

n

Number of observations. If length(*n*) > 1, the length is taken to be the number required.

prob

Probability of success on each trial.

log, log.p

logical; if TRUE, probabilities *p* are given as log(*p*).

lower.tail

logical; if TRUE (default), probabilities are $P[X \leq x]$, = "" otherwise, = "" $p[x = "" > x]$.

Details

The Bernoulli distribution with prob = p has density $p(x) = p^x(1 - p)^{1-x}$ for $x = 0$ or 1.

If an element of x is not 0 or 1, the result of dbern is 0, without a warning; $p(x)$ is computed using Loader's algorithm; see the reference below.

The quantile is defined as the smallest value x such that $F(x) \geq p$, where F is the distribution function.

Value

dbern gives the density, pbern gives the distribution function, qbern gives the quantile function, and rbern generates random deviates.

Finally, let's compute the expected value, mean, and variance for the Bernoulli distribution. Recalling from the last section that these values, the mean, and variance of a discrete random variable are given by:

$$E(X) = \sum_x x * f(x)$$

$$V(X) = \sum_x (x - \mu)^2 * f(x) = E\left(X^2\right) - E^2(X)$$

For more complicated distributions, we will simply give the results. Since there are only two values possible for x, there are only two terms in the sum. The probability is provided with the formula for the distribution.

$$\mu = E[X] = \sum xp(X = x) = p$$

It is easy to show that $E[X^2]$ is the same since both 0 and 1 when squared are still the same. Thus, the variance is:

$$V(X) = p - p^2 = p(1 - p)$$

5.3 Binomial Distribution

Now, the idea behind the Bernoulli distribution is that the experiment is repeated only once. But what happens if we run more than one trial, under the assumption that trials are independent among each other?

For example, consider an experiment made up of a repeated number of independent and identical trials having only two outcomes, like tossing a fair coin {Head, Tail}, or a {red, green} stoplight. These experiments with only two possible outcomes on each trial are called *Bernoulli trials*. Often, they are found by assigning either an S (success) or F (failure) or a 0 or 1 to an outcome. Something either happened (1) or did not happen (0).

A binomial experiment is found by counting the number of successes in n Bernoulli trials, where n is known in advance).

For a binomial experiment, the following assumptions must hold:

(a) Consists of n trials, where n is fixed in advance.
(b) Trials are identical and can result in either a success or a failure (Bernoulli trials).
(c) Trials are independent.
(d) Probability of success, p, is constant from trial to trial.

We have defined the binomial distribution for such an experiment as:

$$b(x;n,p) = p(X = x) = \binom{n}{x} p^x (1-p)^{n-x} \quad \text{for } x = 0,1,\ldots,n.$$

Notice several things about this formula. First, we write $b(x; n, p)$ rather than just $b(x)$ to provide the "parameters" of the distribution. These are values that must be known in order to compute probabilities. The formula is only good for values of x from 0 to n. We cannot have a negative number of successes. Also, if there are n trials, we cannot have more than n successes. Thus, the probability for all other values of x is zero.

The odd-looking notation in the formula, $\binom{n}{x}$, is just another way to write a combination. Thus, as shown in a previous chapter this is computed:

$$\binom{n}{x} = nCx = \frac{n!}{x!(n-x)!}.$$

Finally, let's confirm that this formula works for a single trial, $n = 1$. Such an experiment, like a single flip of a coin, is simply the Bernoulli distribution. Recall that $0! = 1$ so $\binom{1}{1} = \binom{1}{0} = 1$ and therefore:

$$b(x;n=1,p) = \binom{1}{x} p^x (1-p)^{1-x} = p^x (1-p)^{1-x} \quad \text{for } x = 0,1$$

This is the Bernoulli distribution, confirming the Bernoulli is a special case of the Binomial with $n = 1$.

Additional key features of distributions include computing the CDF values, mean and variance. For most discrete distributions, there is no special formula for computing cumulative probabilities; we simply sum as illustrated for unnamed distributions:

$$p(X \le x) = B(x;n,p) = \sum_{y=0}^{x} \binom{n}{y} p^y (1-p)^{n-y} \quad \text{for } x = 0,1,\ldots,n.$$

The mean and variance are found in the manner illustrated for the Bernoulli. The computations require some mathematical tools from sequences and series we do not cover. The results are:

$$\mu = E[X] = np$$
$$\sigma^2 = V[X] = np(1-p)$$

Again, note that for $n = 1$, we get the values for the Bernoulli distribution.

Example 3: Two flips of a fair coin.

Consider a coin flip that we presented earlier; we will flip a fair coin twice and count the number of heads. We see that the coin-flip experiment follows the rules for and is a binomial experiment with $n = 2$ and $p = 1/2$. We will define a "success" as flipping a head, and the random variable is then X = the number of heads on two coin flips, $x = 0, 1, 2$. We say that X is distributed as binomial with the parameters given written as:

$$X \sim \text{binom}(n = 2, p = 1/2)$$

We can compute the probabilities for the experiment. For example, the probability that we get one head in two flips is:

$$b(1; n = 2, p = 1/2) = \binom{2}{1} 0.5^1 (1-0.5)^{2-1} = \frac{2!}{1!1!} 0.5(0.5) = 0.5.$$

This is the result we obtained "by hand." Notice what the formula is giving us. The two 0.5 values represent the probabilities for any sequence of flips with one H and one T. Since this is a fair coin, we have the same probabilities. The combination is the number of such paths: in this case, two of them: HT and TH. We read the combination "2 choose 1"; we are finding paths from two flips with exactly one head.

The mean and variance are:

$$\mu = np = 2\left(\frac{1}{2}\right) = 1$$

$$\sigma^2 = np(1-p) = 2\left(\frac{1}{2}\right)\left(\frac{1}{2}\right) = 0.5$$

The mean makes sense. If you flip a fair coin twice, you would expect, on average, one head.

More complicated examples require more difficult computation, so we pause to see how R can help. The binomial example is similar to other distributions, so we will discuss available commands in some detail. Searching for help on any of the commands (e.g., dbinom) will produce help for all.

There are four commands for each distribution. They have first letters "d," "p," "q," and "r" followed by the name of the distribution. For the binomial, the name is shortened to "binom," and the four commands (with their arguments) are:

```
dbinom(x, size, prob, log = FALSE)
pbinom(q, size, prob, lower.tail = TRUE, log.p = FALSE)
qbinom(p, size, prob, lower.tail = TRUE, log.p = FALSE)
rbinom(n, size, prob)
```

The first argument for each command is the primary input. The next argument(s) is (are) the parameters of the distribution. For the binomial the parameter, "size" is n, and the parameter "prob" is p. An option is provided to convert to the log scale; the default is to not do so, which is what we generally want, so we will not use the option. The other option, appropriate for two of the commands, is "lower.tail." The default is "TRUE." This, essentially, means if we are computing probabilities, they are as we would for the CDF, less than or equal to the given value of X. We will illustrate this option in several examples.

The first command, with "d" as the first letter, is the "density." For a discrete distribution, this is the PMF. In other words, it produces $P(X = x)$. Notice that the primary input is named "x" in the command. For the example we just completed, we can compute the probability of one head in two tosses as well as zero heads in two tosses:

```
> dbinom(1, size = 2, p = 1/2)
[1] 0.5
> dbinom(0, 2, 1/2)
[1] 0.25
```

The "p" in front of the command denotes "probability." This command is the CDF; it provides $P(X \leq x)$ by default. For example, $P(X \leq 1)$ should be 0.75 (0.25 + 0.5):

```
> pbinom(1, size = 2, p = ½)
[1] 0.75
```

If we change the "lower.tail" from the default to FALSE, we get the probabilities above the x value or $P(X > x)$. For example, $P(X > 1)$:

```
> pbinom(1, 2, 1/2, lower.tail = FALSE)
[1] 0.25
```

Be sure you pay attention to whether there is an equal sign for discrete distributions!

The "q" produces quantiles for the distribution. The input is a probability and the command produces the x value with that value for its CDF by default. It is the "inverse" of the "p" command. For example, if we input 0.75, we get the x value of 1:

```
> qbinom(0.75, 2, 1/2)
[1] 1
```

The "r" stands for "random." The command produces random samples from the given distribution. The input, "n," is the number of such samples to generate (a bit confusing for the binomial; it is NOT the parameter of the distribution n). We can, for example, run our two-coin-flip experiment 10 times randomly as:

```
> rbinom(10, 2, 1/2)
 [1] 1 0 1 1 1 2 1 2 1 1
```

On the first random experiment, there was one head. The second had zero heads and so forth. Note that if you run the command a second time, you will obtain different results.

$P(X = 1) = \binom{2}{1} .5^1(1-.5)^{2-1} = 0.50$

If we wanted 5 heads in 10 flips of a fair coin, then we can compute:

$$P(X = 5) = \binom{10}{1} .5^5(1-.5)^{10-5} = 0.2461$$

Example 4: Light bulbs working.

Light bulbs are manufactured in a small local plant. In testing the light bulbs, prior to packaging and shipping, they either work, S, or fail to work, F. The company cannot test all the light bulbs but does test 1 in a random batch of 100 light bulbs per hour produced in the first hour. In this batch, they found 2% that did not work, but all batches were shipped to distributors.

As a distributor, you are worried about the past performance of these light bulbs that you sell individually off the shelf. If a customer buys 20 light bulbs, what is the probability that all work?

Problem Identification: We wish to predict the probability that x light bulbs out of 20 work.

Assumptions: We have a set of n trials, where n is fixed in advance as 20 bulbs. The trials are identical and can result in either a success or a failure (Bernoulli trials) of the bulb. It is reasonable to assume the trials are independent. In other words, whether one bulb fails or not does not depend on whether others fail. Finally, we can also reasonably assume the probability of success, p, is constant from trial to trial.

We can model this situation with a binomial distribution. We define X as the number of bulbs that do not fail out of $n = 20$ bulbs. The probability of success, p, is 0.98 since the trial batch had a 2% failure rate. Finally, we want to compute $P(X = 20)$:

$$b(20; n = 20, p = 0.98) = \binom{20}{20} 0.98^{20}(1-0.98)^{20-20} = 0.98^{20} = 0.667$$

The light bulbs follow the binomial distribution rules stated earlier.

Model: Formula: $b(x; n, p) = p(X = x) = \binom{n}{x} p^x(1-p)^{b-x}$ for $x = 0, 1, 2, \ldots n$

We can use R and the following commands in R:

```
> dbinom(20, 20, 0.98)
[1] 0.667608
```

$\boldsymbol{P(X = x) \rightarrow \text{dbinom}(x, n, p)}$

Is this result surprising? The mean of the distribution is 20(0.98) = 19.6. We can use R to examine the entire PMF. R allows us to get the probabilities for all possible x values. We create a vector of values from 0 to 20 and input to the "dbinom" command. We show only the first and last probabilities; note that the last is for $X = 20$ and is the value we computed, 0.667608.

```
> x = c(0:20)
> pmf = dbinom(x, 20, 0.98)
> pmf
 [1] 1.048576e-34 . . . 6.676080e-01
> plot(x, pmf, type = "h", lwd = 12, ylab = "Probability")
```

$P(X \leq x) \rightarrow$pbinom(x, n, p)

$P(X > x) \rightarrow$ 1–pbinom(x, n, p)

$P(X \geq x) \rightarrow$ 1–pbinom($x - 1$, n, p)

$P(a \leq X \leq b)$ = pbinom(b, n, p) – pbinom($a - 1$, n, p)

$P(a \leq X < b)$ = pbinom(b – 1, n, p) – pbinom($a - 1$, n, p)

$P(a < X \leq b)$ = pbinom(b, n, p) – pbinom(a, n, p)

$P(a < X < b)$ = pbinom($b - 1$, n, p) – pbinom(a, n, p)

If we have discrete data that follows a binomial distribution, then its histogram might look like the plot of the PMF shown in Figure 5.1. We see it is heavily skewed with much higher probabilities for the larger values. This makes sense since the probability of success is high. If $p = 0.5$, the distribution is symmetric. By far, the two most likely outcomes are 19 or 20 successful bulbs. Note the probabilities for most outcomes are essentially 0. $P(X = 0)$ from the partial output above is 1.048e–34. It is extremely unlikely all 20 bulbs fail.

It is symmetric. The keys are the assumptions for the binomial as well as it being discrete.

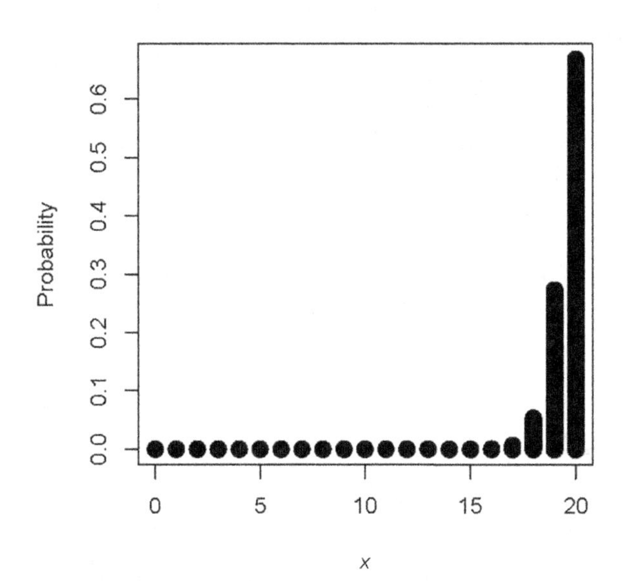

FIGURE 5.1
Binomial PMF for 20-light-bulb example.

Example 5: Weapon accuracy at a range.

A weapon has a 93% accuracy on average. If we fire 10 shots at a target, what is the probability that we hit the target exactly 5 times, at most 5 times, at least 5 times?

Solution:

We choose to use a binomial distribution model because as we can reasonably assume shots are fired independently (although it is possible missing one shot might impact performance on the next); the probability of success is known (93%); and we know in advance the number of shots fired, n (10 shots fired). Note we do assume that the probability is constant from shot to shot, which could be problematic in some situations. So, assume $X \sim \text{binom}(10, 0.93)$.

a. Exactly 5 successes means we want $P(X = 5)$.

```
> dbinom(5, 10, 0.93)
[1] 0.0002946493
```

This is a probability density function (PDF) value that we extract from $n = 5$, under PDF information. The value is 0.0003. $P(X = 5) = 0.0003$. Interpretation: If we fired 10 shots at a target, the probability that exactly 5 of the 10 hit the target is 0.0003.

b. To determine the probability that at most 5 hit the target, we need to compute $P(X \leq 5)$. This is a CDF value that we extract from $n = 5$, since we include 5, and we compute this in R:

```
> pbinom(5, 10, 0.93)
[1] 0.0003139487
```

under the CDF, $P(X \leq 5) = 0.0003$.

Interpretation:

If we fire 10 shots at a target, the probability that 5 or fewer hit the target is $P(X \leq 5) = 0.0003$. This value is not much larger than the probability of exactly 5 hits. This is because the probability of hitting only 5 was already quite small. Probabilities for fewer hits are even smaller.

c. At least 5 hit the target: this is $P(X \geq 5)$. There are several ways to compute this value in R.–This is NOT one of our known forms. We can convert the probability to its complement. $P(X \geq 5) = 1 - P(X < 5) = 1 - P(X \leq 4)$. Using this result, the R command is:

```
> 1-pbinom(4, 10, 0.93)
[1] 0.9999807
```

Alternatively, we change the "lower.tail" option in R. Be careful in doing so; the upper tail option does not have the equal sign in R. $P(X \geq 5) = P(X > 4)$, so the command is:

```
> pbinom(4, 10, 0.93,-lower.tail = FALSE)
[1] 0.9999807
```

We obtain $P(X \leq 4)$ from the CDF table and obtain 0.000 (to four decimal places). $1 - 0.0000 = 1$.

Interpretation:

We have almost certain probability (i.e., probability of 1 time) of expecting 5 or more rounds to hit the target.

5.4 Poisson Distribution

A ***Poisson process*** is a random process that generates occurrences over a certain period of time at a given rate. The Poisson distribution that varies over time (generally its time) is a model of the counts of these occurrences. The parameter for the distribution is λ, which is the *rate* for the given period of time (or area if the setting is spatial rather than temporal). There is a similarity between the binomial and Poisson distributions in that they both are modeling a number of outcomes. However, the binomial setting has a set number of trials, while the Poisson has a set time interval. There exists a rate, called α, for a short time period. Over a longer period of time, λ becomes αt, where t represents time.

A random variable is said to have a Poisson distribution, which is denoted as:

$$X \sim \text{Poisson}(\lambda)$$

The probability distribution function of X is:

$$p(x;\lambda) = \frac{e^{-\lambda}\lambda^{x}}{x!} \quad \text{For } \lambda > 0 \text{ and } x = 0,1,2,3....$$

$p(x;\lambda) = \frac{e^{-\lambda}\lambda^{x}}{x!}$ 0, for $x = 0, 1, 2, 3\ldots$ for some $\lambda > 0$. We see another difference between the Poisson and binomial in the definition. The binomial has a finite number of values that can have probability above 0 due to the set number of trials. There are theoretically infinite values that can have positive probability for the Poisson, as we can have any number of occurrences in a given interval of time (or space). In both cases, negative values have probability 0.

A ***Poisson distribution*** has the mean and variance as:

$$\mu = \sigma^{2} = \lambda.$$

In other words, the mean and variance are the same and equal to the rate. This is a key assumption of a Poisson distribution model. Examples of rates (means) in Poisson models are things like flaws per item, scratches per inch, people arriving per minute, calls received per hour, and so on. The mean, μ, of λ and variance σ^2 of λ.

We consider λ as a *rate per unit time or per unit area.* A key assumption is that with a Poisson distribution the mean and the variance are the same. The mean is a rate such as flaws per item, scratches per inch, per person, per minute, calls received per hour, and so on.

Example 6: Let X represent the number of minor flaws on the surface of a randomly selected airplane. It has been found that on average, there are five flaws per airplane surface unit. Find the probability that a randomly selected airplane surface unit has exactly two flaws.

Solution:

We assume the flaw rate is constant over the surface and model this situation as a Poisson:

$$X \sim \text{Poisson}(\lambda = 5/\text{unit}).$$

We can compute $P(X = 2)$ using R as well as using the formula. The distribution is "Poisson," and the only parameter is lambda:

```
> dpois(2, lambda = 5)
[1] 0.08422434
```

We can also use R.

$P(X = x) = \text{dpos}(x, \lambda)$
$P(X \leq x) = \text{ppos}(x, \lambda)$
$P(X > x) = 1 - \text{ppos}(x, \lambda)$
$P(X \geq x) = 1 - \text{ppos}(x - 1, \lambda)$
$P(a \leq X \leq b) = \text{ppos}(b, \lambda) - \text{ppos}(a - 1, \lambda)$
$P(a \leq X < b) = \text{ppos}(\text{b} - 1, \lambda) - \text{ppos}(\text{a} - 1, \lambda)$
$P(a < X \leq b) = \text{ppos}(b, \lambda) \text{ -ppos}(a, \lambda)$
$P(a < X < b) = \text{ppos}(\text{b} - 1, \lambda) - \text{pposax}, \lambda)$

The probability of exactly two flaws is:

$$p(X = 2) = \frac{e^{-5}5^2}{2!} = .084$$

Poisson data usually at least is slightly positively skewed. The PMF for this example, produced in the same fashion as for the binomial example earlier, is shown in Figure 5.2, and we see the skew. The smaller the rate, the more skew will exist, as smaller counts become much more likely.

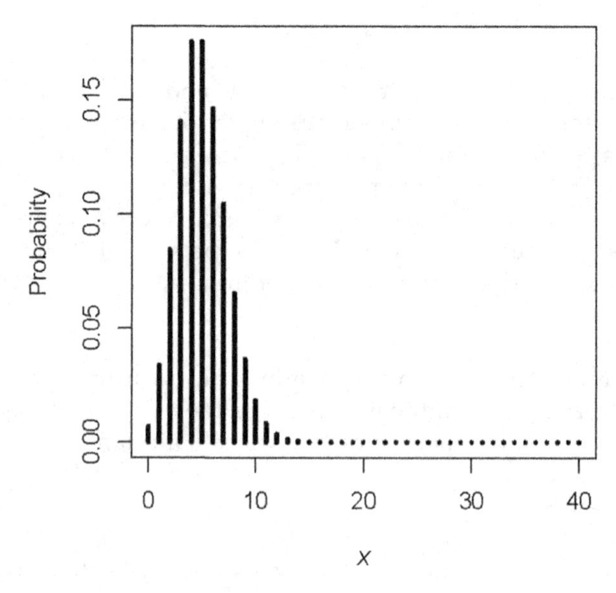

FIGURE 5.2
Poisson PMF for five flaws per unit example.

Example 7: Pulse rates.

Suppose your pulse is read by an electronic machine at a rate of five times per minute. Find the probability that your pulse is read 15 times in a four-minute interval.

Solution:

The units do not agree for the rate and desired probability. We want the probability for a four-minute interval but were given a rate over a one-minute interval. We define the given rate as:

$$\cdot\alpha = 5/\text{minute}$$

The desired rate for our Poisson model is then computed:

$$\lambda = \alpha \ t = 5 \cdot 4 = 20$$

$\lambda = \alpha$, t = 5 times 4 minutes = 20 pulses in a four-minute period. The value makes sense. If we average five per minute, over four minutes we expect 5 + 5+5 + 5 on average. We now model the four-minute-interval situation with $X \sim$ Poisson(20) and compute:

$$p(X = 15) = \frac{e^{-20} 20^{15}}{15!} = .052$$

```
> dpois(15, lambda = 20)
[1] 0.05164885
```

5.5 Chapter 5 Exercises

5.1 If 75% of all purchases at Walmart are made with a card and X is the number among 10 randomly selected purchases made with a card, then find the following:

a. $P(X = 5)$

b. $P(X \leq 5)$

c. μ, and σ^2

5.2 Russell Stover's produces fine chocolates, and it is known from experience that 10% of its chocolate boxes have flaws and must be classified as "seconds."

a. Among six randomly selected chocolate boxes, how likely is it that one is a second?

b. Among the six randomly selected boxes, what is the probability that at least two are seconds?

c. What is the mean and variance for "seconds?"

5.3 Consider the following TV ad for an exercise program: 17% of the participants lose 3 pounds, 34% lose 5 pounds, 28% lose 6 pounds, 12% lose 8 pounds, and 9% lose 10 pounds. Let X = the number of pounds lost on the program.

a. Give the probability mass function of X in a table.

b. What is the probability that the number of pounds lost is at most 6? At least 6?

c. What is the probability that the number of pounds lost is between 6 and 10?

d. What are the values of μ and σ^2, mean, and variance?

5.4 A machine fails on average 0.4 times a month (30 consecutive days). Determine the probability that there are 10 failures in the next year.

5.5 Atlanta's Hartsfield-Jackson International Airport is the busiest airport in the world. On average there are 2,500 arrivals and departures each day.

a. How many airplanes arrive and depart the airport per hour?

b. What is the probability that there are exactly 100 arrivals and departures in one hour?

c. What is the probability that there are at most 100 arrivals and departures in one hour?

5.6 According to Google's Gmail manager, an email management company, an email user gets, on average, 147 emails per day. Let X = the number of emails an email user receives per day. The discrete random variable X takes on the values $x = 0, 1, 2 \ldots$ and follows a Poisson distribution.

a. What is the probability that an email user receives exactly 160 emails per day?

b. What is the probability that an email user receives at most 160 emails per day?

c. What is the standard deviation?

5.7 Leah's iPhone receives about six telephone calls between 8 a.m. and 10 a.m. What is the probability that Leah receives more than one call in the next 15 minutes?

5.8 Last National Bank is concerned about the level of service at its drive-in window during the weekday noon hour from 12 p.m. to 1 p.m. A historical study of customer arrivals during the window's busy period revealed that, on average, 30 customers per hour arrive, with a Poisson distribution. Find the probability that from 12:30 p.m. to 12:45 p.m. exactly 11 customers arrive. Find the probability that between 8 a.m. and 11 a.m. (both inclusive) arrive.

5.9 An operator receives an average of 20 calls per hour in accordance with a Poisson process. Let X be the random variable for calls. What is the probability that she gets 9 calls in 30 minutes? What is the probability of receiving at least 9 calls in 30 minutes?

5.10 Your basketball team is playing a series of five games against your opponent. The winner is the team that wins more games (out of five). Let's assume that your team is much more skilled and has 75% chances of winning a game. It means there is a 25% chance of losing. What is the probability of your team getting three or more wins?

5.11 A box of candies has many different colors in it. There is a 15% chance of getting a pink candy. What is the probability that exactly 4 candies in a box are pink out of 10?

5.12 A coin is flipped four times. Calculate the probability of obtaining more heads than tails.

5.13 An agent sells **life insurance** policies to five equally aged, **healthy** people. According to recent data, the probability of a person living in these conditions for 30 years or more is 2/3. Calculate the probability that after 30 years:

a. All five people are still living.

b. At least three people are still living.

c. Exactly two people are still living.

5.14 If from six to seven in the evening 1 telephone line in every 5 is engaged in a conversation, what is the probability that when 10 telephone numbers are chosen at random, only 2 are in use?

5.15 The probability of a man hitting the target at a shooting range is 1/4. If he shoots 10 times, what is the probability that he hits the target exactly 3 times? What is the probability that he hits the target at least once?

5.16 There are 10 red and 20 blue balls in a box. A ball is chosen at random and it is noted whether it is red. The process repeats, returning the ball 10 times. Calculate the expected value and the standard deviation of this game.

5.17 It has been determined that 5% of drivers checked at a road stop show traces of alcohol, and 10% of drivers checked do not wear seat belts. In addition, it has been observed that the two infractions are independent from one another. If an officer stops five drivers at random:

a. Calculate the probability that exactly three of the drivers have committed any one of the two offenses.

b. Calculate the probability that at least one of the drivers checked has committed at least one of the two offenses.

5.18 A pharmaceutical lab states that a drug causes negative side effects in 3 of every 100 patients. To confirm this affirmation, another laboratory chooses 5 people at random who have consumed the drug. What is the probability of the following events?

a. None of the five patients experience side effects.

b. At least two experience side effects.

c. What is the average number of patients that the laboratory should expect to experience side effects if they choose 100 patients at random?

5.19 Abe has a hot dog stand at the college. Usually, Abe sells about 70% of his hot dogs each day. If a football weekend has Abe expect 50 customers, what is the probability of the following events?

a. No one buys a hot dog.

b. Between 23 and 35 buy hot dogs.

c. Between 23 and 35 do not buy hot dogs.

d. Compute the mean and variance.

5.20 An insurance company insures its customers against a very rare disease. The data of the last year show that the number of cases per year among all customers was roughly 1.2. Assume that the number of customers, as well as the other circumstances, has not changed.

a. What is the probability that there will be no case among the customers of the insurance in 2021?

b. What is the probability that the number of cases will be larger than 27?

c. What is the probability that the number of cases is between 23 and 27?

5.21 Given Y is a binomial random variable with $n = 20$ trials and $P(s) = 0.65$, compute the following:

a. $P(Y = 12)$, $P(Y \leq 12)$, $P(Y < 12)$, $P(Y > 12)$, $P(Y \geq 12)$

b. $P(Y \leq 6)$, $P(Y < 6)$

c. $P(6 < Y < 12)$, $P(6 < Y \leq 12)$, $P(6 \leq Y < 12)$, $P(6 \leq Y \leq 12)$

5.22 Given the following data table:

X	1	2	3	4	5	6	7	8	9
$P(X = x)$	0.05	0.09	0.013	0.088	0.25	0.3	0.1	0.08	0.02

a. Is this a distribution? Explain why or why not.

b. Find the CDF.

c. Compute the mean, variance, and standard deviation.

d. Compute the following probabilities:

$$P(1 < X < 4),\ P(1 \leq X < 4),\ P(1 < X \leq 4),\ P(1 \leq X \leq 4)$$

5.23 Given the following data table:

X	10	12	13	14	15	16	17	18	20
$P(X = x)$	0.04	0.10	0.03	0.08	0.25	0.2	0.1	0.15	0.05

a. Is this a distribution? Explain why or why not.

b. Find the CDF.

c. Compute the mean, variance, and standard deviation.

d. Compute the following probabilities:

$$P(12 < X < 15),\ P(12 \leq X < 15),\ P(12 < X \leq 15),\ P(12 \leq X \leq 15)$$

5.24 Given the following data table, is this a valid distribution> Explain your answer.

X	1	2	3	4	5	6	7	8	9
$P(X = x)$	−0.05	0.14	0.013	0.088	0.25	0.3	0.1	0.08	0.02

Is this a distribution? Explain why or why not.

5.6 Other Discrete Distributions: Hypergeometric, Geometric, Negative Binomial

In this section, we present a few additional discrete distributions: the hypergeometric, negative binomial, and geometric distributions. We present the characteristics of each, parameters of each and examples for computing probabilities in R.

5.6.1 Hypergeometric Distribution

The hypergeometric distribution deals with models with sampling without replacement in a finite population. As with the binomial distribution, there are two possible outcomes for each sample, a success or a failure. X is then the number of successes in the samples taken. An example is dealing cards from a single deck. Once the ace of spades is dealt, no one else can receive that card.

Note that there are several different ways to define the parameters for this distribution. We will use a "parameterization" consistent with R to avoid confusion in using the software. The probability distribution depends on three parameters (regardless of how they are defined):

m = the number of successes available in the population

n = the number of failures available in the population from N

k = the number of objects sampled

Returning to the cards example, suppose we wish to draw a spades and will pick 5 cards randomly without replacement. N would define $m = 13$ "successes" in the deck, the spades; $n = 39$ "failures" in the deck, not spades; and $k = 5$ cards selected. X would then be the number of spades actually observed in the cards selected. We write:

$$X \sim \text{hypergeometric}(m, n, k)$$

The probability function for the distribution involves several combinations:

$$h(x;m,n,k) = P(X = x) = \frac{\binom{m}{x}\binom{n}{k-x}}{\binom{m+n}{k}}.$$

There are constraints here such as x cannot be greater than m, $(k - x)$ greater than n, or k greater than $(m + n)$.

The formula looks complicated but is interpretable when we think about the probability lessons in previous chapters. The denominator is the total number of ways to select k objects from those available. This is the sample space. In the numerator we have the events of interest. The first combination gives the total ways to select successes from those available. For each of these, there are a number of ways to select the failures, which is why we multiply by the second combination.

The mean of the distribution is also intuitive; the number sampled times the proportion of successes in the population or:

$$\mu = k\left(\frac{m}{m=n}\right).$$

$$f(x) = \frac{\binom{a}{x}\binom{N-a}{n-x}}{\binom{N}{n}}, \quad \max\{0, n - N + a\} \le x \le \min\{n, a\}$$

In R, we can–compute the probability density for $P(X = x)$ using

```
> dhyper(x, a, N – a,)
```

To compute $P(X \le x)$ we use

> phyper(x, a, $N - a$, n)

The mean, $\mu = np$, and the variance is less intuitive:

$$\sigma^2 = k \cdot m \cdot n\left(\frac{m+n-k}{(m+n)^2(m+n-1)}\right)$$

In R, the commands use the abbreviation "hyper" after the letters described earlier. $[na(N-a)(N-n)]/(N^2(N-1))$, which simplifies to $np(1-p)\,(N-n)/(N-1))$.

Example 8: iPhone defectives.

A shipment of 25 Apple 11 iPhones contains 5 that are found defective. Ten of these 25 are selected at random. Compute the expected value and the variance. What is the probability that exactly 2 are defective? What is the probability that at most 2 are defective? What is the probability that at least 2 are defective? What is the probability that between 2 and 5 are defective (inclusive)?

Solution:

We note that we are sampling without replacement from a finite population with two possible outcomes, defective or not. Thus, the hypergeometric model is appropriate. The parameters here are:

$m = 5$ defective (since this is the interest in the probability questions)

$n = 20$ not defective

$k = 10$ sampled from the population

We will define X as the number in the sample that are not defective (i.e., successes).

The mean and variance are computed using the formulas:

$$\mu = 10\left(\frac{5}{20+5}\right) = 2$$

$$\sigma^2 = 10 \cdot 5 \cdot 20\left(\frac{5+20-10}{(5+20)^2(5+20-1)}\right) = 1000\left(\frac{15}{15000}\right) = 1$$

Again, note that the mean makes sense: 20% of the iPhones are defective, so we would expect 2 of a random sample, on average, to be defective.

Checking the R helps; we can obtain the order of the parameters. The order actually makes sense with the number of successes, m; then failures, n; and then the sample size:

```
dhyper(x, m, n, k, log = FALSE)
```

$P(X = 2)$ is then computed as below. In a situation like this, with three parameters, it may be a good idea to actually use the parameter name in the command. Doing so avoids the need to order them as expected by default. We illustrated this with a second version of the same command:

```
> dhyper(2,5,20,10)
[1] 0.3853755
```

```
> dhyper(2,m = 5,k = 10,n = 20)
[1] 0.3853755
```

The remaining problems are answered: $E[X] = (5/25)\ 10 = 2$.

a. $V[X] = [10(5)\ (25 - 5)\ (25 - 10)]/25^2(24) = 1$

   $P(X = 2)$

   > *dhyper*(2, 5, 20, 10)

   0.385

   $P(X \leq 2)$

```
> phyper(2,m = 5,n = 20,k = 10)
[1] 0.6988142
```

phyper(2, 5, 20, 10)
[1] 0.6988142

b. $P(X \geq 2) = 1 - P(X \leq 1)$; or using lower.tail = FALSE, note this is $P(X \geq 1)$

```
> 1-phyper(1,m = 5,n = 20,k = 10)
[1] 0.6865613
> phyper(1,m = 5,n = 20,k = 10,lower.tail = FALSE)
[1] 0.6865613
```

c. 1 – phyper(1,5,20, 10)

   [1] 0.6865613

   $P(2 \leq X \leq 5) = P(X \leq 5) - P(X \leq 1)$

```
> phyper(5,5,20,10)-phyper(1,5,20,10)
[1] 0.6865613
```

5.6.2 Geometric Distribution

The geometric distribution consists of a number of Bernoulli trails. This time we do not fix the number of trails in advance as we did with the binomial. Instead, we perform trial after trial stopping only when a success is obtained. The random variable (RV), X, is then the number of trials before the first success (note: another parametrization defines the RV as the number of rolls until the first success). We opt for this definition to agree with R. The assumptions for the trials are the same as for the binomial—namely, independent trials, each with a constant probability of success and two outcomes. Since we do not fix the number of trials, we do not have n and the only parameter is p. We write:

$$X \sim \text{geometric}(p)$$

The probability distribution function for a geometric random variable is:

$$g(x;p) = P(X = x) = p(1-p)^x \ \textit{for}\ x = 0,1,2,\ldots$$

Note that x must be at least 0; we need at least one trial to get a success or 0 before the first success. The formula makes sense, as it simply multiplies the probabilities for–a given number of trials. One (the last) will be a success with probability p. The other x will be failures. $F(x) = p\ (1 - p)^{x-1}$, $x = 1, 2,$

A geometric random variable has the following properties.

i. Each trial has a success or a failure.
ii. The trials are independent.
iii. $P(\text{Success}) = p$.
iv. The process continues until the first success is obtained.

The parameters, mean and variance, of this distribution are:

$$E[X] = 1/p \text{ and } V[X] = (1 - p)/p^2$$

Example 9: Favorite number on a die.

Consider rolling a fair die until your favorite number appears. What is the probability that the first 5 (your favorite number) appears in the sixth roll? On or before the sixth roll?

Solution:

We have $X \sim \text{geometric}(p = 1/6)$ and define X as the number of rolls before the first success. We can compute the first probability by noting that to have the number appear on the sixth roll, we must have $X = 5$ rolls prior. Using the formula, or in R:

$$P(X = 6) = (1/6)\ (5/6)^5 = 0.06697$$

```
> dgeom(5, p = 1/6)
[1] 0.0669796
```

For the second question, we need to add the probabilities of 0, 1, 2, 3, and 4 rolls before we see a 5. Or, we can use pgeom in R:

$$P(X \leq 5)$$

```
> pgeom(5, p = 1/6)
[1] 0.665102
```

Example 10: Consider the COVID-19 pandemic. Consider observing patients with COVID-19 until one recovers. What is the probability that there are 10 patients until we see the first recovery if the $P(s) = -1/10$?

The R commands are for the geometric distribution are as follows:

$P(X = x)$

> dgeom($x - 1$, p)

$P(X \leq x)$

> pgeom($1 - x$, p)

We will illustrate with the parameters from Example 4.

$P(X = 10)$

dgeom(9, 0.1)

[1] 0.03874205

So,

$$P(x = 10) = 0.1^1(0.9^9) = 0.0387$$

5.6.3 Negative Binomial Distribution

A distribution somewhat similar to the geometric, the negative binomial models the number X of" trials needed to produce k n successes in a series of Bernoulli trials in a negative binomial experiment is a negative binomial random variable and has a negative binomial distribution. Again, the assumptions include the following.

If repeated independent trials, each with two outcomes, results in a constant success with probability p and a failure with probability $(1 - p)$, then the probability distribution of the random variable X, defined as in R, is the number of trial failures that occur before, on which the kth success that occurs is written as:

$$X \sim \text{NB}(n, p)$$

with distribution function defined as:

$$f(x) = \binom{x+n-1}{nk-1} p^n p^k (1-p)^{x=k}$$

$$\text{for } x = k, k + 1, k + 2, \ldots 0, 1, 2, \ldots$$

In R we can use the following commands:

$$P(X = x)$$

```
dnbinom(x, size, prob, mu, log = FALSE)
```

The parameters are given by size = n and prob = p. The parameter mu is for an alternative parameterization, so we can simply ignore it.

$$P(X \leq x)$$

pnbinom(q, size, prob, mu, lower.tail = TRUE, log.p = FALSE)

Example 11: Oil drilling.

An oil company has a probability of success of 25% of striking oil when drilling a new oil well on a certain Ewing property. Find the probability that a company drills $x = 7$ wells in order to strike oil three times?

Solution:

We define X as the number failed wells to have three successes. So, $n = 3$ and $p = 0.25$.

We want to know the probability of drilling seven wells. That would mean four failed wells, so $P(X = 4)$:

```
> dnbinom(4, size = 3, prob = 0.25)
[1] 0.07415771
```

> $r = 3$
> $p = 0.20$
> $n = 7 - r$
> # exact
> dnbinom($x = n$, size = r, prob = p)
[1] 0.049152

Example 12: NBA championship.
Since the 2021 NBA season has been cut short, we will examine the hypothetical information from the 2021 NBA championship between the Golden State Warriors and Boston Celtics. The championship is the best of seven, so the winning team must obtain four wins. Let's state that the odds makers consider the Golden State Warriors, who have acquired some new players, the favorites and assign a 0.60 chance of winning over the Boston Celtics for each game.

a) Find the probability that the Warriors will win the series in a sweep, four games.
b) Find the probability that the Warriors will win the series in six games.
c) Find the probability that the Warriors win the series.

Solution:
Here X is the number of losses by the Warriors before they win $n = 4$ and $X \sim NB(n = 4, p = 0.6)$.

a) To win in a sweep, four games, would mean $X = 0$ losses. $P(X = 0)$ is:

```
> dnbinom(0, size = 4, prob = 0.6)
[1] 0.1296
```

b) A six-game series would mean $X = 2$ losses. $P(X = 2)$ is:

```
> dnbinom(3, size = 4, prob = 0.6)
[1] 0.165888
```

$P(X = 6)$
> $r = 4$
> $p = 0.60$
> $n = 7 - r$
> dnbinom($x = n$, size = r, prob = p)
[1] 0.165888

c) The Warriors can win the series with $X = 0, 1, 2$ or 3 losses. $P(X \leq 3)$.

```
> pnbinom(3, size = 4, prob = 0.6)
[1] 0.710208pnbinom(x, size = r, prob = p)
[1] 0.3369600 0.5443200 0.7102080 0.8263296 0.9006474 0.9452381 0.9707185
```

Note we use the "p" here as in pnbinom.
We see the $P(X < 6) = 0.94523$

5.7 Discrete Distribution Summary of Known Distributions

5.7.1 Binomial Distribution

A binomial distribution can be thought of as simply the number of successes in a certain number of independent trials when there are two possible outcomes on each trial (success or failure). The parameters are n, the number of trials that must be set in advance, and p, the probability of a success, which is assumed to be constant for all trials. SUCCESS or FAILURE outcome in an experiment or survey that is repeated multiple times.

An example would be the number of free throws in basketball out of three shots with a probability of making one shot of 0.9. The first variable in the binomial formula, n, stands for the number of times the experiment runs.

The second variable, p, represents the probability of one specific outcome. For example, let's suppose you wanted to know the probability of getting a 1 on a die roll. if you were to roll a die 20 times, the probability of rolling a one on any throw is 1/6. Roll 20 times, and you have a binomial distribution of (n = 20, p = 1/6). SUCCESS would be "roll a one," and FAILURE would be "roll anything else." If the outcome in question was the probability of the die landing on an even number, the binomial distribution would then become (n = 20, p = 1/2). That's because your probability of throwing an even number is one-half.

5.7.2 Poisson Distribution

A Poisson distribution is a tool that helps to predict the probability of certain events happening when you know how often the event has occurred. It gives us the probability of a given number of events happening in a fixed interval of time (or space). The parameter of the distribution is the rate, λ, that events occur on average for the period of time in question.

Poisson distributions are valid only for integers on the horizontal axis; λ (also written as μ) is the expected number of event occurrences.

For example, if we know the mean number rate (or mean) number of flaws in a single blown piece of glass is two, then we can compute the probability that the next random piece of blown glass has three flaws.

5.7.3 Geometric Distribution

The geometric distribution represents the number of failures before you get a success in a series of independent Bernoulli trials. The only parameter is p, the probability of a success on–each trial. This discrete probability distribution is represented by the probability density function:

$$f(x) = (1 - p)^{x - 1}p$$

For example, you ask people outside a polling station who they voted for until you find someone who voted for the independent candidate in a local election. The geometric distribution would represent the number of people who you had to poll before you found someone who voted independent. You would need to get a certain number of failures before you got your first success.

5.7.4 Hypergeometric Distribution

The hypergeometric distribution is a probability distribution that's very similar to the binomial distribution, as it models the number of successes in a certain number of trials. In fact, the binomial distribution is a very good approximation of the hypergeometric distribution as long as you are sampling 5% or less of the population.

Therefore, in order to understand the hypergeometric distribution, you should be very familiar with the binomial distribution. Plus, you should be fairly comfortable with the combinations formula. The hypergeometric has two main properties, however: (1) a random sample of size n k is selected without replacement. The other parameters of the model are the number of available successes, m, and failures, n, to choose from. An example would be the number of red balls selected if 10 are taken out of a bowl with 30 red and 20 white balls from N items, and (2) of the N items, k are classified as successes and $N-k$ are failures (not successes). The number of successes X is a hypergeometric random variable.

5.7.5 Negative Binomial Distribution

The negative binomial experiment is almost the same as a binomial experiment with one difference: a binomial experiment has a fixed number of trials, and in a negative binomial experiment, it is not fixed. If repeated independent trials can result in a success with probability p and a failure with probability $1 - p$, then the distribution of the random variable is a negative binomial.

The negative binomial is similar to the binomial with two differences (the number of trials, n, is not fixed).

A random variable models Y = the number of failures, x, needed in trials needed to make achieve nr successes if each trial has success probability of p.

An example is the NBA championship, which is a best of seven games. If we have an estimated probability that one team wins for each game, time is favored, let's see the probability of winning 0.55 a game over the other team; then we can compute the probability of that team winning a specific number of games. {4, 5, 6, 7} follows a negative binomial distribution since we need $n = 4$ successes. $X = 3$ losses, for example, would be the team winning the series in seven games.

5.8 Chapter 5 Exercises

5.25 A shipment of 35 Apple 11 iPhones contains 7 that are found defective. Twelve of these 35 are selected at random. Compute the expected value and the variance. What is the probability that exactly 3 are defective? What is the probability that at most 3 are defective? What is the probability that at least 3 are defective? What is the probability that between 3 and 7 are defective (inclusive)?

5.26 Consider rolling a fair die until our favorite number appears. What is the probability that the first 3 appears in the sixth roll?

5.27 Consider rolling a pair of fair die until our favorite number appears. What is the probability that the first 7 appears in the seventh roll?

5.28 Consider the COVID-19 pandemic. Consider observing patients with COVID-19 until one recovers. What is the probability that there are 100 patients until we see the first recovery if $P(s) = 1/100$?

5.29 An oil company has a probability of success of 35% of striking oil when drilling a new oil well on J. R. Ewing's Texas property. Find the probability that a company drills $x = 6$ oil wells to strike oil three times?

5.30 An oil company has a 45% probability of success of striking oil when drilling a new oil well on J. R. Ewing's Texas property. Find the probability that a company drills $x = 6$ oil wells to strike oil four times?

5.31 Since the NBA season was been cut short in 2020, we will examine the hypothetical information from the 2020 NBA championship. The championship is the best of seven, so the winning team must obtain four wins. Let's consider the Golden State Warriors who have acquired some new players have a 0.65 chance of winning over the Boston Celtics.

a) Find the probability that the Warriors will win the series in a sweep, four games.

b) Find the probability that the Warriors will win the series in six games.

c) Find the probability that the Warriors win the series.

5.32 In the 2020 presidential election, almost 52% of the country preferred the Democratic candidate, 47% voted for the Republican candidate, and the rest for Independent or other candidates. If we randomly select 100 eligible voters, what is the probability that 60 will prefer the Democratic candidate?

5.33 A fair die is rolled 10 times. What is the probability that 1, 2, and 4 occurred once each, 3 and 6 occurred twice, and 6 occurred thrice?

5.34 A certain vaccine is only 94.2% effective against a virus. Suppose the drug is administered to 1,000 students at a college. Answer the following:

a. Find the approximate probability that no one gets the virus.

b. Find the approximate probability that one gets the virus.

c. Find the approximate probability that more than two get the virus.

5.9 Chebyshev's Inequality

For any set of data and any constant $k > 1$, the proportion of the data that MUST lie within k standard deviations of the mean is at least

$$P(|X - \mu| < k\sigma) \geq 1 = (1 - 1/k^2)$$

This is the Chebyshev's inequality.

Example 13: Tire tread process is supposed to produce a mean tire tread of 8 mm with a standard deviation of 0.025 mm. With what minimum probability can we assert that the tire tread will be between 7.9 and 8.1 mm?

$$7.9 = \mu - 0.1, 8.1 = \mu + 0.1 \longrightarrow X - \mu| < 0.1$$

$$0.1 = k\sigma = 0.025\kappa \rightarrow \kappa = .1/0.025 = 4$$

$$P(|X - \mu| < k\sigma) \geq 1 = (1 - 1/4^2) = 1 - 1/16 = 15/16$$

Example 14: We are provided the following data set,

3, 3, 4, 4, 5, 5, 6, 6, 6, 7, 8, 8, 8, 8, 11, 12, 12

for false alarms to a local fire station.

Find the interval for which 82% of false alarms fall.

Solution:

Sample–mean = 0.824

Sample standard deviation = 2.856

At least 82% of false alarms must fall between ___ and ___ false calls.

$(1 - 1/k^2) = 0.82$

$-0.18 = -1/k^2$

$k^2 = 1/0.18 = 5.5555$

$k = 2.3570$

So, 82% must lie between 2.357 standard deviations of the mean. From the above, S_x = 2.856, so 2.357 standard deviations is equivalent to (2.856 times 2.357) or 6.73165.

The mean is 6.824. Thus, 82% fall between (6.824 – 6.73165) and (6.824 + 6.73165), which is

[0.09234, 13.55565].

Let's look at the empirical results; 100% of the data lie in this interval, and 100% is greater than 82% as it should be.

Example 15: What % falls between 3.824 and 9.824?

Mean end point = half interval size

6.824 – 3.824 = 3

half interval size/standard deviation = k

3/2.856 = 1.0504

Since, $k > 1$, we may proceed.

$(1 - 1/k^2) = 9.36\%$

So, at least 9.36% of the data falls in this interval.

Empirically, we count 12 of the 17 that fit in this interval.

Example 16: Coffee cans are filled to have a mean weight of 16 oz.–with a standard deviation of 0.02 oz., at least what percent of cans contain–between 15.80–and 16.20 oz. of coffee?

16.20 – 16 = 0.20 (half interval)

$k = 0.20/0.02 = 10$

Since $k > 1$ we proceed.

$(1 - 1/k^2) = 1 - 1/100 = 0.99$, at least 99% of the cans fall into this interval.

5.10 Chapter 5 Exercises

5.35 Tire tread process is supposed to produce a mean tire tread of 8 mm with a standard deviation of 0.025 mm. With what minimum probability can we assert that the tire tread will be between 7.8 and 8.2 mm?

5.36 We are provided the following data set:

3, 3, 4, 4, 5, 5, 6, 6, 6, 7, 8, 8, 8, 8, 11, 12, 12

for false alarms to a local fire station.

Find the interval for which 85% of false alarms fall.

5.37 We are provided the following data set:

3, 3, 4, 4, 5, 5, 6, 6, 6, 7, 8, 8, 8, 8, 11, 12, 12

for false alarms to a local fire station.

Find the interval for which 90% of false alarms fall.

5.38 What percent of the data in exercises 5.37 falls between 3.824 and 9.824?

5.39 Coffee cans are filled to have a mean weight of 16 oz. with a standard deviation of 0.02 oz.; at least what percent of cans contain between 15.85 and 16.15 oz. of coffee?

5.40 What percentage is guaranteed to fall within two standard deviations of the mean? Within three standard deviations? Within five standard deviations?

6

Continuous Probability Models

Objectives

1. **Know the continuous distributions.**
2. **Know how to solve probability problems with continuous distributions.**
3. **Understand the use of the normal distribution.**
4. **Be able to interpret, use, and compute inverses.**

6.1 Introduction

Some random variables do not have a discrete range of values. In the previous chapter, we saw examples of discrete random variables and discrete distributions. What if we were looking at time as a random event? Time has a continuous range of values and, thus, as a continuous random variable requires a continuous probability distribution as a model. We define a continuous random variable as any random variable measured on continuous scale. Other examples include altitude of a plane, the percent of alcohol in a person's blood, net weight of a package of frozen chicken wings, the distance by which a round misses a designated target, or the time to failure of an electric light bulb. We cannot list the sample space because the sample space is infinite. We need to be able to define a distribution as well as its domain and range.

For any continuous random variable, X, we can define a probability density function (PDF), $f(x)$. Unlike for a discrete random variable, the function itself doesn't give the probability of a certain value of X. Instead, the probabilities must be computed for a range of values.

For those that have seen calculus, we obtain probabilities for a range of x with the integral of the PDF of $f(x)$. For example, the probability that X is a value between a and b is defined as:

$$P(a \le x \le b) = \int_a^b f(x)dx$$

We know, from calculus, that this computation is the area under the curve of the PDF.

DOI: 10.1201/9781003317906-6

In similar fashion, we can define the CDF, $F(x)$. As with discrete distributions, this is the probability of values at or below a given x. Thus, the cumulative distribution function (CDF) is:

$$F(x) = P(X \leq x) = \int_{-\infty}^{x} f(x)dx$$

$F(b) = P(X \leq b)$ or the area under the PDF up to point x. A technical point for those who have had calculus is that the we could just write $P(X < x)$ here; for continuous distributions, there is no probability of an individual x value.

For those without a calculus background, another way to think of the integral is as a summing up of continuous values. If you look at the CDF for discrete distributions, the expressions are similar to the one above with a sum instead of the integral.

To be a valid probability density function (PDF):

a) $f(x)$ must be greater than or equal to zero for all x in its domain. In other words, we cannot have the possibility of negative probability, and

b) the integral $\int_{-\infty}^{\infty} f(x)dx = 1$. $\int_{-\infty}^{\infty} x \cdot f(x)dx = 1 =$ the area under the entire graph of $f(x)$ must be 1. In other words, the total probability for all values of X possible, cannot be more than 1.

The expected value or average mean value of a random variable x, with PDF defined as above, is also defined using integrals:

$$E[X] = \int_{-\infty}^{\infty} x \cdot f(x)dx$$

Again, this looks like the weighted average sum we used for discrete distributions with the integral in place of a sum. The variance is similarly computed using the same "shortcut" formula we used for discrete distributions.

In this chapter, we will see some modeling applications using many continuous distributions such as the uniform, the exponential distribution, and the normal distribution. For each of these distributions, we will not have to use calculus to get our answers to probability questions.

Since we do not require calculus, we will discuss only a few of these distributions that we obtain results with R. There are many other situations where we might need calculus; we point these out but do not demonstrate the methods, and these are marked appropriately.

6.2 Uniform Distribution

The uniform probability distribution models situations where there is a constant probability over a range of possible values of X. A graphical depiction of the distribution is shown in Figure 6.1. The general formula for the probability density function for a uniform distribution is

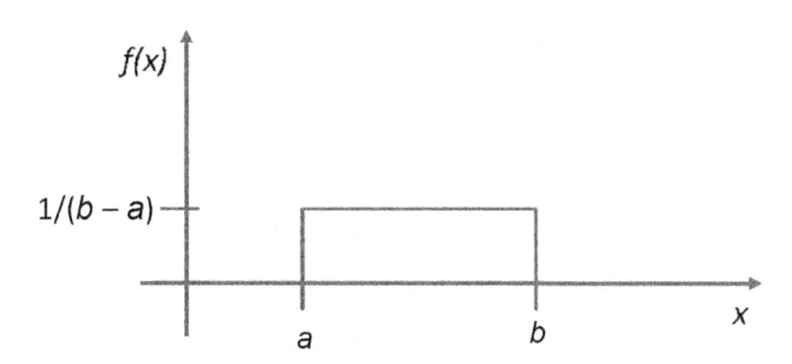

FIGURE 6.1
PDF of the Uniform (*a*, *b*) Distribution.

The general formula for the probability density function for a uniform distribution is

$$f(x) = \frac{1}{b-a} \text{ for } a \le x \le b.$$

$f(x) = \frac{1}{B-A}$ for A ≤ X ≤ B where A is a location parameter and (B – A) is a scale parameter. The parameters of the distribution are a and b, the values between which there is positive probability, and we write $X \sim$ Unif(a, b). The special case where $A = 0$ and $B = 1$ is called the standard uniform distribution.

The cumulative uniform distribution function is found as the integral of the density function from *a to x*:

$$F(x) = \int_a^x \frac{1}{b-a} dx = \frac{x-a}{b-a}$$

For those students without calculus, we do not need to use an integral since we now have a general form for the CDF as shown above. We may just use the $F(x)$ formula: $F(x) = P(X < x) = (x - a) / (b - a)$.

Some common statistics and properties of the uniform distribution are:

$$\mu = E(X) = \frac{a+b}{2}.$$

Because the distribution is symmetric, the median equals the mean = (A + B)/2.

$$\sigma^2 = V(X) = \frac{(b-a)^2}{12}.$$

Median = (A + B)–/ 2

As with discrete distributions, the standard deviation is the square root of the variance.

Range = B – A

Variance = $(B - A)^2 / 12$

Standard deviation = $\sqrt{\frac{(B-A)^2}{12}}$

The uniform distribution defines equal probability over a given range for a continuous distribution. For this reason, it is important as a reference distribution. It is intuitive to compute probabilities from this distribution as we will demonstrate in examples. We will use those examples to introduce R for continuous distributions.

One of the most important applications of the uniform distribution is in the generation of random numbers. That is, almost all random number generators generate random numbers on the (0, 1) interval. For other distributions, some transformation is applied to the uniform random numbers. We will see this in the chapter on Monte Carlo simulations.

Example 1: Old faithful eruptions.

It is common knowledge the old faithful erupts every 91 minutes. You have waited 20 minutes already to watch for an eruption. What is the probability you will see of an eruption?

Solution:

If we assume an eruption occurs with equal probability over 91-minute intervals, we can model this situation using the Uniform distribution. We can define the interval from 0 to 91 minutes by starting the "clock" with time 0 at the last eruption. Thus, we assume the model for probability of an eruption $X \sim \text{Unif}(0, 91)$.

The PDF for the distribution is then:

$$f(x) = \frac{1}{b-a} = \frac{1}{91-0} = \frac{1}{91}.$$

Figure 6.2 describes the situation in terms of the PDF. We wait for a 20-minute interval depicted by the dashed lines:

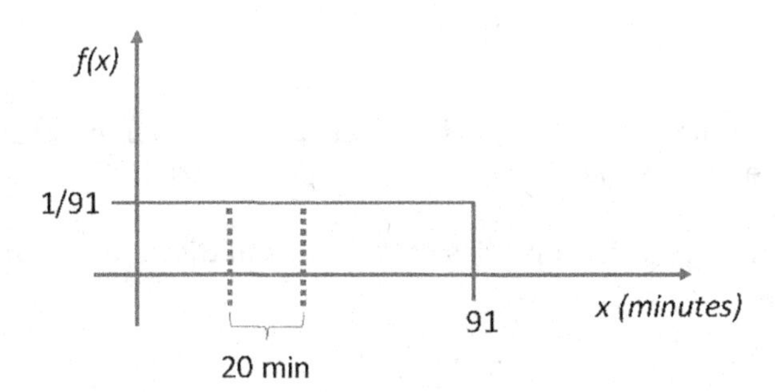

FIGURE 6.2
Old faithful PDF.

Let's take a minute to look at the area under $f(x)$, in Figure 6.2. The function forms a rectangle, so the area is just the length times the height or:

$$\text{Length} \times \text{height} = 91 \times (1/91) = 1$$

We have a valid PDF.

To compute the probability we observe an eruption, we just compute the area of the box formed by the two dashed lines or:

$$P(\text{eruption after in our 20 minutes wait}) = 20 \times (1/91) = \mathbf{0.21978}$$

We will not draw graphs for the following examples but recommend doing so as you work through each problem.

Using the uniform distribution, Area = (1/91) × (a + 20 – a)

= (1/91) × 20
= 20/91
= **0.21978**

Example 2: Math quiz.

Suppose in a mathematics quiz there are 30 participants. A question is given to all 30 participants, and the time allowed to answer it is 25 seconds. Find the probability of a random participant responding within 6 seconds?

Solution:

Again, we assume an equal probability of an answer at any point in the interval and model the probability for a single participant answer time as $X \sim$ Unif(0, 25). The PDF is then given the interval of probability distribution = [0 seconds, 25 seconds]

$$f(x) = \frac{1}{b-a} = \frac{1}{25-0} = \frac{1}{25}.$$

We can also write a formula for the CDF using the result in the previous section:

$$F(x) = P(X \le x) = \frac{x-a}{b-a} = \frac{x}{25}.$$

Density of probability = 1/ (B-A) = 1 /(25 – 0)

We are interested in $P(X < 6)$. We can compute this directly from the CDF:

$$F(6) = P(X \le 6) = \frac{6}{25}$$

Note we can also solve the problem in the same manner as Example 1, thinking about the interval of probability distribution of successful event = interest [which is from 0 seconds to 6 seconds. The area under the curve for this interval can be computed with length times height:]

6 × (1/25) = 6/25 = 0.24

We will use this example to examine the R commands for a continuous distribution. They are the same as for discrete distributions with a "d," "p," "q," or "r," followed by the name of the distribution. However, because the density (PDF) does not give probabilities for a given x, the "d" functions are not used in the same way as in the discrete case.

The description of the "dunif" command is below. We see the two parameters are the min and the max (with defaults the standard Uniform(0,1) distribution).

```
dunif(x, min = 0, max = 1, log = FALSE)
```

For our example, $X \sim$ Unif(0, 25), we compute "dunif" for three values of X, 10, 15 and 20:

```
> dunif(10, min = 0, max = 25)
[1] 0.04
> dunif(15, 0, 25)
[1] 0.04
> dunif(20, 0, 25)
[1] 0.04
```

Why are all three results the same? The "d" is the distribution or density, which for a continuous distribution is NOT a probability but rather the height of the PDF, $f(x)$, curve at that particular x value. For the uniform, the height is constant. In our example the height of the curve is $1/25 = 0.04$.

This means that to compute probabilities for continuous distributions, we must always use the "p" function, the CDF. In our example, we wanted $P(X < 6)$ which equals $F(6)$, so we can compute this using "punif" as:

```
> punif(6, 0, 25)
[1] 0.24
```

Let's look at how to apply this in a more complicated probability calculation in the next example. The probability $P(x < 6)$

The probability ratio = 6/25

There are 30 participants in the quiz.

Hence the participants likely to answer it in 6 seconds = $6/25 \times 30 \approx 7$

Example 3: Airline flight.

Suppose a flight is about to land and the announcement says that the expected time for the plane to will definitely land is in the next 30 minutes. What is the average time before the flight lands. Find the probability of getting the flight landing between 25 and 30 minutes and between 10 and 20 minutes using R.

Solution:

Once again, we assume constant probability of landing over the interval and model the probability of landing times as $X \sim$ Unif(0, 30). The mean (average) time for this distribution is:

$$\mu = E(X) = \frac{a+b}{2} = \frac{0+30}{2} = 15\,minutes$$

This answer may seem obvious; that is good, as the formula for the mean should provide a result that passes the "common sense test."

Given

Interval of probability distribution = [0 minutes, 30 minutes], you should be able to compute the two probabilities fairly quickly using the methods in the previous examples. The interval from 25 to 30 has a width of 5, so the probability is 5/30 = 0.16667. The interval from 10 to 20 is 10, so the probability is 10/30 = 1/3 = 0.333. This will help us understand how to use the R commands for continuous distributions.

For the interval 25–30, we want $P(25 < X < 30)$ (note we will not put the equal sign in, since for continuous distributions, the expressions are the same). Since there are no values above 30, this is equivalent to:

$$P(25 < X < 30) = 1 - P(X < 25) = 1 - F(25).$$

Drawing a picture helps in determining this relationship. So, in R we compute:

```
> 1 - punif(25, 0, 30)
[1] 0.1666667
```

Recall the "lower.tail" option in R also allows us to compute:

$$P(25 < X < 30) = P(X > 25)$$

```
> punif(25, 0, 30, lower.tail = FALSE)
[1] 0.1666667
```

The second probability is $P(10 < X < 20)$. We can write this in terms of CDF expressions as:

$$P(10 < X < 20) = P(X < 20) - P(X < 10) = F(20) - F(10).$$

The R code is then:

```
> punif(20, 0, 30) - punif(10, 0, 30)
[1] 0.3333333
```

We first get all the probability below 20 minutes and then remove the probability under 10 minutes. Drawing a picture and first writing appropriate probability expressions will be critical to solving problems involving continuous distributions.

Density of probability = $1/\ (30-0) = 1/30$

Interval of probability distribution of successful event = [0 minutes, 5 minutes]

The probability $(25 < X < 30)$

The probability ratio = $5/30 = 1/6$

Hence the probability of getting flight land between 25 and 30 minutes = 0.16666

Example 4: Random numbers

Suppose a random number N is taken from 690 to 850 in uniform distribution. Find the probability number N is greater than the 790?

Solution:

Given the interval of number in probability distribution = [690, 850]

Density of probability = $1/\ (850-690) = 1/160$

Now the probability is $(790 < X < 850)$

Interval of probability distribution of successful event = [790, 850] = 60

The probability ratio = $60/160 = 0.375$

Hence the probability of N number greater than 790 = 0.375

Example 5: Trains

Suppose a train is delayed by approximately 60 minutes. What is the probability that train will reach the destination within 57–60 minutes?

Solution:

Given interval of probability distribution = [0 minutes, 60 minutes]

Density of probability = 1/ (60 – 0) = 1/60

Interval of probability distribution of successful event = [0 minutes, 3 minutes]

Now the probability is $(57 < X < 60)$

The probability ratio = 3/60

The probability of train to reach between 57 and 60 minutes = 3/60 = 0.05

Example 6: Percentiles and random numbers.

Assume we have a uniform distribution with CDF as $F(x) = 1/23$ from $0 \le X \le 23X \sim \text{Unif}(0, 23)$. Determine the 90th percentile and generate 3 random numbers from the distribution.

Solution:

The first problem is an "inverse" problem. We have the probability 0.9 and want to find the x value that makes the expression $P(X < x) = 0.9$ true. A percentile is known as a quantile, so we use the "q" functions in R to solve this problem:

```
> qunif(0.9, 0, 23)
[1] 20.7
```

We can check this answer. If correct, $P(X < 20.7)$ should be 0.9:

$20.7\ (1/23) = 0.9$

```
> punif(20.7, 0, 23)
[1] 0.9
```

Functions with "r" for distributions produce random numbers. Three random values from the distribution are:

```
> runif(3, 0, 23)
[1] 19.46160 11.57077 19.68403
```

$P(X < k) = 0.90$

$(k - 0) / 23 = 0.90$

We solve for k, $k = 20.7$

6.3 Exponential Distribution

In probability theory and statistics, the **exponential distribution** (a.k.a. negative exponential distribution) is a commonly occurring family of continuous probability distributions. As an example, the exponential distribution occurs naturally when describing the lengths of the inter-arrival times in a homogeneous Poisson process. It describes the time between events in a Poisson process (described in the previous chapter)—that is, a process in which events occur continuously and **independently at a constant average rate.**

The exponential distribution occurs naturally when describing the lengths of the inter-arrival times in a homogeneous Poisson process.

In real-world scenarios, the assumption of a constant rate (or probability per unit time) is rarely satisfied. For example, the rate of incoming phone calls differs according to the time of day. But if we focus on a time interval during which the rate is roughly constant, such as peak periods from 12 to 2 p.m. during workdays, the exponential distribution can be used as a good approximate model for the time until the next phone call arrives. Similar caveats apply to the following examples, which yield approximately exponentially distributed variables:

- The time until a radioactive particle decays, or the time between clicks of a Geiger.
- The time it takes before your next telephone call.
- The time until default (on payment to company debt holders) in reduced form credit risk modeling.

Exponential variables can also be used to model situations where certain events occur with a constant probability per unit length, such as the distance between mutations on a DNA strand or between roadkills on a given road.

In queuing theory, the service times of agents in a system (e.g., how long it takes for a bank teller to serve a customer) are often modeled as exponentially distributed variables. (The inter-arrival of customers, for instance system is typically modeled by the Poisson distribution in most management science textbooks.)

6.3.1 Reliability

Reliability theory and reliability engineering also make extensive use of the exponential distribution, to model the time until failure of products.

The exponential distribution has a single parameter, the rate λ. Thus we write $X \sim \exp(\lambda)$ for a model using this distribution. The PDF and CDF for the distribution are defined:

$$\text{PDF: } PDFf(x) = \lambda e^{-\lambda x} \text{ for } x \geq 0$$
$$\text{CDF: } F(x) = P(X < x) = 1 - e^{-\lambda x} \text{ for } x \geq 0$$

Note that the values of x must be positive; the times between arrivals, for example, cannot be negative. Further, the range of possible values with positive probability is infinite. We could, theoretically, have a very large time between arrivals. As we will see in a typical graph for the distribution in one of the examples, however, the probability quickly becomes very small as x becomes really large.

Let's look at the continuous distribution of a random variable X. The properties of the distribution include equality of the mean and standard deviation:

$$\mu = 1/\lambda,$$
$$\text{Variance} = \sigma^2 = 1/\lambda^2$$
$$\text{Standard deviation} = \sigma = 1/\lambda$$

Another important property of the distribution is known as the "memoryless" property. This means that the probability of waiting four minutes is the same regardless of how long we have already waited. We will illustrate this property in an example. If assuming these properties does not make sense in a given setting, the exponential distribution is not a good model. This is an exponential distribution, where λ is the rate.

CDF = $P(X < x) = 1 - e^{-lx}$ for $x \geq 0$ (represents the area under the curve).

In solving a problem like $P(X < 2)$, we could set up the integral as

$$\int_2^X \text{»} e^{-\lambda x} dx$$

However, we have already found the CDF, $P(X < x) = F(x) = 1 - e^{-\lambda x}$ for $x \geq 0$. So we can take advantage of that result.

In R, the command for $P(X < x)$ = pexp(x, μ)

Reliability = 1 – Failure

Recall from our chapter on classical probability the rules for intersection and union as well as independence.

Series -----(A)-----(B)----

$P(A$ and $B)$ must work. A and B are independent, so

$$P(A \cap B) = P(A) * P(B)$$

Parallel events

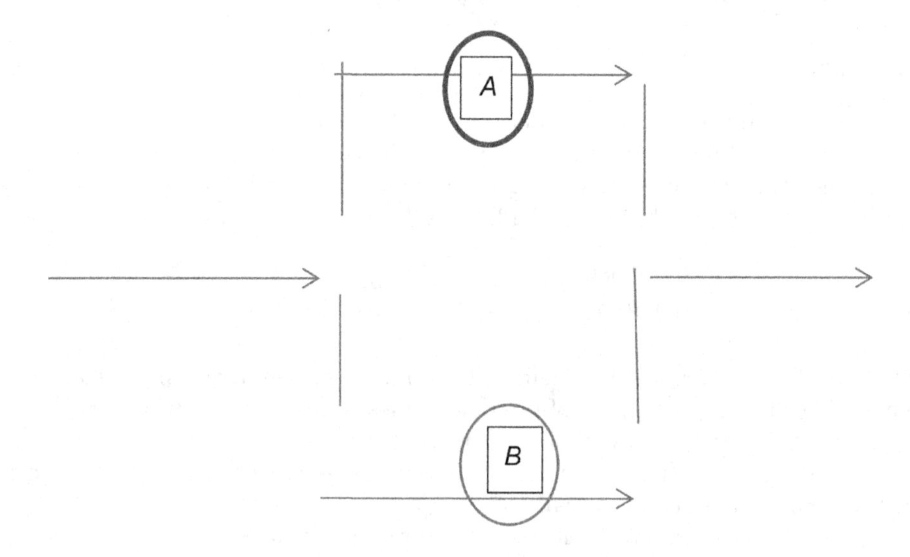

$$P(A \text{ or } B) = P(A) + P(B) - P(A \text{ and } B)$$
$$P(A \text{ or } B) = P(A) + P(B) - P(A) * P(B)$$

Example 7: Postal clerk.

Let X = amount of time (in minutes) a postal clerk spends with his or her customer. The time is known assumed to have an exponential distribution with the average amount of time equal to four minutes. Determine an appropriate model for the situation and define the PDF and CDF; then find the probabilities the clerk spends less than five minutes, more than five minutes and between two and six minutes with a customer.

Solution:

We are told the exponential distribution is reasonable, so we just need to determine the rate parameter, λ. We know that $\mu = 1/\lambda$ for the distribution, and we are given the mean, so we solve:

$$\mu = 4 = 1/\lambda$$
$$\lambda = 1/4$$

The rate is one customer every four minutes or 1/4 of a customer per minute.

X is a continuous random variable since time is measured. It is given that $\mu = 4$ minutes. To do any calculations, you must know μ, the decay parameter.

$$\lambda = 1/\mu$$
$$\text{Therefore, } \lambda = 1/4 = 0.25$$

The standard deviation, σ, is the same as the mean. $\mu = \sigma$

The distribution notation for our model is then X~Exp(λ). Therefore, $X \sim$ Exp(0.25). The PDF and CDF are:

The probability density function is $f(X) = \lambda \cdot e^{-\lambda x}$. The number e = 2.71828182846 . . . . It is a number that is used often in mathematics.
$f(x) = 0.25 \cdot e^{-0.25x}$ where $x > 0$ and $\lambda = 0.25$.
CDF $F(x) = P(X < x) = 1 - e^{-\lambda x} = 1 - e^{-0.25}x$ for $x > 0$

The graph of the PDF is shown in Figure 6.3. We see there is very little area under the curve (probability) for ranges of x values beyond 20.

$$\mu = -4.$$

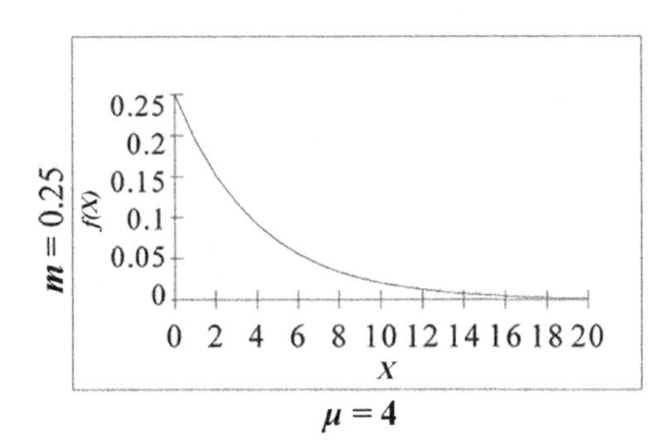

FIGURE 6.3
Exponential distribution PDF with mean rate $\lambda = 1/4$.

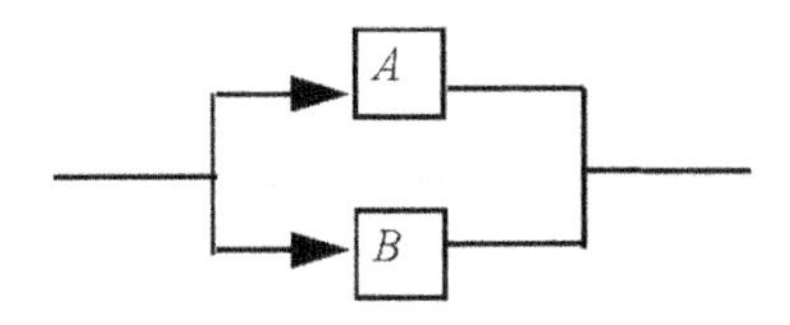

FIGURE 6.4
Parallel system of two components A and B.

Notice the graph is a declining curve. When $X = 0$,

we can compute the probabilities by writing probability expressions in terms of the CDF. We illustrate using both the formula and R. The distribution is "exp," and the parameter is the "rate" for the R functions.

The probability of spending less than five minutes is $P(X < 5)$, which is exactly in the form of the CDF. $P(X > 5)$, $P(2 < X < 6)$

$$P(X < 5) = F(5) = 1 - e^{-0.25 * 5} = 0.713495$$

```
> pexp(5, rate = 1/4)
[1] 0.7134952
```

The probability of spending more than 5 minutes is $P(X > 5)$. We use the complement to write an expression using the CDF. Note we can compute this in R using the "lower.tail = FALSE" option as well.

$$P(X > 5) = 1 - P(X < 5) = 1 - 0.713495 = 0.2865$$

```
> 1 - pexp(5, 1/4)
[1] 0.2865048

> pexp(5, 1/4, lower.tail = FALSE)
[1] 0.2865048
```

The last probability is of time between two and six minutes or $P(2 < X < 6)$. Here we first get the probability of time less than six minutes and then subtract the time less than two minutes. We used this approach in the uniform distribution examples previously.

$$P(2 < X < 6) = P(X < 6) - P(X < 2) = F(6) - F(2) = 0.7768698 - 0.393469 = 0.3834008$$

```
> pexp(6, 1/4) - pexp(2, 1/4)
[1] 0.3834005
```

Example 8: Postal clerk "memoryless" property.

For the model in the previous problem, find the probability that a clerk spends more than four to five minutes with a randomly selected customer. Then find the probability the clerk spends more than four minutes with a randomly selected customer given the clerk already spent two minutes with the customer.

The first probability is $P(X > 4)$:

```
> pexp(4, 1/4, lower.tail = FALSE)
[1] 0.3678794
```

Solution:

Find $P(4 < X < 5)$.

The second is a conditional probability, so we need to review the definition of conditional probability in a previous chapter. Doing so, we recall:

$$P(A|B) = \frac{P(A \cap B)}{P(B)}.$$

For our example, since the clerk already spent two minutes and we want the chance of spending more than an additional four; event A is the probability of more than six minutes. B is "given," which is the probability of spending more than two minutes (since that is what has already occurred):

$$P(X > 6 | X > 2) = \frac{P(X > 6 \cap X > 2)}{P(X > 2)}.$$

We know how to compute the denominator. Let's look at the numerator more closely. The event is that the clerk spends more than six minutes with a customer AND spends more than two minutes with a customer. If the clerk does spend more than six minutes, then he or she had to have spent more than two minutes with the customer. Thus,

$$P(X > 6 \cap X > 2) = P(X > 6)$$

The desired probability is then:

$$P(X > 6 | X > 2) = \frac{P(X > 6)}{P(X > 2)}.$$

```
> pexp(6,1/4,lower.tail = FALSE) / pexp(2,1/4,lower.tail = FALSE)
[1] 0.3678794
```

This is the same probability of just spending more than four minutes with the customer. A remarkable result known as the "memoryless" property!

We use the CDF.

The cumulative distribution function (CDF) gives the area to the left.

$$P(X < x) = 1 - e^{-\lambda x}$$
$$P(X < 5) = 1 - e^{-0.25 \cdot 5} = 0.7135$$
$$\text{and } P(X < 4) = 1 - e^{-0.25 \cdot 4} = 0.6321$$

Note:

You can do these calculations easily on a calculator.

The probability that a postal clerk spends four to five minutes with a randomly selected customer is

$$P(4 < X < 5) = P(X < 5) - P(X < 4) = 0.7135 - 0.6321 = 0.0814$$

Example 9: Postal worker quantiles.

For the postal worker example, determine the time at which half of all customers are finished or within how long. Compare this to the average (mean) time.

Solution:

The time when half the customers finish is the 50th percentile, or the median. This is the inverse problem where we want to find the quantile q to satisfy the equation:

$$P(X < q) = 0.5$$

Using the formula for the CDF and solving for q:

$$P(X < q) = 0.5 = 1 - e^{-0.25q}$$

$$0.5 = e^{-0.25q}$$

$$\ln(0.5) = -0.25q$$

$$q = 2.77$$

In R we use "qexp":

```
> qexp(.5, 1/4)
[1] 2.772589
```

Find the 50th percentile.

$P(X < k) = 0.50$, $k = 2.8$ minutes (calculator or computer)

Half of all customers are finished within 2.8 minutes.
You can also do the calculation as follows:

$P(X < k) = 0.50$ and $P(X < k) = 1 - e^{-0.25 \cdot k}$
Therefore, $0.50 = 1 - e^{-0.25 \cdot k}$ and $e^{-0.25 \cdot k} = 1 - 0.50 = 0.5$
Take natural logs: $\ln(e^{-0.25 \cdot k}) = \ln(0.50)$. So, $-0.25 \cdot k = \ln(0.50)$
Solve for: $k = \ln(0.50)/-0.25 = 2.8$

Example 10: Which is larger, the mean or the median?
The mean of the distribution is four minutes (given, but also 1/0.25), while the median is only 2.877. The mean is larger due to the right skew in the exponential distribution, which impacts the mean more than the median.

Example 11: Exponential distribution.
Twenty units were reliability tested with the following results of a histogram as shown in Figure 6.5.

Number of Units in Group	Time-to-Failure
7	100
5	200
3	300
2	400
1	500
2	600

FIGURE 6.5
Histogram of data

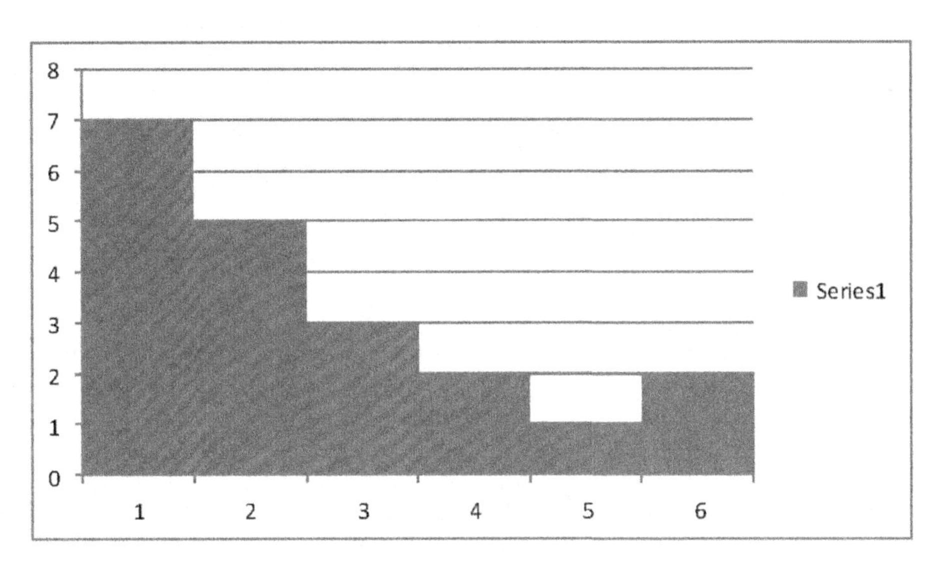

FIGURE 6.5
(Continued)

	Column 1
Mean	255
Standard error	37.32856
Median	200
Mode	100
Standard deviation	166.9384
Sample variance	27868.42
Kurtosis	−0.10518
Skewness	0.959154
Range	500
Minimum	100
Maximum	600
Sum	5100
Count	20

OK, now what?
Assume an exponential distribution with $\mu = 255$ hours or
$\lambda = 1/255 = 0.00392156$ or 0.00392156 failures per hour.
So the average lifetime is 255 hours.

$$P(X > 3) = 1 - P(X < 3) = 1 - e^{(0.0039 * 3)} = 0.0116957$$

This about a 1.1% chance of having more than 3 failures in a given hour.
So what if we want the following:
P(more than 3 failures in a day)
λ is now $(0.00392156 * 24) = 0.094117$ per day

$$P(X > 3) = 1 - P(X < 3) = 1 - 0.754 = 0.24599$$

6.4 The Normal Distribution

A continuous random variable X is said to have a normal distribution with parameters μ and σ (or μ and σ^2), where $-\infty < \mu < \infty$ and $\sigma > 0$, if the PDF of X is:

$$f(x;\mu,\sigma)=\frac{1}{\sqrt{2\pi\sigma}}e^{\frac{-(x-\mu)^2}{(2\sigma^2)}},-\infty \le x \le \infty$$

The plot of the normal distribution is our bell-shaped curve; see Figure 6.6. The mean and variance of the distribution are given by the parameters, μ and σ. Computing probabilities with this function is mathematically difficult, so we will take advantage of R for all computations.

The normal distribution is the most "famous" and used of all statistical distributions. The reason is twofold. One reason is that a surprising number of things tend to follow the bell-shaped curve. The second is that even if a given experiment does not follow the normal distribution, if we form new random variables that are either sums or averages of other random variables, the new random variables are normally distributed. This remarkable result is known as the Central Limit Theorem (CLT) and will be very important to much of the statistical inference covered in later chapters.

Normal distribution calculations:

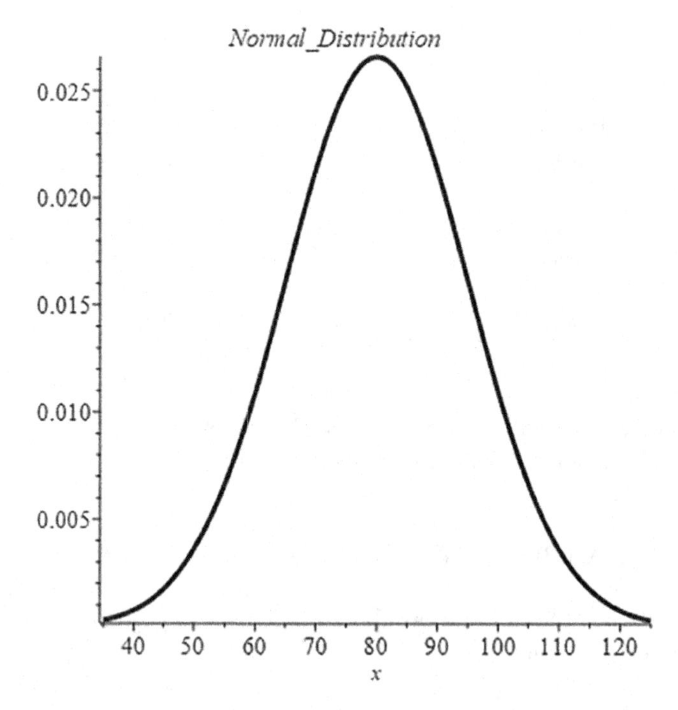

FIGURE 6.6
Bell-shaped curve of the normal distribution.

To compute $P(a < x < b)$ when x is a normal random variable, with parameters μ and σ, we must evaluate

$$\int \frac{1}{\sqrt{2\pi\sigma}} e^{\frac{-(x-\mu)^2}{(2\sigma^2)}} dx$$

Since none of the standard integration techniques can be used to evaluate this integral, the standard normal random variable Z with parameters $m = 0$ and $\sigma = 1$ has been numerically evaluated and tabulated for certain values. Since most applied problems do not have parameters of $m = 0$ and $\sigma = 1$, "standardizing" transformation can be used $Z = \frac{x - \mu}{\sigma}$.

Example 12: Suppose x is distributed as normal with mean 10 and standard deviation 2. Find the probability that x is less than 9, greater than 11, between 9 and 11. Also, determine the median and 90th percentile of the distribution. Finally, determine the probability of being within 1 standard deviation, within 2 standard deviations and within 3 standard deviations of the mean. For example, the amount of fluid dispensed into a can of diet coke is approximately a normal random variable with mean 11.5 fluid ounces and a standard deviation of 0.5 fluid ounces. We want to determine the probability that between 11 and 12 fluid ounces, $P(11 < x < 12)$ are dispensed.

Solution:

a. Probabilities: less than 9, greater than 11, and between 9 and 11

In R, the two parameters required are the mean and the standard deviation. We can compute the probabilities:

$$P(X < 9)$$

```
> pnorm(9, mean = 10, sd = 2)
[1] 0.3085375
```

$$P(X > 11)$$

```
> pnorm(11, 10, 2, lower.tail = FALSE)
[1] 0.3085375
```

$$P(9 < X < 11)$$

```
> pnorm(11, 10, 2) - pnorm(9, 10, 2)
[1] 0.3829249
```

Can you explain why the first two probabilities are the same? We recommend drawing pictures and shading the area under the bell-shaped curve represented by each probability statement. Doing so quickly reveals that the reason is the symmetry of the normal distribution. These probability statements are "mirror images" of one another.

Notice also that these three probabilities add up to 1. The regions divide the bell-shaped curve into three parts and represent all the possible probability.

b. Median and 90th percentile

The "q" for quantile command in R yields:

```
> qnorm(0.5, 10, 2)
[1] 10
> qnorm(0.9, 10, 2)
[1] 12.5631
```

The median is equal to the mean since the distribution is symmetric.

c. Probabilities of being within 1, 2, and 3 standard deviations of the mean.

To solve this problem, we need to interpret "within *x* standard deviations." Since the standard deviation is 2, we are within 1 standard deviation of the mean 10 if between 8 and 12 (10 – 2 and 10 + 2). For two standard deviations, we add/subtract 2 * 2 = 4 and for three standard deviations 3 * 2 = 6. The probability statements and calculations are: $Z_1 = (11 - 11.5)/0.5 = -1$

$$P(8 < X < 12) \text{ is within 1 standard deviation of the mean}$$

```
> pnorm(12, 10, 2) - pnorm(8, 10, 2)
[1] 0.6826895Z2 = (12-11.5)/.5 = 1
```

$$P(6 < X < 14) \text{ is within 2 standard deviations of the mean}$$

```
> pnorm(14, 10, 2) - pnorm(6, 10, 2)
[1] 0.9544997
```

$$P(4 < X < 16) \text{ is within 3 standard deviations of the mean}$$

```
> pnorm(16, 10, 2) - pnorm(4, 10, 2)
[1] 0.9973002
```

These probabilities are worth noting, as they hold for all normal distributions and can be useful for estimation and also "common sense" checks of results. For the normal distribution:

- Approximately 2/3 of the probability is within 1 standard deviation of the mean.
- Approximately 95% of the probability is within 2 standard deviations of the mean.
- Approximately 99% of the probability is within 3 standard deviations of the mean.

6.5 Checking Normality

Later in the course, we will present methods to test things like whether data is distributed normally (hypothesis testing). Since the normal distribution is used often for modeling, checking that data is actually distributed normally is important. We can use histograms and boxplots to look for a bell-shaped curve and symmetry. Without the hypothesis testing tools, there is one additional plot that we will present here and return to in later chapters known as a Q–Q plot.

A Q–Q plot is a scatterplot of the quantiles estimated from data and the theoretical quantiles of a given distribution. If the data follows the distribution, these values should be roughly equal, so plotting them should lead to a straight line $y = x$. We will illustrate the Q–Q plot in an example.

Example 13: Normal distribution Q–Q plot.

Produce a random sample of 100 observations from a normal distribution and test that the sample is normally distributed using graphs. Repeat the experiment using a sample from an exponential distribution.

Solution:

We can use "rnorm" to obtain random samples from a normal distribution. In our solution, we decided to use the normal distribution from the previous example with mean 10 and standard deviation 2. You can use any values you like. We then produce first a histogram of the 100 values and the Q–Q plot.

```
> normsamp <- rnorm(100, 10, 2)
> hist(normsamp)
> qqnorm(normsamp)
```

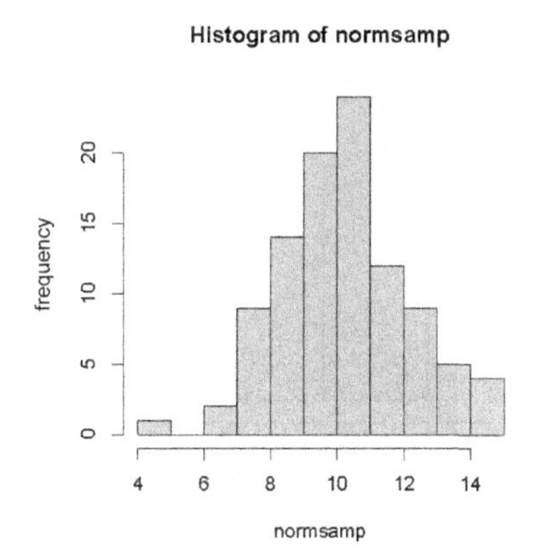

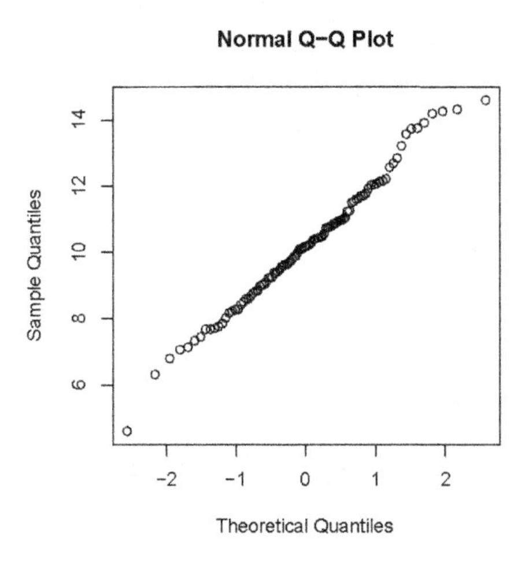

The histogram is reasonably a bell-shaped curve. 100 is actually a fairly small sample and a larger sample should look increasingly normal. The Q–Q plot does appear to follow a straight line. The slight departures from the line are not enough to question the assumption that the data is from a normal distribution.

We next use "rexp" to obtain random samples from an exponential distribution. In our solution, we decided to use a rate parameter of 1/10, which is a mean of 10 (see Section 6.3 on the exponential distribution), the same as the mean of the normal distribution we used. This is not necessary, and you should try this for a variety of exponential distributions.

```
> expsamp <- rexp(100, rate = 1/10)
> hist(expsamp)
> qqnorm(expsamp)
```

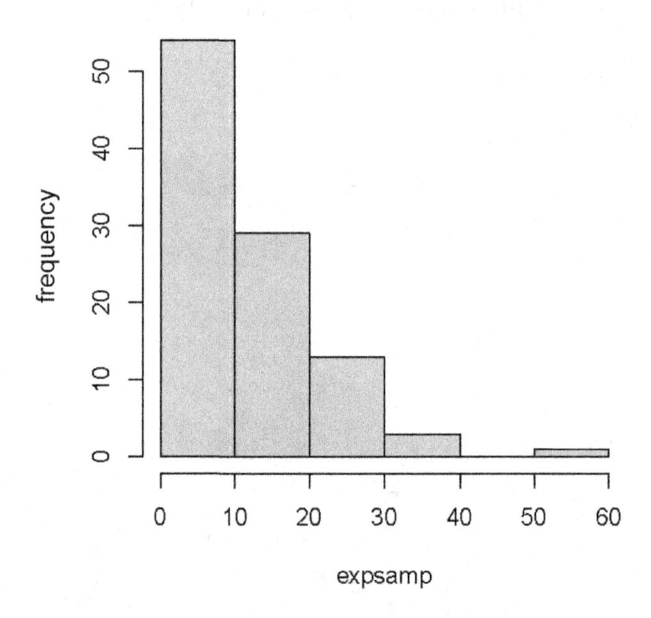

This probability statement $P(11 < x < 12)$ is equivalent to $P(-1 < Z < 1)$. If we use the tables, we can compute this to be 0.8413 – 0.1587 = 0.6826. However, we can use R to compute the area between 11 and 12. A graphical depiction is given in Figure 6.7.

The R command to find $P(X < x)$ and $P(X \leq x)$-→pnorm(x, μ, σ)

And for $P(Z < z) \rightarrow$ pnorm(z)

$P(a \leq X \leq b) = P(a \leq X < b) = P(a < X \leq b) = P(a < X < b) \rightarrow$ pnorm(b, μ, σ) – pnorm(a, μ, σ)

And in Z, which

$P(z1 < Z < z2) \rightarrow$nnorm(z1) – pnorm(z2)

To find the $P(a < x < b)$ or in our case $P(11 < x < 12) = 0.682689$.

Therefore, 68.26% of the time the cans are filled between 11 and 12 fluid ounces.

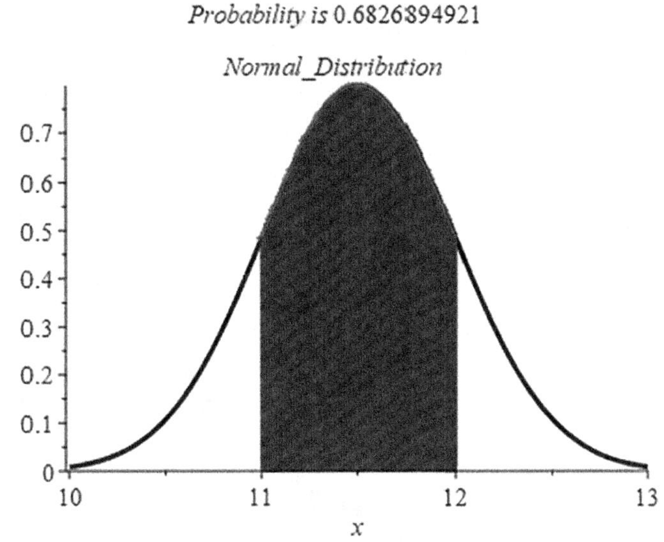

A left Riemann sum approximation of $\int_{11}^{12} f(x)\,dx$, where $f(x) = 0.5641895835\, e^{-2.000000000\,(x-11.5)^2}\sqrt{2}$ and the partition is uniform. The approximate value of the integral is 0.6826733603. Number of subintervals used: 100.

FIGURE 6.7
Normal Distribution area from 11 to 12.

6.5.1 Inverse of Normal Distribution

Often in either the RV X or Z we need the value of x or z that corresponds to a specific probability.

In R the command is

> qnorm, "probability", "mean", "stdev")

For example, if we need the value of $Z \sim N(0,1)$ that corresponds with a 0.95 probability,

> qnorm(0.95, 0, 1)

[1] 1.6448536

For example, if we need the value of $X \sim N(100, 15)$ that corresponds with a 0.95 probability,

> qnorm(0.95, 100, 15)

[1] 124.672804

The histogram is nowhere near a bell-shaped curve. This is not surprising, as we know the exponential distribution is right skewed. The Q–Q plot reflects the lack of normality, Figure 6.8. Instead of following a straight line, there is a clear "U" shape. Skewed distributions will exhibit this shape; if left skewed, as an "upside down" U.

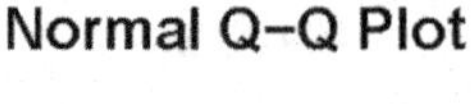

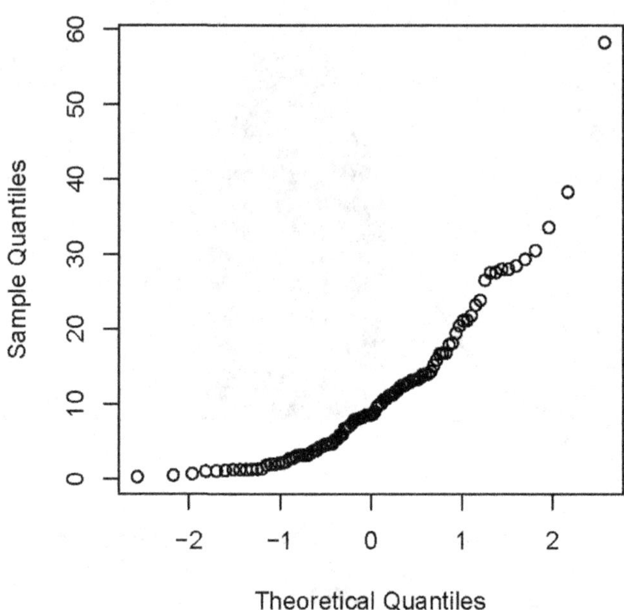

FIGURE 6.8
Q–Q plot of data.

6.6 Chapter 6 Exercises

6.1 Find the following probabilities:

a. $X \sim N(\mu = 10, \sigma = 2)$, $P(X > 6)$

b. $X \sim N(\mu = 10, \sigma = 2)$, $P(6 < x < 14)$

6.2 Determine the probability that lies within one standard deviation of the mean, two standard deviations of the mean, and three standard deviations of the mean for a standard normal (mean 0 and standard deviation 1) distribution. Draw a sketch of each region.

6.3 A tire manufacturer thinks that the amount of wear per normal driving year of the rubber used in their tire follows a normal distribution with mean = 0.05 inches and standard deviation 0.05 inches. If 0.10 inches is considered dangerous, then determine the probability that $P(X > 0.10)$.

6.4 $X \sim N(0,1)$. Compute $P(X > 2)$.

6.5 $X \sim N(0,1)$. Compute $P(X > 1)$.

6.6 $X \sim N(0,1)$. Compute $P(X > 3)$.

6.7 $X \sim N(0,1)$. Compute $P(0 < X < 2)$.

6.8 Find the value of X if $X \sim N(\mu = 10, \sigma = 2)$ that represents the 99th and 95th percentile.

6.9 Find the value of X if $X \sim N(\mu = 20, \sigma = 4)$ that represents the 95th percentile.

6.10 Find the value of X if $X \sim N(\mu = 20, \sigma = 4)$ that represents the 99th percentile.

6.11 Find the value of X if $X \sim N(\mu = 0, \sigma = 1)$ that represents the 90th, 95th, and 99th percentiles.

6.12 Find the value of X if $X \sim N(\mu = 0, \sigma = 1)$ that represents the 97.5th percentile.

6.13 Find the value of X if $X \sim N(\mu = 0, \sigma = 1)$ that represents the 99.5th percentile.

6.14 X is a normally distributed variable with mean $\mu = 30$ and standard deviation $\sigma = 4$. Find:

a. $P(X < 40)$

b. $P(X > 21)$

c. $P(30 < X < 35)$

6.15 A radar unit is used to measure speeds of cars on a motorway. The speeds are normally distributed with a mean of 90 km/hour and a standard deviation of 10 km/hour. What is the probability that a car picked at random is traveling at more than 100 km/hour?

6.16 For a certain type of computer, the length of time between charges of the battery is normally distributed with a mean of 50 hours and a standard deviation of 15 hours. John owns one of these computers and wants to know the probability that the length of time will be between 50 and 70 hours.

6.17 Entry to a certain university is determined by a national test. The scores on this test are normally distributed with a mean of 500 and a standard deviation of 100. Tom wants to be admitted to this university, and he knows that he must score better than at least 70% of the students who took the test. Tom takes the test and scores 585. Will he be admitted to this university?

6.18 The lengths of similar components produced by a company are approximated by a normal distribution model with a mean of 5 cm and a standard deviation of 0.02 cm. If a component is chosen at random,

a. what is the probability that the length of this component is between 4.98 and 5.02 cm?

b. what is the probability that the length of this component is between 4.96 and 5.04 cm?

6.19 The length of life of an instrument produced by a machine has a normal distribution with a mean of 12 months and standard deviation of 2 months. Find the probability that an instrument produced by this machine will last a) less than 7 months b) between 7 and 12 months.

6.20 The time taken to assemble a car in a certain plant is a random variable having a normal distribution of 20 hours and a standard deviation of 2 hours. What is the probability that a car can be assembled at this plant in a period of time a) less than 19.5 hours? b) between 20 and 22 hours?

6.21 A large group of students took a test in physics, and the final grades have a mean of 70 and a standard deviation of 10. If we can approximate the distribution of these grades by a normal distribution, what percent of the students:

a. scored higher than 80?

b. should pass the test (grades ≥ 60)?

c. should fail the test (grades < 60)?

6.22 The annual salaries of employees in a large company are approximately normally distributed with a mean of $50,000 and a standard deviation of $20,000.

a. What percent of people earn less than $40,000?

b. What percent of people earn between $45,000 and $65,000?

c. What percent of people earn more than $70,000?

6.23 X is a normally distributed variable with mean $\mu = 30$ and standard deviation $\sigma = 4$.

a) $P(x < 40)$

b) $P(x > 21)$

c) $P(30 < x < 35)$

6.24 The amount of time spouses shop for anniversary cards can be modeled by an exponential distribution with the average amount of time equal to 8 minutes. Write the distribution, state the probability density function, and graph the distribution.

6.25 Let X = amount of time (in minutes) a postal clerk spends with his or her customer. The time is known to have an exponential distribution with the average amount of time equal to 4 minutes. Let X be a **continuous random variable** since time is measured. It is given that $\mu = 4$ minutes. To do any calculations, you must know m, the decay parameter. Graph the PDF and CDF

6.26 For the information presented in exercise 6.25, find the probability that a clerk spends 4 to 5 minutes with a randomly selected customer.

6.27 The number of days ahead travelers purchase their airline tickets can be modeled by an exponential distribution with the average amount of time equal to 15 days. Find the probability that a traveler will purchase a ticket fewer than 10 days in advance. How many days do half of all travelers wait?

6.28 On the average, a certain computer part lasts 10 years. The length of time the computer part lasts is exponentially distributed.

a) What is the probability that a computer part lasts more than 7 years?

b) On the average, how long would five computer parts last if they are used one after another?

6.29 The number of miles that a particular car can run before its battery wears out is exponentially distributed with an average of 10,000 miles. The owner of the car needs to take a 5000-mile trip. What is the probability that he will be able to complete the trip without having to replace the car battery?

6.30 Scores on a test have a mean of 77.7 and a standard deviation of 15, find the probability of scores that fall below 85. Assume a normal distribution.

6.31 Each month, an American household generates an average of 28 pounds of newspaper for recycling. Assume a standard deviation of 2 pounds. Find the probability that:

the family generates between 27 and 31 pounds.

More than 30.2 pounds per month.

6.32 The American Automobile Association (AAA) reports that the average time it takes to respond to an emergency call is 25 minutes. Assume the random variable is approximately normally distributed with a standard deviation of

4.5 minutes. If 80 calls are randomly selected, approximately how many will be answered in less than 15 minutes? Less than 20 minutes?

6.33 An exclusive college desires to only accept the top 10% of all graduating seniors based upon results of a national placement test. The test has a mean of 600 and a standard deviation of 100. Find the cutoff for the exam. Assume the random variable is normally distributed.

6.34 The average hourly wage of workers is $11.76. Assume the RV is approximately normally distributed. If the standard deviation is $2.72, find these probabilities:

a. Worker earns more than $12.55.

b. Worker earns less than $8.

c. Worker earns between $10-$13.

6.35 Determine the 90% percentile of a normal RV whose mean is 89 and standard deviation is 11.2.

6.36 The amount of time spouses shop for anniversary cards can be modeled by an exponential distribution with the average amount of time equal to 8 minutes. Write the distribution, state the probability density function, and graph the distribution.

6.37 Let X = amount of time (in minutes) a postal clerk spends with his or her customer. The time is known to have an exponential distribution with the average amount of time equal to 6 minutes. What is the probability the clerk spends less than 3 minute with a customer? More than 10 minutes?

6.38 Scores on a test have a mean of 87.7 and a standard deviation of 15.3. Find the probability of scores that fall below 85. Assume a normal distribution.

6.39 The average hourly wage of workers is $11.76. Assume the RV is approximately normally distributed. If the standard deviation is $2.72, find these probabilities:

a. Worker earns more than $12.55.

b. Worker earns less than $8.

c. Worker earns between $10–$13.

6.40 Determine the 90% percentile of a normal RV whose mean is 89 and standard deviation is 11.2.

7

Other Continuous Distribution (Some Calculus Required): Triangular, Unnamed, Beta, Gamma

Objectives

1. **Know the application of probability to continuous probability distribution functions.**
2. **Apply the calculus as needed to solve problems involving continuous distributions.**
3. **Find μ and σ^2 of continuous distribution functions.**

We already discussed the rules for probability in previous chapters to include for some of the more common distributions. In this chapter, we extend those ideas to some additional more complicated and less common distributions and demonstrate how calculus is used with continuous distributions. We also briefly mention two flexible and important distributions for modeling, the gamma and beta distributions.

7.1 Right Triangular Distribution

A right triangular distribution is a triangle whose area under the nonnegative curve from a to b is equal to 1 as shown in Figure 7.1 for $a = 0$ and $b = 2.5$. Notice that we can confirm that this graph does represent a valid PDF using geometric. The area of the triangle is 1/2 base × height. Here area = 1/2 (2.5)(0.8) = 1. Also, the curve is always greater than 0.

The PDF is given by the equation of the line. The intercept is 0.8. The slope is calculated using the "rise" over the "run." We can use the two points where the line intersects the axis (0, 0.8) and (2.5, 0) and compute:

$$slope = \frac{y_2 - y_1}{x_2 - x_1} = \frac{0 - 0.8}{2.5 - 0} = -0.32$$

Thus, the equation of the PDF is:

$$f(x) = 0.8 - 0.32x \text{ for } 0 < X < 2.5$$

DOI: 10.1201/9781003317906-7

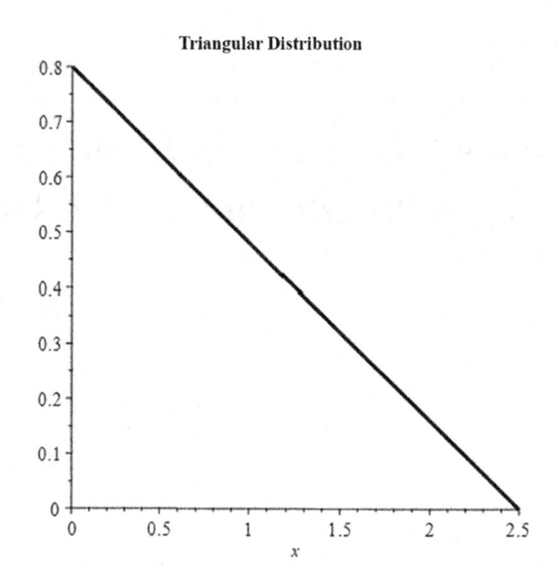

FIGURE 7.1
Right triangular distribution.

From the previous chapter, we know the CDF is defined as:

$$F(x) = P(X < x) = \int_{-\infty}^{x} f(x)dx$$

For $X < 0$, the values of $f(x)$ are 0, so the $F(x)$ is zero for that region. In the region from $x = 0$ to $x = 2.5$, we can compute the CDF:

$$F(x) = \int_{0}^{x} (0.8 - 0.32x)dx = 0.8x - 0.16x^2.$$

Notice that when $x = 2.5$, $F(2.5) = 0.8(2.5) - 0.16(2.5)^2 = 1$. Above $x = 2.5$, there is no additional probability, so the CDF above that value remains one. The final CDF is defined as:

$$F(x) = P(X < x) = \begin{cases} 0 & x < 0 \\ -0.16x^2 + 0.8x & 0 \le x \le 2.5 \\ 1 & x > 2.5 \end{cases}$$

Example 1: Computation with right triangular distributions.
Using the right triangular distribution from Figure 7.1 compute the following:

(a) $P(X < 0.5)$
(b) $P(X > 1)$
(c) $P(0.5 < X < 1)$
(d) Find the mean and standard deviation of the distribution

Solution:

(a) $P(X < 0.5) = F(0.5) = -0.16\,(0.5^2) + 0.8\,(0.5) = 0.36$

Note we could compute from the PDF as $\int_0^{0.5} f(x)dx$. This calculation, however, was already done in computing the CDF.

(b) $P(X > 1) = 1 - P(X < 1) = 1 - F(1) = 1 - (-0.16 + 0.8) = 1 - 0.64 = 0.36$

Again, we could compute using PDF as $\int_1^{2.5} f(x)dx$

(c) $P(0.5 < X < 1) = F(1) - F(0.5) = 0.64 - 0.36 = 0.28$

(d) Mean = $\int_0^{2.5} x * f(x)dx = 0.83333$, variance = $\int_0^{2.5} (x - 0.83333)^2 f(x)dx = 0.347$, and standard deviation = $\sqrt{0.347} = .58925$

7.2 General Triangular Distributions

A more general triangle like the one is Figure 7.2, a more flexible model. We will use this graph as an example to illustrate how such a probability distribution is created and utilized.

Example 2: Given the triangular distribution in Figure 7.2, (a) determine the PDF and CDF and (b) show it is a valid distribution.

Finding the PDF:

The PDF is built as a piecewise function. We have four regions for values of x: x below 1, from 1 to 6, from 6 to 9, and then greater than 9. For the regions $x < 1$ and $x > 9$, the PDF values are 0. For the other two regions, the PDF is described by equations of lines.

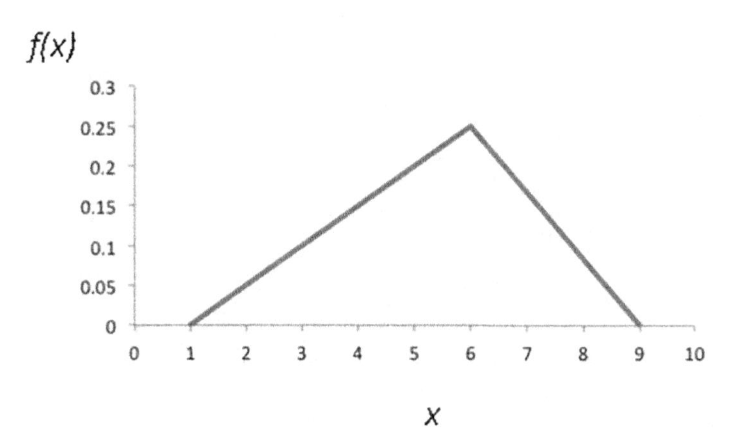

FIGURE 7.2
Triangular distribution for Example 2.

In the region $1 < x < 6$, we use the two known points (1, 0) and (6, 0.25). The slope of the line is:

$$slope = \frac{y_2 - y_1}{x_2 - x_1} = \frac{0.25 - 0}{6 - 1} = \frac{1}{20}.$$

The point slope form for the equation of the line is:

$$slope = \frac{y - y_1}{x - x_1}$$

We will use (1, 0) as the point in this equation giving us:

$$\frac{1}{20} = \frac{y - 0}{x - 1}.$$

Solving the equation for y produces: $y = \frac{1}{20}x - \frac{1}{20}$.

In the region $6 < x < 9$, we use the two known points (6, 0.25) and (9, 0) to first get the slope:

$$slope = \frac{0 - 0.25}{9 - 6} = \frac{-1}{12}.$$

Using the point (9, 0) in the point slope form of the line and solving for y, the equation of the line $y = \frac{-1}{12}x + \frac{9}{12}$.

The PDF is then:

$$f(x) = \begin{cases} 0 & x < 1 \\ \frac{1}{20}x - \frac{1}{20} & 1 \le x \le 6 \\ \frac{-1}{12}x + \frac{9}{12} & 6 \le x \le 9 \\ 0 & x > 9 \end{cases}$$

Finding the CDF:

When $X < 1$, the CDF is 0, as the PDF is 0. We begin accumulating probability at $X = 1$. The CDF for the first line from $X = 1$ to $X = 6$ is:

$$F(x) = P(X < x) = \int_1^x \left(\frac{1}{20}x - \frac{1}{20} \right) dx = \frac{1}{40}x^2 - \frac{1}{20}x + \frac{1}{40}$$

The CDF for the region from $x = 6$ to $x = 9$ is tricky. The cumulative probability must include all the probability for $x < 6$ as well. We can obtain that amount by using the result for the previous region or:

$$P(X<6)=F(6)=\frac{1}{40}6^2-\frac{1}{20}6+\frac{1}{40}=\frac{36-12+1}{40}=\frac{5}{8}.$$

After $x = 6$, the CDF is the 5/8 plus the accumulation of probability under the second line up to x or:

$$F(x)=P(X<x)=\frac{5}{8}+\int_6^x\left(\frac{-1}{12}x+\frac{9}{12}\right)dx.$$

After computing the integral, noting that the cumulative probability above $x = 9$ does not increase, so the CDF should be 1 for those values, we can write the CDF as:

$$F(x)=P(X<x)=\begin{cases} 0 & x<1 \\ \frac{1}{40}x^2-\frac{1}{20}x+\frac{1}{40} & 1\le x\le 6 \\ \frac{-1}{24}x^2+\frac{9}{12}x-\frac{19}{8} & 6\le x\le 9 \\ 1 & x>9 \end{cases}$$

Confirm this is a valid distribution:

There are several ways to do this. The integral of the PDF should be 1 (the area under the curve). We already computed the area under the first line as 5/8. Thus, the area under the second should be 3/8. That is:

$$\int_6^9\left(\frac{-1}{12}x+\frac{9}{12}\right)dx=\frac{3}{8}.$$

If you start computing this integral, you will realize you did it before, in finding the CDF. So, a "shortcut" is to use the CDF. If we compute the CDF for $x = 9$, the result should equal 1, as we have accumulated all the probability at that point:

$$F(9)=P(X<9)=\frac{-1}{24}9^2+\frac{9}{12}9-\frac{19}{8}=\frac{-81+162-57}{24}=1.$$

A check can also be done by computing the areas of the two right triangles formed by the graph.

Example 3: Given the triangular distribution in Example 2 find the following:

(a) $P(X < 5)$
(b) $P(X < 7)$
(c) $P(X > 7)$
(d) $P(5 < X < 7)$

Solutions:

(a) $P(X<5)=F(5)=\frac{1}{40}5^2-\frac{1}{20}5+\frac{1}{40}=\frac{25-10+1}{40}=0.4$

(b) Be sure to use the correct equation from the CDF:

$$P(X<7)=F(7)=\frac{-1}{24}7^2+\frac{9}{12}7-\frac{19}{8}=\frac{-49+126-57}{24}=\frac{5}{6}$$

(c) Using the complement, $P(X>7)=1-P(X<7)=1-\frac{5}{6}=\frac{1}{6}$

(d) $P(5<X<7)=F(7)-F(5)=\frac{5}{6}-0.4=\frac{13}{30}$

7.3 General Continuous Distributions

We can create probability distributions to model a variety of situations other than just the triangular distribution using functions. The key is that the function is never negative and the area under the curve is 1. We will look at several examples of such functions.

Example 4: A quadratic function.

Consider the function, $f(x) = x^2/3$ from $-1 < x < 2$.

First, let's check to ensure this is a valid probability distribution. The function $x^2/3$ is always greater than or equal to 0 over our interval, as we can confirm from the graph in Figure 7.3.

We next use integration to ensure the area under the curve is one:

$$\int_{-1}^{2}\frac{x^2}{3}dx=\frac{x^3}{9}\bigg|_{-1}^{2}=\frac{8}{9}-\frac{-1}{9}=1.$$

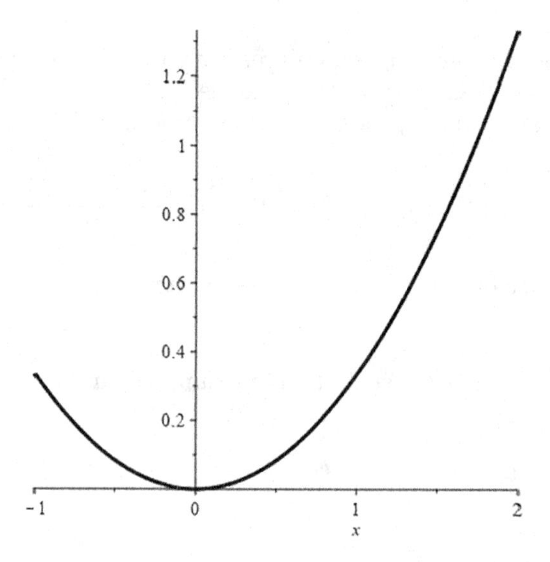

FIGURE 7.3
Plot of $x^2/3$ that is always greater than or equal to 0 in our interval.

Our function is a valid probability distribution. The calculation above actually provides us the form of the function for the CDF:

$$F(x) = P(X < x) = \begin{cases} 0 & x < -1 \\ \frac{x^3}{9} & -1 \le x \le 2 \\ 1 & x > 2 \end{cases}$$

We can compute the expected value (mean), variance, and standard deviation for our valid probability distribution function.

Recall that the expected value is computed for continuous distributions as:

$$E[X] = \mu = \int_{-\infty}^{\infty} x f(x) dx.$$

For our example, the integral is 0 except over the region –1 to 2 (known as the "support"), so:

$$E[X] = \mu = \int_{-1}^{2} x \frac{x^2}{3} dx = \frac{x^4}{12}\bigg|_{-1}^{2} = \frac{5}{4}.$$

The variance is found as the expected value of the square difference from the mean or:

$$V[X] = \int_{-\infty}^{\infty} (x-\mu)^2 f(x) dx.$$

We have computed the mean, which can be substituted into this integral. However, it is often easier to use the shortcut formula and first compute the expected value of X^2:

$$E\left[X^2\right] = \int_{-1}^{2} x^2 \frac{x^2}{3} dx = \frac{x^5}{15}\bigg|_{-1}^{2} = \frac{33}{15}.$$

Then we obtain the variance:

$$V[X] = E\left[X^2\right] - (E[X])^2 = \frac{33}{15} - \frac{25}{16} = \frac{51}{80}.$$

The standard deviation, σ, is $\sqrt{51/80} = 0.7984$

Example 5: Confirm that the following is a valid probability density function, and find the probability that $X < b$.

$$f(x) = \begin{cases} \frac{5}{8b} & \frac{2b}{5} \le x \le 2b \\ 0 & else \end{cases}$$

Solution:

If we integrate the PDF over the support, result should equal 1:

$$\int_{2b/5}^{2b} \frac{5}{8b} dx = \frac{5x}{8b}\bigg|_{2b/5}^{2b} = \frac{10b}{8b} - \frac{2b}{8b} = 1$$

Thus, this is a valid PDF for all values of b.

$$P(X < b) = \int_{2b/5}^{b} \frac{5}{8b} dx = \frac{5x}{8b}\bigg|_{2b/5}^{b} = \frac{5b}{8b} - \frac{2b}{8b} = \frac{3}{8}.$$

7.4 The Gamma Distribution

There are several named distributions that are particularly flexible for modeling. They are useful, as one need not develop a function to describe the situation assuming they produce a graph that fits the data. We will look at two very briefly: the gamma and beta distributions.

Graphs of the gamma distribution for various values of its two parameters are shown in Figure 7.4. We can see that the distribution is useful for modeling situations where the

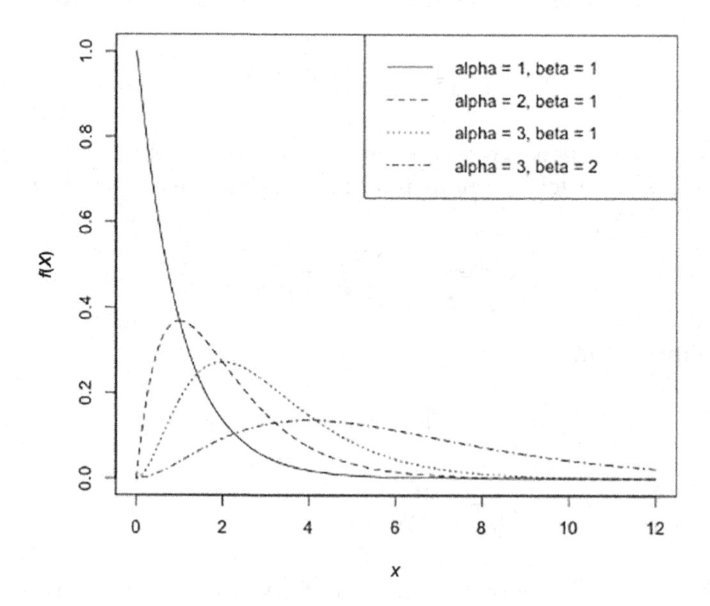

FIGURE 7.4
Plot of the gamma distribution for various parameter values.

random variable, X, is always nonnegative. The parameters of the distribution can change the shape quite dramatically, which makes it a good choice for modeling many such random variables.

We also note that the exponential distribution previously covered in Chapter 5, is a special case of the gamma distribution with $\alpha = 1$. Another special case is the chi-square distribution, χ^2 with parameter v > 0, which is a gamma distribution with parameters $\alpha = v/2$ and $\beta = 2$. We will see that the χ^2 is used for inference in many situations in later chapters.

If $X \sim Gamma(\alpha, \beta)$, the PDF is:

$$f(x) = \begin{cases} \frac{1}{\beta^{\alpha}\Gamma(\alpha)} x^{\alpha-1} e^{-x/\beta} & x \geq 0 \\ 0 & else \end{cases}$$

Both parameters must be positive. The alpha parameter is known as the shape parameter, as it changes the degree of curvature of the graph. The beta parameter is known as the scale parameter, as it shifts the distribution along the x-axis.

There is a strange looking function you may not recognize in the PDF. It is known as the gamma function and, for $\alpha > 1$, is defined as:

$$\Gamma(\alpha) = \int_0^{\infty} x^{\alpha-1} e^{-x} dx.$$

However, it is almost always computed using factorials, as it can be shown that:

1. For any $\alpha > 1$, $\Gamma(\alpha) = (\alpha\text{-}1)\Gamma(\alpha\text{-}1)$
2. For any positive integer, n, $\Gamma(n) = (n - 1)!$
3. $\Gamma(1/2) = \sqrt{\pi}$

The mean and variance of the distribution are:

$$E[X] = \alpha\beta$$
$$Var(X) = \alpha\beta^2$$

Probabilities and quantiles are best computed using R with commands similar to those previously covered for various distributions. The two parameters are referred to as the "shape" (α) and the "scale" (β) parameters:

```
pgamma(q, shape, scale, lower.tail = TRUE)
```

Example 6: Gamma computations.

The curve in Figure 7.4 with the highest peak is gamma with $\alpha = 1$ and $\beta = 1$. The lowest peak has parameters $\alpha = 3$ and $\beta = 2$. Compare these two distributions further by computing their mean, standard deviation, and median and finding the probability that X is less than 1/2 and greater than 3/2.

Solution:

The mean and standard deviation for each distribution:

$X \sim \text{Gamma}(\alpha = 1, \beta = 1)$

$$E[X] = \alpha\beta = 1$$
$$\text{Var}(X) = \alpha\beta^2 = 1$$
$$SD(X) = \sqrt{1} = 1$$

$X \sim \text{Gamma}(\alpha = 3, \beta = 2)$

$$E[X] = \alpha\beta = 3(2) = 6$$
$$\text{Var}(X) = \alpha\beta^2 = 3(2^2) = 12$$
$$SD(X) = \sqrt{12} = 3.46$$

The curve with the higher peak has the lower variability, as expected since the graph is much "tighter." The mean is also lower. The distribution is shifting to lower values of x.

We can compute the median in R using the "q" with the distribution:

```
> qgamma(0.5, shape = 1, scale = 1)
[1] 0.6931472
> qgamma(0.5, shape = 3, scale = 2)
[1] 5.348121
```

The median is less affected by skew, so is lower than the mean for both distributions.

Probabilities are similarly computed in R:

$$P(X < ½):$$

```
> pgamma(0.5, shape = 1, scale = 1)
[1] 0.3934693
> pgamma(0.5, shape = 3, scale = 2)
[1] 0.002161497
```

$$P(X > 3/2):$$

```
> pgamma(1.5, shape = 1, scale = 1, lower.tail = FALSE)
[1] 0.2231302
> pgamma(1.5, shape = 3, scale = 2, lower.tail = FALSE)
[1] 0.9594946
```

The probabilities make sense. The flatter distribution has very few small x values.

7.5 Beta Distribution

Very often, we deal with proportions. A PDF that is often applied to random variables that are restricted to values between 0 and 1 is the beta distribution. Like the gamma

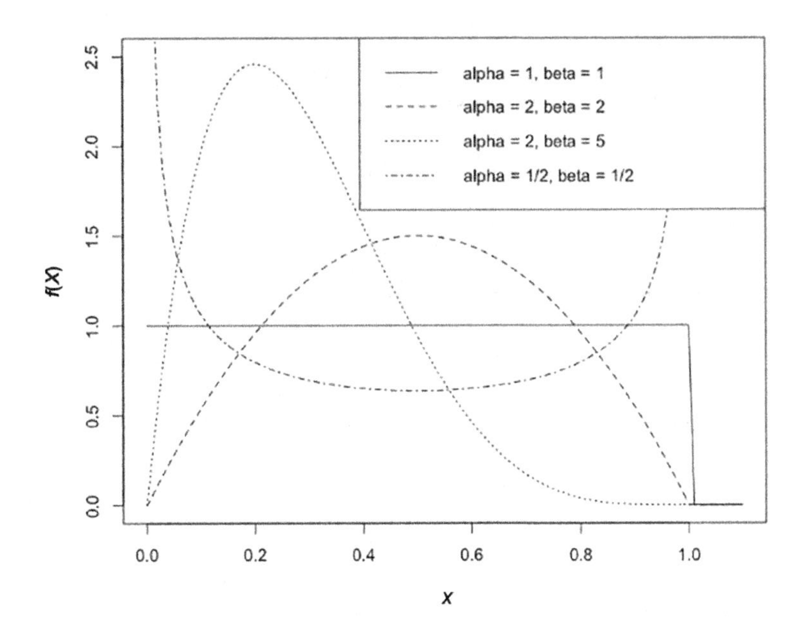

FIGURE 7.5
Plot of the beta distribution for various parameter values.

distribution, the beta distribution has two positive parameters, α and β, and can produce a wide variety of curves as shown in Figure 7.5. In fact, if both parameters equal one, we have the uniform distribution.

The beta PDF is, like the gamma, a complicated looking function:

$$f(x) = \begin{cases} \dfrac{\Gamma(\alpha+\beta)}{\Gamma(\alpha)\Gamma(\alpha)} x^{\alpha-1}(1-x)^{\beta-1} & 0 \le x \le 1 \\ 0 & else \end{cases}$$

For a Beta(α,β) distribution, the mean and variance are known:

$$E[X] = \alpha/(\alpha+\beta)$$

$$Var(X) = \frac{\alpha\beta}{(\alpha+\beta)^2(\alpha+\beta+1)}$$

We can compute various probabilities using R. The command for CDF values is "pbeta." Since the scale of the distribution is always between 0 and 1, the two parameters are both referred to as shape parameters. Looking at the help in R, we can confirm that "shape1" refers to alpha and "shape2" is beta. The command with primary options is:

```
pbeta(q, shape1, shape2, lower.tail = TRUE).
```

To obtain quantiles, the "qbeta" command is used as we have seen for other distributions.

Example 7: Premium gas.

At a local gas station, the proportion of customers who use premium gasoline is a random variable with Beta(1,4). If the gas station serves on average 850 customers every week, how many customers are expected to use premium gasoline over the next week? Additionally, find the probability that more than 25% of the customers will purchase premium gasoline over the next week.

Solution:

The expected proportion is found using the mean of the beta distribution:

$$E[X] = \alpha/(\alpha + \beta) = 1/(1+4) = 1/5$$

Since we have 850 customers, we would expect 850 * 0.2 = 170 customers to purchase premium gasoline.

Using R to obtain the probability that more than 25% purchase premium gasoline, we can use the complement:

$$P(X > 0.25) = 1 - P(X \leq 0.25) = 1 - F(0.25)$$

```
> 1 - pbeta(0.25, shape1 = 1, shape2 = 4)
[1] 0.3164063
```

Or, as we have seen in previous examples, we can change the "lower.tail" option:

```
> pbeta(0.25, 1, 4, lower.tail = FALSE)
[1] 0.3164063
```

7.6 Chapter 7 Exercises

7.1 In Example 7, compute the expected value if $n = 1000$.

7.2 In Example 7, if X is Beta(2, 5), then compute $P(X > 0.30)$.

7.3 Suppose a system contains a certain type of electrical component whose time to failure, measured in years, is given by a random variable T. We assume the T is exponentially distributed, a gamma distribution with $\alpha = 1$ and the mean time to failure is $\beta = 5$. If we installed four components, what is the probability that at least three are functioning at the end of 10 years.

7.4 Suppose a system contains a certain type of electrical component whose time to failure, measured in years is given by a random variable T. We assume the T is exponentially distributed, a gamma distribution with $\alpha = 1$ and the mean time to failure is $\beta = 6$. If we installed four components, what is the probability that at least two are functioning at the end of nine years.

7.5 The length of time between customer's complaints about a specific type of chair follows a gamma distribution with $\alpha = 2$ and $\beta = 4$. Changes were made in dealing with the quality control of the product. It has been 15 months since the last complaint; has quality control worked?

7.6 The total number of hours, measured in units of 100 hours, that a family runs their new Shark vacuum over a period of one year is a continuous random variable X with the following density function:

$$f(x) = \begin{cases} x & 0 < x < 1 \\ 2 - x & 1 \le x < 2 \\ 0 & else \end{cases}$$

(a) Verify that this is a valid density function.

(b) Find the probability that over a period of one year, a family runs the vacuum less than 110 hours and between 65 and 105 hours.

(c) Find the expected amount of time they use the vacuum over a year.

7.7 Assume we have a possible distribution for an item's size (measured in micrometers) given by:

$$f(x) = \begin{cases} 3x^{-4} & x > 1 \\ 0 & else \end{cases}$$

(a) Verify that this is a valid density function.

(b) Compute $F(x)$, the CDF.

(c) Compute the probability that a random particle exceeds three micrometers.

(d) Compute the probability that a random particle is between two and four micrometers.

7.8 Consider the function:

$$f(x) = \begin{cases} \frac{2(x+2)}{5} & 0 < x < 1 \\ 0 & else \end{cases}$$

(a) Graph the function, and use geometry to validate that this is a valid PDF. If it is not valid, change one value in the function to create a valid PDF.

(b) Verify the function is a valid PDF using calculus (either the original or the new function if you had to modify in part a).

(c) Compute the probability $P(1/4 < x < 1/2)$, and find the mean and standard deviation of the distribution.

(d) Find the median of the distribution. How does it compare to the mean? Explain why this is the case by considering the graph.

8

Sampling Distributions

Objectives

1. **Know the central limit theorem (CLT).**
2. **Know how to apply the central limit theorem.**
3. **Know how CLT applies to means and totals.**

This chapter is somewhat of a "bridge" from probability to statistics. The central limit theorem (CLT) acts as that bridge. In statistics, it is critical to remember the difference between a population and a sample. Let's define again a population and a sample.

Population: The entire collection of all the observations of a particular type: population for the entire United States, height and weight of every student in your college, and so forth.

Sample: The data set that is acquired through the process of observation that is a random selection from the population.

Probability models and random variables and their probability distributions are generally models for the population. Very often in analysis, it is not practical to obtain an entire population. It might be easier and more useful to obtain a random sample. Therefore, we will want to infer the characteristics of the population from the sample's statistics.

In statistics, we are interested in using a sample to make inferences about the population or characteristics of the population such as the mean or standard deviation. In order to do this, we need to be able to identify distributions for the statistics from samples. The CLT is an important such tool for identifying a "sampling distribution."

8.1 The Sampling Distribution of the Mean

The Central Limit Theorem is one of the most important theorems in statistics. Applications abound, and it is the primary tool for much of statistical inference: using data to make statements about hypotheses of interest.

To illustrate the CLT, Let's assume we start with a random sample of $n = 49$ observations from a population with an unknown distribution. The histogram for the random variable, X, is shown in Figure 8.1.

DOI: 10.1201/9781003317906-8

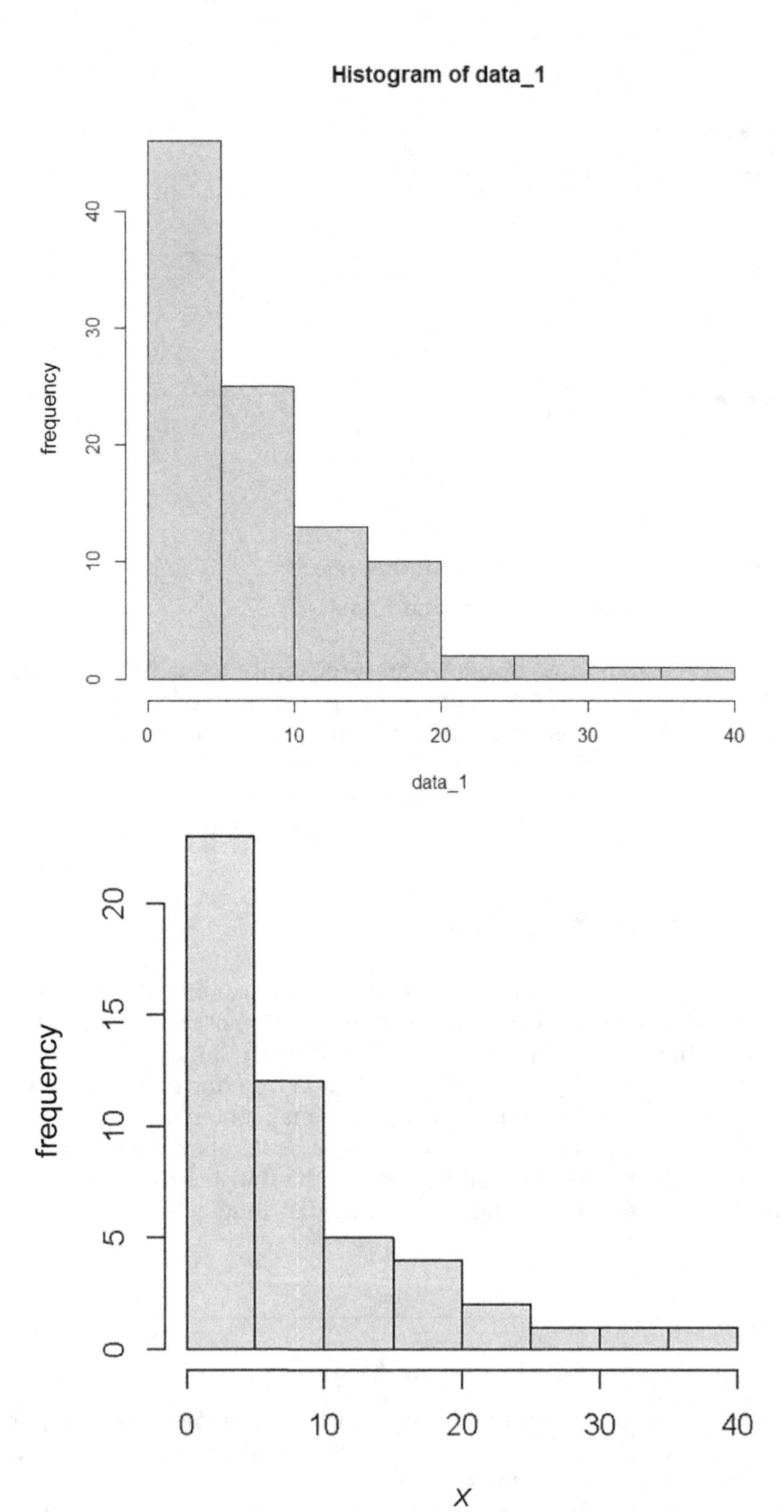

FIGURE 8.1
Histogram for sample from an unknown distribution (n = 49) random variable X.

We clearly see this, as the data is skewed right. We might even recognize the shape as very similar to an exponential distribution. Suppose further that we take this sample of size 49 and compute the mean of the sample as 9.75845 and the standard deviation as 9.69856. The fact that these are close to being the same is further evidence that the underlying population, X, might follow an exponential distribution.

For sure, the distribution is heavily skewed, with only positive values of x, so the distribution is almost certainly not normal. In fact, we see from the normal Q–Q plot in Figure 8.2 that any assumption of normality is not reasonable. We discussed in a previous chapter that this plot should follow a straight line if the distribution is normal. In this case we have the "U" shape associated with skewed distributions.

We have obtained the mean and standard deviation for the sample. However, these values might be different if we collected a second sample of $n = 49$ from the population. Thus, the question for us is how much would the mean differ in a second or third sample? How can we account for the uncertainty in our estimate of the mean because we are using a sample?

Generally, it is not possible to sample repeatedly from a population. We have a single sample and must use it alone for inference. Since the data in Figure 8.1 was generated randomly from the population on the computer (we used R), we have the ability to repeat the process and compute the mean for a 2nd, 3rd, 4th, . . . , nth sample. Such a procedure results in what is known as a sampling distribution of the mean.

Very often in analysis, it is not practical to obtain an entire population. It might be easier and more useful to obtain a random sample. Therefore, we will want to infer the characteristics of the population from the sample's statistics.

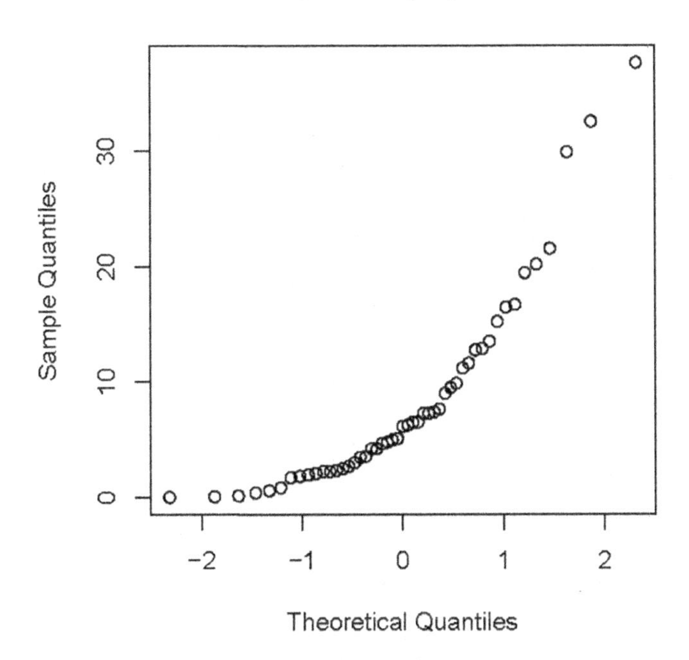

FIGURE 8.2
Q–Q plot for sample from an unknown distribution ($n = 49$).

A ***sampling distribution*** of a statistic:

The sampling distribution is the probability distribution for all possible values of the statistics computed from a sample of size n. For example, the sampling distribution of the sample mean, $\overline{x}$ is the probability distribution of all possible values of the random variable, $\overline{x}$, computed from a sample size n from a population with mean μ and standard deviation σ.

The idea behind this is as follows:

Step 1. Obtain a random sample of size n.

Step 2. Compute the sample mean, $\overline{x}$.

Step 3. Assuming that we are sampling from a finite population, repeat steps 1 and 2 until many distinct random samples of size n have been obtained.

Again, we do not really want (and, in fact, cannot) to repeat steps 1 and 2 over and over. We would prefer to infer information about the population's parameters from a single sample. Let's do this experiment, and see what the sampling distribution of the mean looks like.

8.1.1 Sampling Distribution of the Mean of n = 49 from Skewed Distribution

We repeated the experiment 10,000 times, each time taking a sample of n = 49 from our right skewed distribution of Figure 8.1 (which we will now confess is not unknown; it was, in fact, an exponential distribution with a mean of 10) and computed the sample mean for each of the 10,000 samples. The 10,000 means are plotted in the histogram in Figure 8.3. This picture is an estimate of the sampling distribution of the mean.

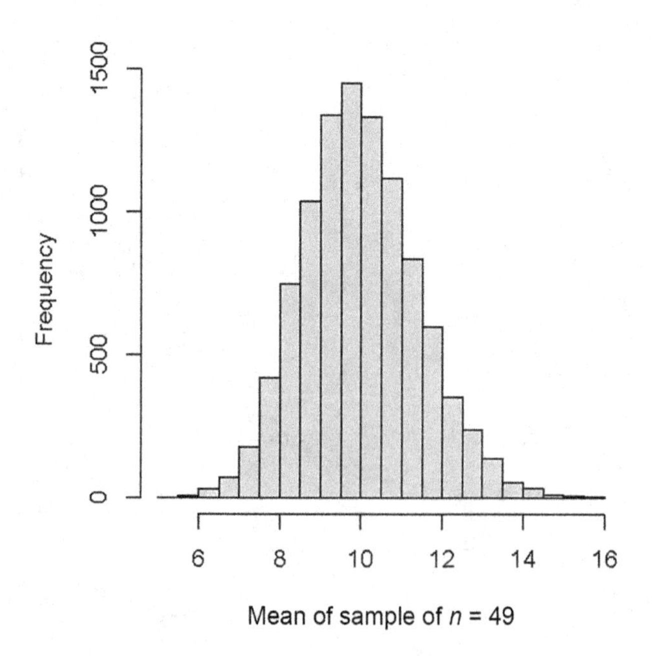

FIGURE 8.3
Histogram of 10,000 means of samples from an exponential distribution (n = 49).

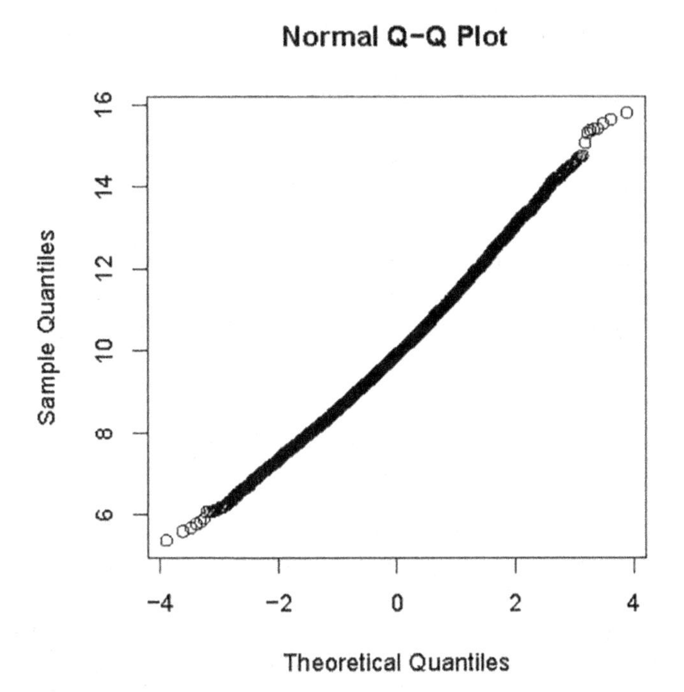

FIGURE 8.4
Q–Q plot of 10,000 means of samples from an exponential distribution ($n = 49$).

The result is remarkable. The samples all came from a highly skewed distribution. However, the distribution of the means from this distribution looks relatively normal! The Q–Q plot, shown in Figure 8.4, confirms that the assumption of a normal distribution is reasonable.

This result is the primary tenet of the CLT:

> Regardless of the distribution of the population, the **sampling distribution of the sample means is approximately normal** if the sample size is large enough (generally $n > 30$).

Note what this does NOT say: the sample itself will **not** follow a normal distribution, unless the underlying population is normal.

8.2 Central Limit Theorem (CLT)

Looking at average or total results, not individual random variables.

The Central Limit Theorem (CLT) is one of the most important theorems in probability. It more formally states that if X_1x_1, X_2x_2, . . . , X_nx_n are a random sample from a distribution with a mean μ and a standard deviation σ and n is sufficiently large **($n > 30$)**, then the distribution of the sample average, $\bar{X}$, or TT (the total) has a normal distributions defined as:

$$\bar{X} \sim N\left(\mu_{\bar{x}} = \mu, \sigma_{\bar{X}}^2 = \frac{\sigma^2}{n}\right).$$

The CLT actually applies not just to the sample mean, but also to the sample total, *T*. The sample total is just the sum of the sample observations and is computed in obtaining the sample mean, so they are related. The formal distribution for the sample total is:

$$T \sim N\left(\mu_T = n\mu, \sigma_T^2 = n\sigma^2\right).$$

We demonstrated, at least with a simulated example, how the sampling distribution appears normally distributed. We will not prove the result. A text on mathematical statistics contains such proofs. Before moving on, however, we want to also look, at least with examples, at the parameters of the normal distribution. We will do this for the sample mean as before. The sample total is shown in similar fashion.

We see from the CLT definition that the sample mean is distributed normally with a mean that is the same as the mean of the distribution of the population from which the sample was obtained. Refer to our example and the sampling distribution of the mean shown in Figure 8.3. The samples for each mean came from an exponential distribution with mean 10. We see from the histogram the "mean of the means" appears close to 10. In fact, the exact value came out to 9.98.

The variance of the sample means is more difficult to "see." According to the CLT, the variance is:

$$\sigma_{\bar{X}}^2 = \frac{\sigma^2}{n}.$$

The variance of the exponential distribution is the mean squared, so in our example 10^2 = 100. Each sample has $n = 49$, so the variance theoretically is:

$$\sigma_{\bar{X}}^2 = \frac{\sigma^2}{n} = \frac{100}{49} = 2.04.$$

For the 10,000 means in the histogram, the variance is 1.99, very close to the theoretical value.

The formula for the variance has important consequences. The larger the sample, the less variability there is in the sampling distribution of the mean: in other words, we are less likely to obtain a sample mean that is far from the actual mean.

We illustrate this principle by rerunning the experiment that produced Figure 8.3 but using samples of $n = 99$ each time so that the theoretical variance is approximately 1. The histogram is in Figure 8.5. Once again, the center of the histogram is around 10 (the actual mean of the sample means is 10.008 in this simulation). The variance is clearly smaller. For samples of $n = 49$ in Figure 8.3, the spread of the histogram is from around 6 to 14. With the larger sample, the histogram in Figure 8.5 ranges from about 8 to 12. The variance of these means is, in fact, 1.02, again quite close to the CLT claimed value.

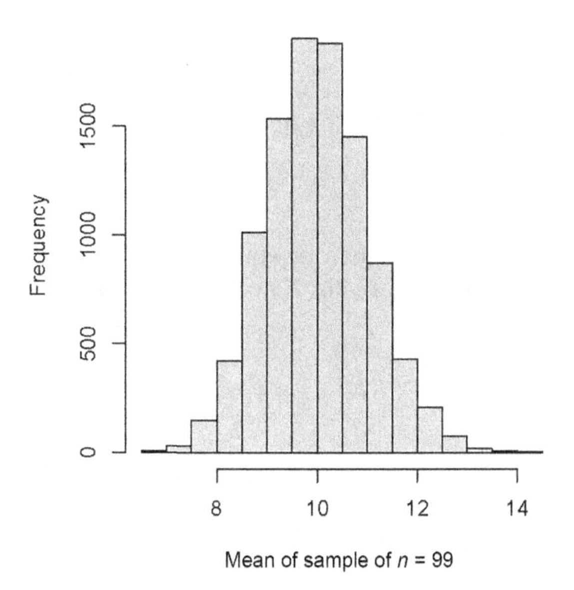

FIGURE 8.5
Histogram of 10,000 means of samples from an exponential distribution (n = 99).

8.3 CLT Applications

The central limit theorem (when applicable) will be used extensively in many applications to follow in the book involving statistical inference: confidence intervals and hypothesis testing. For now, we look an example that helps ensure understanding of the CLT itself. We have the following parameters:

$$\mu_{\bar{X}} = \mu \quad \sigma_{\bar{X}}^2 = \frac{\sigma^2}{n} \quad \mu_T = n \cdot \mu$$

$$\sigma_T^2 = n \cdot \sigma$$

Example 1: Aloe.

For example, when a batch of a certain pharmaceutical is prepared, the amount of the natural substance aloe found is measured. A random variable with mean value 4.0 grams and standard deviation 0.35 grams that is modeled using a Gamma(2, 2) distribution (see Chapter 7 to review this distribution). What is the mean and standard deviation of the amount of aloe? What is the probability that the aloe content in a batch is between 3.5 and 4.5 grams? If a sample of n = 50 batches are prepared, what is the mean and variance of the sample average? What is the probability that the ***sample average*** of the aloe is between 3.5 and 4.5 grams?

Solution:

a) The mean and standard deviation of aloe in a batch are given by the mean and variance of the gamma distribution:

$E[X] = \alpha\beta = 2(2) = 4$

$\text{Var}(X) = \alpha\beta^2 = 2(2^2) = 8$

$\text{SD}(X) = 2.83$

b) We can use R to obtain the $P(3.5 < X < 4.5)$:

```
> pgamma(4.5, 2, 2) - pgamma(3.5, 2, 2)
[1] 0.006060958
```

c) Since $n = 50$, which is greater than or equal to 30, we evoke the CLT; then the sample average aloe random variable, $\bar{X}$, follows a normal distribution with mean 4.0 and variance $8/50$:

$$\bar{X} \sim N\left(\mu_{\bar{X}} = 4, \sigma_{\bar{X}}^2 = \frac{8}{50}\right).$$

The standard deviation is, $\frac{0.35}{\sqrt{50}}\ \sqrt{8/50} = 0.04950$

d) We can now compute the probability:

$$P(3.5 < \bar{X} < 4.5)$$

Recall earlier, using the normal distribution from part (c) and R, we transformed a normal random variable X into a Z random variable:

```
> pnorm(4.5, mean = 4, sd = 0.4) - pnorm(3.5, 4, 0.4)
[1] 0.7887005
```

$Z = \frac{x - \mu}{\sigma}$, now we will use a variation with the parameters from the CLT

$$Z = \frac{(\bar{x} - \mu)}{\frac{\sigma}{\sqrt{n}}}$$

$$P\left(3.96 < \bar{X} < 4.02\right) = P\left(\frac{(3.96 - 4.0)}{0.0495} < Z < \frac{(4.02 - 4)}{0.0495}\right) = 0.447386$$

Using R commands

pnorm(0.404, 0, 1)-pnorm(–0.8080, 0, 1)

The probability of the sample average in the interval is quite high while the probability for a single batch is low. The variance of the sample average is much smaller (0.4 compared to 8), so the probability of a sample mean near the population mean of 4 is fairly high. We see this in Figure 8.6.

The normal distribution and the central limit (when applicable) are used in many applications of confidence intervals and hypothesis testing.

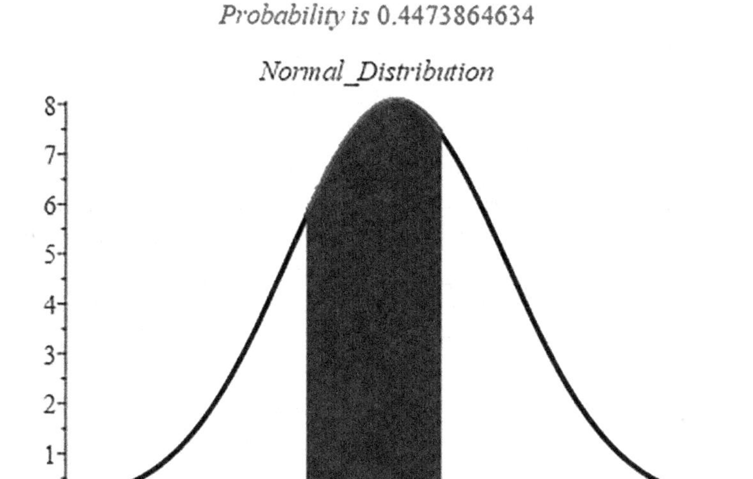

A left Riemann sum approximation of $\int_{3.96}^{4.02} f(x)\, dx$, where $f(x) = 5.698884681\, e^{-204.0608101\, (x-4)^2}\, \sqrt{2}$ and the partition is uniform. The approximate value of the integral is 0.4468978149. Number of subintervals used: 100.

FIGURE 8.6
Normal curve with mean = 4, standard deviation = 0.04950 from 3.96 to 4.02.

8.4 Chapter 8 Exercises

Find the following probabilities: assume σ given is for *x*-bar and is already divided by *n*.

8.1 $\bar{X} \sim N(\mu = 10, \sigma = 2)$, $n = 64$, $P(\bar{X} > 6)$

8.2 $\bar{X} \sim N(\mu = 10, \sigma = 2)$, $n = 64$, $\mathbf{P}(6 < \bar{X} < 14)$

8.3 A tire manufacturer thinks that the amount of average wear per normal driving year of the rubber used in their tire follows a normal distribution with mean = 0.05 inches and standard deviation 0.05 inches. A sample of 100 tires is taken. If an average of 0.10 inches is considered dangerous, then determine the probability that $P(X > 0.10)$.

8.4 A sample of 100 items with mean $\mu = 30$ and standard deviation $\sigma = 4$. Find:

a) $P(\bar{X} < 31.5)$

b) $P(\bar{X} > 30.4)$

c) $P(30.4 < \bar{X} < 33)$

8.5 A sample of 100 items with mean $\mu = 30$ and standard deviation $\sigma = 4$. Find

a) $P(\bar{x} < 31.5)$

b) $P(\bar{x} > 30.4)$

c) $P(30.4 < \bar{x} < 33)$

8.6 A radar unit is used to measure speeds of cars on a motorway. A sample of 81 cars is made with a mean of 90 km/hour and a standard deviation of 10 km/hour. What is the probability that the average car speed is more than 95 km/hour?

8.7 A sample of 144 Dell laptops is taken of battery life with a mean of 50 hours and a standard deviation of 15 hours. John owns one of these computers and wants to know the probability that the average length of time will be between 49 and 55 hours.

8.8 Entry to a certain university is determined by a national test. A sample of 1,000 students gave us a mean of 500 and a standard deviation of 100. What is the probability that the average score will be greater than 550?

8.9 A sample of 64 wire products yields a mean of 5 cm and a standard deviation of 0.02 cm.

a) What is the probability that the average length of this component is between 4.98 and 5.02 cm?

b) What is the probability that the average length of this component is between 4.96 and 5.04 cm?

8.10 A random sample of 100 machines yields the life of an instrument produced by a machine with a mean of 12 months and standard deviation of 2 months. Find the probability that an average instrument produced by this machine will last:

a) less than 7 months.

b) between 7 and 12 months.

8.11 We sampled 144 cars being assembled at a plant and found the mean of 20 hours and a standard deviation of 2 hours. What is the probability that the average period of time for car assembly will be:

a) less than 19.5 hours?

b) between 20 and 22 hours?

8.12 A large group of students (36) took a test in physics, and the final grades have a mean of 70 and a standard deviation of 10. What is the probability that the average score will be scored higher than 80?

8.13 According to the National Center for Health Statistics, 15% of all Americans have hearing trouble. (a) In a random sample of 120 Americans, what is the probability that at most 12% have hearing trouble? (b) Suppose a random sample of 120 students who use headsets results in 26 students with hearing problems, what might you conclude?

8.14 X is a normally distributed variable with mean $\mu = 30$ and standard deviation $s = 4$ from a sample of size, $n = 64$. Find:

a) $P(\bar{X} < 31)$

b) $P(\bar{X} > 31)$

c) $P(30.5 < \bar{X} < 32)$

8.15 A radar unit is used to measure the average speeds of cars on a motorway. A sample of 112 car speeds are normally distributed with a mean of 90 km/hour and a standard deviation of 10 km/hour. What is the probability that a car picked at random has an average speed of more than 92 km/hour?

8.16 For a certain type of computer, the average length of time between charges of the battery is normally distributed. A sample of 10 computers is made with a mean of 50 hours and a standard deviation of 15 hours. What is the probability that the length of time will be between 49.5 and 51 hours?

8.17 Entry to a certain university is determined by a national test. The scores on this test are normally distributed with a mean of 500 and a standard deviation of 100 based upon a sample of 100 students. Tom wants to be admitted to this university, and he knows that he must score better than at least 70% of the students who took the test. Tom takes the test and scores 585. Will he be admitted to this university?

8.18 The length of similar components produced by a company are approximated by a normal distribution model with a sample of size, $n = 48$, having a mean of 5 cm and a standard deviation of 0.02 cm. If a component is chosen at random:

a) what is the probability that the length of this component is between 4.98 and 5.02 cm?

b) what is the probability that the length of this component is between 4.96 and 5.04 cm?

8.19 The length of life of an instrument produced by a machine has a normal distribution from a sample of size, $n = 32$, with a mean of 12 months and standard deviation of 2 months. Find the probability that an instrument produced by this machine will last:

a) less than 11 months

b) between 11 and 12 months

8.20 The average time taken to assemble a car in a certain plant is a random variable having a normal distribution of 20 hours and a standard deviation of 2 hours. What is the probability that a car can be assembled at this plant in an average period of time if our sample size is 150 cars?

a) Less than 19.5 hours?

b) Between 20 and 22 hours?

c) If our sample is 100 cars, resolve parts (a) and (b).

8.21 A large group of students took a test in physics and the final grades of 35 students have a mean of 70 and a standard deviation of 10. If we can approximate the distribution of these grades by a normal distribution, what percent of the students:

a) scored higher than 80?

b) should pass the test (grades ≥ 60)?

c) should fail the test (grades < 60)?

8.22 The annual salaries of employees in a large company are approximately normally distributed with a mean of $50,000 and a standard deviation of $20,000 based upon a sample of 200 people.

a) What percent of people earn less than $40,000?

b) What percent of people earn between $45,000 and $65,000?

c) What percent of people earn more than $70,000?

9

Estimating Parameters

Objectives

1. **Understand the differences between point and interval estimates.**
2. **Apply interval estimates to population means and proportions.**
3. **Know how to use formulas for sample size for margins of error, E.**
4. **Apply interval estimates to variances and standard deviations.**
5. **Know and be able to apply the *t* distribution as appropriate.**

When we compute the mean of a sample, $\bar{X}$, we obtain what is known as a point estimate. This is a single number that is our "best guess" about the true mean of the population, μ. The problem with a point estimate is that it varies from sample to sample. If we use a second sample and compute the mean, it is very unlikely to be the same as the mean from the first sample. In the previous chapter, we saw that there is a sampling distribution of values for the statistic.

Given there is variability in the point estimate, we would like to also compute a range of values that we think likely contain the true parameter of interest. In this way, we capture some of the uncertainty in estimating from a sample. The range of values is known as an interval estimate. We will explore interval estimates known as confidence intervals for a variety of situations in this chapter.

9.1 Confidence Interval for the Mean, Known Variance

The basic concepts and properties of confidence intervals can be understood by starting with the result of the CLT. If the sample size n is sufficiently large (generally $n > 30$) from a distribution with mean μ and variance σ^2, we learned that:

$$\bar{X} \sim N\left(\mu, \frac{\sigma^2}{n}\right)$$

DOI: 10.1201/9781003317906-9

This result is also true if the distribution of the population is itself normal, regardless of the sample size. Thus, we will initially consider confidence intervals using two assumptions:

1) the population distribution is normal or the sample size is large enough ($n > 30$)
2) the standard deviation σ is known

In its simplest form, we are trying to find a region for μ (and thus a confidence interval) that will contain the value of the true parameter of interest with some degree of certainty. The degree of certainty, or the level of confidence, is set as $1 - \alpha$. α is the type I error or probability our interval does not contain the true parameter. Thus, α is set to a fairly low value: 0.1, 0.05, or even 0.01. The most common choice is $\alpha = 0.05$ in which case the level of confidence is 1 – 0.05, or 0.95. This is known as a 95% confidence interval.

Example 1: Finding critical values.

a) If we set $\alpha = 0.05$, find $z_{\alpha/2}$ that is defined as the quantile of a standard normal distribution, $Z \sim N(0, 1)$ so that $P\left(-z_{\alpha/2} < Z < z_{\alpha/2}\right) = 1 - \alpha$.
b) confirm the value you found in a) for any arbitrary choice of mean and variance for a normally distributed random variable X makes the following true: $P\left(\mu - z_{\alpha/2}\sigma < X < \mu + z_{\alpha/2}\sigma\right) = 1 - \alpha$ (again, with $\alpha = 0.05$)

Solution:

a) If we set $1 - \alpha = 1 - 0.05$, then the probability expression tells us we want quantile values with 0.95 of the probability between them. The situation is easier if you draw a graph like the one in Figure 9.1. With 0.95 probability between the values, we see that half of 0.05 (α) is in the tails outside of the quantiles. We can find these quantiles in R. We only need to find one of them; let's look at the positive value.

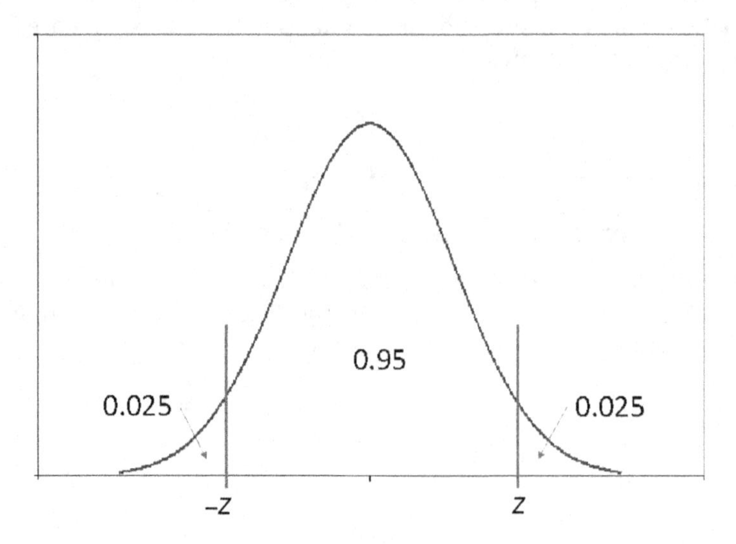

FIGURE 9.1
Graph of normal quantiles for $\alpha = 0.05$.

We notice from the graph that there is 0.025 + 0.95 = 0.975 probability to the left of this value. Thus, we solve:

$$P(Z < z_{\alpha/2}) = 0.975$$

```
> qnorm(0.975, mean = 0, sd = 1)
[1] 1.959964
```

b) We can do this for any number of normal distributions. We want:

$$P\left(\mu - z_{\alpha/2}\sigma < X < \mu + z_{\alpha/2}\sigma\right)$$

With $z_{\alpha/2} = 1.96$ from part (a). Here are two examples in R; you should do a few of your own:

$\mu = 0, \sigma = 1$:
```
> pnorm(0 + 1.96*1, mean = 0, sd = 1) - pnorm(0-1.96*1, 0, 1)
[1] 0.9500042
```
$\mu = 20, \sigma = 4$:
```
> pnorm(20 + 1.96*4, 20, 4) - pnorm(20-1.96*4, 20, 4)
[1] 0.9500042
```

The example is extended to the case where the random variable is the sample mean with assumed distribution:

$$\bar{X} \sim N\left(\mu, \frac{\sigma^2}{n}\right).$$

If this distribution is correct, then we know that there is $1 - \alpha$ probability that the sample average is in the interval:

$$\mu - z_{\alpha/2}\sqrt{\frac{\sigma^2}{n}}, \mu + z_{\alpha/2}\sqrt{\frac{\sigma^2}{n}}.$$

If our sample mean falls into in this interval then:

$$\bar{X} - z_{\alpha/2}\sqrt{\frac{\sigma^2}{n}}, \bar{X} + z_{\alpha/2}\sqrt{\frac{\sigma^2}{n}}$$

Must include the true mean, μ. If $\bar{X}$ is within a certain distance of μ, then the reverse must be true!

Since there is $1 - \alpha$ probability of obtaining such a sample mean, there is that same probability this interval actually includes the true mean.

The discussion above leads to the proper interpretation for any confidence interval. If we took 100 experiments of n random samples each and calculated the 100 confidence intervals in the same manner:

$$\bar{X} \pm z_{\alpha/2}\sqrt{\frac{\sigma^2}{n}}$$

we would expect, on average, 95 of the 100 confidence intervals to contain the true mean, μ, if we set $1 - \alpha = 0.95$, for example. We do not know which of the confidence intervals contain the true mean and, of course, the flip side is we would expect 5% of the intervals to NOT include the true mean. Thus, to a modeler, each confidence interval built will either contain the true mean or it will not contain the true mean. We would say something like: "We are 95% confident the interval formed from our sample contains the true mean."

Note a subtle but critical point: the interval changes with the sample. The true mean, μ, does not change! Beware of interpreting the interval in a way that suggests the true mean is random or variable.

Example 2: Amount of Diet Coke in a can.

The amount of fluid dispensed into a can of Diet Coke is subject to variability. The factory has collected data over many years that suggest the standard deviation in the amount is 0.5 fluid ounces. A sample of 36 Diet Cokes was taken, and a sample mean of $\bar{x} = 11.35$ was found. What are the 95% and 90% confidence intervals for the true mean? Which interval is wider and why?

Solution:

We could assume that the distribution of the amount of fluid is approximately normal. However, we do not need to make this assumption since the sample size is reasonably large (>30), so we can use the CLT to assume the distribution of sample means is normal. We also have knowledge about the true standard deviation, so we can use the confidence interval formula.

We need the critical values for the normal distribution. For the 95% confidence interval, $1 - \alpha = 0.95$, which is the situation of Example 1 and shown in Figure 9.1. We determined the critical value was 1.96 in that example.

For a 90% confidence interval, Figure 1 is modified to have 0.90 probability between the two quantiles so that $\alpha/2 = 0.1/2 = 0.05$ is in each tail. To find the critical value using the positive quantile, we solve:

$$P(Z < z_{0.1/2}) = 0.95$$

```
> qnorm(0.95, mean = 0, sd = 1)
[1] 1.644854
```

We then form the two confidence intervals:

$$\bar{X} \pm z_{\alpha/2}\sqrt{\frac{\sigma^2}{n}}$$

95% confidence interval:

$$11.35 \pm 1.96\sqrt{\frac{0.5^2}{36}} = [11.187, 11.513]$$

90% confidence interval:

$$11.35 \pm 1.645\sqrt{\frac{0.5^2}{36}} = [11.213, 11.487]$$

The 95% confidence interval is wider. The reason is that we want to be more confident we have formed an interval from our sample that actually includes the true mean. Making the interval wider gives us more confidence. An easy way to think of this is to ask what interval would you form if you wanted to be 100% confident. The obvious answer is to make the interval from negative to positive infinity!

9.2 Confidence Interval for the Mean, Unknown Variance

When forming the confidence interval in the previous section, we needed assumptions that the sample mean was normally distributed. There was one other assumption: that we knew the value of the true variance. Unfortunately, this is rarely the case. The obvious solution is to replace the population variance with the sample variance and form the confidence interval as:

$$\bar{X} \pm z_{\alpha/2}\sqrt{\frac{s^2}{n}}$$

The problem is that this introduces additional variability in our estimates, and the distribution of the sample mean is no longer normal. We now need a new distribution, the Student t distribution.

9.2.1 The Student's *t* Distribution

W. S. Gosset, working for Guinness breweries, found this problem in 1908. His confidence intervals designed to help with quality control were consistently not wide enough. He explored the distribution of the sample averages and saw that they looked normal, a bell shape, but with a larger variance. He uncovered and found the "Student's *t*" or just "*t*" distribution modeled the sample averages. The "student" comes from the fact he used the pseudonym "Student" to avoid getting fired for publishing work that involved Guinness data.

As mentioned, the shape of the *t* distribution is very similar to the shape of the standard normal distribution but just more spread out. Thus, the *t* distribution is a bell-shaped curve centered at 0. The *t* distribution has a (slightly) different shape for each possible sample size. The distribution is somewhat broader than Z, reflecting the additional uncertainty resulting from using s in place of σ. The amount of additional spread depends on the sample size n. As n gets larger and larger, the shape of the *t* distribution approaches the standard normal. For small n, it is much more spread out.

A plot of the distribution for two different sample sizes, $n = 3$ and $n = 30$, and the standard normal distribution are shown in Figure 9.2. The *t* for $n = 30$ and Z distributions are

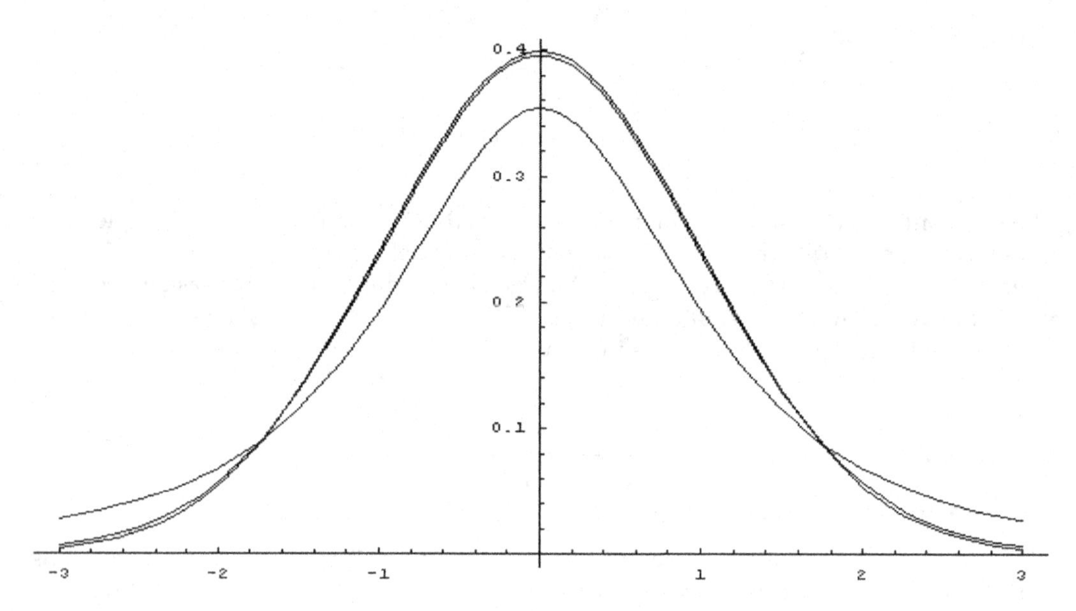

FIGURE 9.2
Graph of the standard normal distribution and t distributions for $n = 3, 30$.

the two curves that are very close. The t distribution is slightly wider. For $n = 3$, the t distribution is noticeably wider.

Since the shape of the curve only depends on the sample size, the t distribution has a single parameter, the degrees of freedom, or df. For the sample mean, df = $n - 1$. The reason the df is not just n is that we "lose a degree of freedom" since we have had to estimate one quantity, the sample mean itself.

As with other distributions, we can use R to compute probabilities and quantiles for the t distribution. The parameter is "df," and the commands are similar to others we have seen as illustrated in the examples to follow.

Example 3: Suppose we have a random sample of size $n = 10$ with a standard deviation of 2.5 from a normal distribution with a mean of 11.5. Find the $P\left(\bar{X} > 13\right)$ using the t distribution. Compare the result to those obtained with the normal distribution.

Solution:

To use the t distribution, we must first use the transformation, to "standardize" the value:

$$T = \frac{\bar{X} - \mu}{\sqrt{s^2/n}} = \frac{13 - 11.5}{\sqrt{2.5^2/10}} = 1.898$$

So, $P\left(\bar{X} > 13\right)$ is equivalent to $P(T > 1.898)$. Now, we can use R. With a sample of $n = 10$, our df = 9:

```
> pt(1.898, df = 9, lower.tail = FALSE)
[1] 0.04508799
```

Had we inappropriately assumed the sample mean was still distributed as normal, the result would have been instead:

$$\bar{X} \sim N\left(11.5, \frac{2.5^2}{10}\right)$$

```
> pnorm(13, 11.5, sd = sqrt((2.5^2)/10), lower.tail = FALSE)
[1] 0.02888979
```

The normal distribution has less spread, so the probability of being above 13 is smaller by quite a bit.

We can also use R to get the inverse value of the t distribution, which will be important for confidence intervals. We demonstrate in the next–example.

Example 4: Find the value of the Student's t distribution with five degrees of freedom that places 95% of the distribution between $-T$ and T in the same fashion as done previously for the normal distribution (Figure 9.1 with T instead of z). How does this compare to the value from the standard normal distribution?

Solution:
We apply the quantile function "qt" of the Student's t distribution using the probability 0.975 to obtain the positive value since 97.5% of the probability is below this value.

```
> qt(0.975, df = 5)
[1] 2.570582
```

The value is quite a bit larger than the 1.96 from the normal distribution. The t distribution is more spread and will therefore make the confidence interval wider and account for the variability due to estimating the standard deviation from the sample.

9.2.2 Confidence Interval Using t Distribution, σ Not Known

We are now ready to form confidence intervals. The formula is familiar:

$$\bar{X} \pm t_{\alpha/2, n-1}\sqrt{\frac{s^2}{n}}$$

The differences are we now use the sample standard deviation, s, which leads to using a t critical value rather than a Z.

Before we proceed to an example, we will first pause to briefly review when to use t, when to use Z, and when neither is appropriate. The flowchart in Figure 9.3 can help. If σ is known, we proceed to the right and can use Z as long as either the sample is from a normal distribution or we have a large enough sample for the CLT to hold ($n > 30$ roughly).

If σ is not known, we move to the left in the flowchart. We can again use t as long as the same assumptions hold.

It is worth noting two things as part of this discussion. One is that as n gets large, the t distribution becomes increasingly similar to the Z distribution. Thus, for large enough

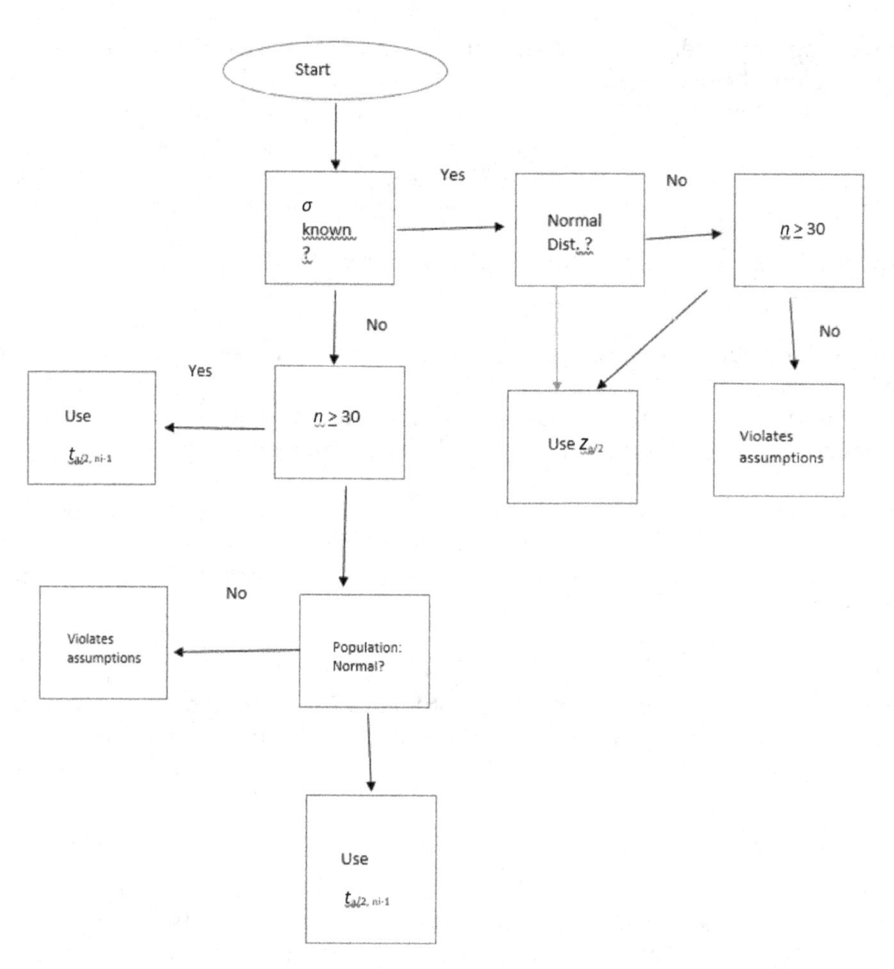

FIGURE 9.3
Flowchart of which confidence interval to use.

samples, a Z will give the same results as a t and therefore could be used even if the sample standard deviation is used. Generally, a t is just used for all, and some software only do a t for sample data.

The second point is about the requirement of normality for smaller samples. Intervals formed with the t (or Z) distributions are said to be "robust." What this means is that the intervals are still approximately correct if the assumption is not perfect. If the distribution is far from normal—highly skewed, for example—these intervals are clearly inappropriate. However, for nonnormal but symmetric distributions, they are usually not bad. Small samples, of course, are problematic for other reasons, as they may not fully represent the population.

Example 5: Comparing confidence intervals from a large sample.

We have a random sample of size $n = 100$, with a sample mean of 35 and a sample standard deviation of 10. Compute the 95% confidence interval for the true mean using both t and Z. What do you observe? Interpret the confidence interval from the t distribution.

Solution:

We first use R to obtain the t value of $t_{\alpha/2,n-1}$ as we did in the previous example. Here the df = 100 − 1 = 99:

```
> qt(.975,99)
[1] 1.984217
```

We then substitute values to form the interval using:

$$\bar{X} \pm t_{\alpha/2,n-1}\sqrt{\frac{s^2}{n}}$$

$$35 \pm 1.9842\sqrt{\frac{10^2}{100}}$$

The confidence interval is [33.0158, 36.9842].

For the normal distribution interval, the critical value is 1.96, and the interval would then be [33.04, 36.96]. We see that the two intervals are similar with the t interval slightly wider. With a large sample size, the difference is not great.

The interpretation of the interval is that if we take numerous samples of size 100 from this population and create 95% confidence intervals in this fashion, then we would expect approximately 95% to include the true population mean. We do not know which ones, to include the one from our sample, would not contain the true mean. A more common way of saying this is we are 95% confident that the interval based on our sample includes the true mean.

Example 6: Comparing intervals from a small sample.

Let's assume we have a random sample from a normal distribution. The sample size is $n = 25$, and we have a sample mean of 35 and a sample standard deviation of 10. Compute a 99% confidence interval using both t and Z and compare.

Solution:

For the t distribution, first compute the critical value:

```
> qt(0.995, df = 24)
[1] 2.79694
```

Using this value in our CI formula:

$$35 \pm 2.79694\sqrt{\frac{10^2}{25}}$$

and we obtain a 99% CI [29.406, 40.594].

For the Z distribution, first compute the critical value:

```
> qnorm(0.995)
[1] 2.575829
```

Substituting this value into the CI formula above produces an interval: [29.848, 40.152]. The intervals differ by more than in the previous example. The normal interval width is almost 1 less. The smaller sample size makes a difference!

9.3 Confidence Intervals for Proportions

The point estimate for a population proportion from a random sample is the number of successes, x, in a sample size of n.

$$\hat{p} = \frac{x}{n}$$

We would prefer an interval estimate for the true population proportion. In order to proceed with building our interval estimate, we must check for the normality of the proportion. Previously, we have mentioned the normality plot, but now we introduce a new check for proportions. The shape of the distribution of all possible sample proportions is approximately normally distributed provided $np(1 - p) \geq 10$ where $p = x/n$.

The confidence interval is then formed as we saw earlier:

$$\bar{X} \pm z_{\alpha/2}\sqrt{\frac{\sigma^2}{n}}$$

The one difference is that since X is binary (success/failure), we know the variance is that of a Bernoulli random variable or:

$$\sigma^2 = p(1-p)$$

Thus, a confidence interval for a proportion becomes:

$$\hat{p} \pm z_{\alpha/2}\sqrt{\frac{p(1-p)}{n}}$$

Of course, once again we do not know the true parameter p that is found in the variance. In fact, that is the very parameter that led us to produce a sample point estimate and to form a confidence interval. Several options are used typically. One is to use the sample proportion. A second is to use $p = 1/2$, which produces the largest possible variance. This latter option then produces a "conservative" or wider interval.

Example 7: Poll reports.

In a news poll conducted, 911 registered voters nationwide were asked, "Have you switched wireless carriers?" Of the 911 respondents, 191 said yes. Obtain a point estimate for the true proportion of registered voters that have switcher wireless services and then find a 95% confidence interval.

Solution:

The point estimate is:

$$\hat{p} = 191/911 = 0.20965$$

The 95% confidence interval may be found provided the proportion is approximately normally distributed. Our informal check is $np(1 - p) \geq 10$:

$$911 * 0.20965 = 190.2 \geq 10$$

So, we can infer that $\hat{p}$ is reasonably normal and proceed with finding the confidence interval.

The critical value for normal 95% confidence interval we have seen previously is approximately 1.96. If we use the sample proportion to estimate the variance we get:

$$\hat{p} \pm z_{\alpha/2}\sqrt{\frac{p(1-p)}{n}}$$

$$0.20965 \pm 1.96\sqrt{\frac{0.20965(1-0.20965)}{911}}$$

producing an interval of [0.18323, 0.23609].

We interpret this as we are 95% confident that the interval includes the true proportion.

If we want a more conservative interval we can use $p = 1/2$ for the variance:

$$0.20965 \pm 1.96\sqrt{\frac{0.5(1-0.5)}{911}}$$

for a wider interval of: [0.177, 0.242].

9.3.1 Margin of Error and Sample Size

You should have noticed that every confidence interval we formed involved adding and subtracting a quantity from the point estimate. The amount we add/subtract is known as the margin of error. So, for example, for the confidence interval of a proportion, we define a margin of error, ME, as:

$$ME = z_{\alpha/2}\sqrt{\frac{p(1-p)}{n}}$$

In our previous example, the ME using the sample proportion is:

$$1.96\sqrt{\frac{0.20965(1-0.20965)}{911}} = 0.0264$$

If we desire to have a smaller ME, say 0.01, instead of 0.0264 we can calculate the sample size we would need, instead of 911, in order to make the confidence interval "tighter" as demonstrated in the next example.

Example 8: Sample size for poll reports example.

Determine a formula to calculate the sample size required to produce a desired ME for the confidence interval for a proportion. Apply your formula to determine the required sample size to reduce the ME to 0.01 for the poll report example.

Solution:

The ME for a proportion is given by:

$$ME = z_{\alpha/2}\sqrt{\frac{p(1-p)}{n}}.$$

We can solve this equation for n:

$$n = \frac{z_{\alpha/2}^2 p(1-p)}{ME^2}$$

For the example with ME = 0.01:

$$n = \frac{1.96 \cdot 0.20965(1-0.20965)}{0.01^2} = 3247.659$$

or 3,248 as the sample size.

9.4 Confidence Intervals for a Population Variance and Standard Deviation

Let $X_1, X_2, \ldots, X_n$ be a random sample from $N(\mu, \sigma^2)$. We know $(n-1)s^2/\sigma^2$ is distributed as a chi-square with $n-1$ degrees of freedom. We can use this fact to obtain confidence intervals on σ^2 as well as do hypothesis tests on σ^2.

The 100(1-α)% confidence interval for σ^2 from a normal population is

$$\left[\frac{(n-1)s^2}{\chi^2_{\frac{\alpha}{2},n-1}}, \frac{(n-1)s^2}{\chi^2_{1-\frac{\alpha}{2},n-1}}\right]$$

Example 9: A random sample of college student heights come from a normal distributed population. We want a 95% confidence interval on the variance.

Data for Math 106 Spring Semester
male = 0,
female = 1

N	Gender	Age	Height	Weight	Class	Time	Family	heights		
1	1	21	65	145	3	7	4	65	66.89286	mean
2	1	18	65	125	1	10	3	65	4.058128	st dev
3	0	19	60	135	1	15	7	60		
4	0	18	71.5	141	1	10	5	71.5		
5	1	19	65	150	1	10	7	65		
6	0	18	73	195	1	5	5	73		
7	0	19	74	189	1	7	5	74		
8	1	19	65	105	1	10	5	65		
9	0	19	69	135	3	8	3	69		
10	0	18	64	103	1	18	6	64		
11	1	20	69	140	2	30	5	69		
12	1	19	62	120	2	10	4	62		

13	1	19	66	150	2	10	5	66
14	1	19	68	132	2	10	3	68
15	1	21	61	185	4	12	8	
16	1	18	60	100	1	8	4	
17	1	18	65	118	1	8	4	
18	1	19	66	240	1	10	4	
19	0	19	75	160	1	2	4	
20	0	18	69	140	1	7	6	
21	0	19	70	165	2	5	3	
22	1	18	69	130	1	10	4	
23	1	18	66	145	1	4	16	
24	1	18	63	145	1	20	4	
25	1	19	72	150	1	5	3	
26	1	18	62	109	1	5	5	
27	1	18	64	130	1	5	5	
28	1	24	63	128	3	30	3	
29	1	18	68	135	1	9	4	
30	1	18	69	125	1	2	3	
31	1	19	64	127	1	15	5	
32	0	19	84	173	1	15	4	
33	1	18	70	180	1	10	4	
34	0	19	66	133	1	10	6	
35	1	18	57	110	1	15	7	
36	1	18	65	125	1	10	5	
37	1	19	64	140	1	12	8	
38	0	18	67	148	1	15	3	
39	1	18	61	115	1	10	5	
40	1	18	69	135	1	12	4	
41	0	20	71	162	2	10	7	
42	1	19	65	127	2	5	9	

We will need $\chi^2_{0.025,13} = 24.7356, \chi^2_{.975,13} = 5.008751$

We obtain these values from either a χ^2 table or directly from R. The commands are

```
> chiupper = qchisq((1-0.95)/2, 13, lower.tail = FALSE)
> chiupper
[1] 24.7356
> chilower = qchisq((1-0.95)/2, 13)
> chilower
[1] 5.008751
```

We substitute into our confidence interval formulas to obtain the interval [7.13, 35.20].

Example 10: The length of a steel rod is only acceptable if the standard deviation is at most 0.6 mm. Use a 0.05 level of significance to find the confidence intervals for the standard deviation of the steel rods. Our sample has 36 rods and the sample standard deviation is 0.75 mm. What does the confidence interval tell us about the acceptable level of steel rods?

$$\left[\sqrt{\frac{(n-1)s^2}{\chi^2_{\frac{\alpha}{2},n-1}}}, \sqrt{\frac{(n-1)s^2}{\chi^2_{1-\frac{\alpha}{2},n-1}}}\right]$$

The two χ^2 values are 53.2033 and 20.569.

$$(n - 1)s^2 = 35*.75^2 = 19.6875$$
$$[0.608, 0.978]$$

Since 0.6 is not in this interval, our rods will not acceptable.

9.5 Chapter 9 Exercises

9.5.1 Find and Interpret the 95% and 99% CI for Problems 9.1–9.5

9.1 Numerous complaints have been made that a certain hot coffee machine is not dispensing enough hot coffee into the cup. The vendor claims that on average the machine dispenses at least 8 ounces of coffee per cup. You take a random sample of 36 hot drinks and calculate the mean to be 7.65 ounces with a standard deviation of 1.05 ounces.

9.2 A machine that produces cylindrical tubes is set so that the average diameter is 0.50 inch. In a sample of 100 tubes, it was found that the sample mean was 0.51 inch. Assume that the standard deviation is 0.05 inch.

9.3 In the midst of labor-management negotiations, the president of a company argues that the company's blue-collar workers, who are paid an average of $30,000 per year, are well paid. That figure is disputed by the union, which does not believe that the mean blue-collar income is less than $30,000. To test the company president's belief, an arbitrator draws a random sample of 350 blue-collar workers from across the country and asks each to report his or her annual income. The sample average is $27,000. If the arbitrator assumes that the blue-collar incomes are distributed with a standard deviation of $8,000, what do your confidence intervals suggest about the company claims?

9.4 A company has put in place new safety regulations. In a test of the effectiveness of the new regulations, a random sample of 50 workers was chosen. The number of person-hours lost in the month prior to and the month after the installation of the safety regulations was recorded. Assume that the population standard deviation is = 5 and the sample average was found as 1.20 hours lost. Interpret your confidence intervals in context of the problem.

9.5 The average, μ, of time spent reading newspapers online is: 8.6 minutes. Do people in leadership positions spend more time than the national average time per day reading newspapers? We sampled 100 leaders and found that they spend 8.66 minutes reading the paper (or from the web) with a standard deviation of 0.1 minutes. Do the confidence intervals suggest this group does spend more time reading on average?

9.6 Find a 95% CI for a sample whose mean filling weight is 16 oz./container, $s = 0.8$ oz., sample size = 16, $\alpha = 0.05$.

9.7 A trial for a new headache medicine found 104 men out of 1,123 men who took the drug still had the headache after three hours. Find both a 95% and 99% confidence interval for the true proportion of men who still have a headache after three hours.

9.8 Compute the margin of error for Exercise 9.7.

9.9 If we want the margin of error to be 1/2 of the value found in Exercise 9.8, then find the sample size required to reduce the margin of error.

9.10 A random sample of 25 male college student heights comes from a normal distributed population. We find the average height is 70 inches with a standard deviation of 5 inches. Find a 95% confidence interval on the variance.

9.11 A random sample of 25 male college student heights comes from a normal distributed population. We find the average height is 70 inches with a standard deviation of 5 inches. Find a 99% confidence interval on the standard deviation.

9.12 A random sample of 16 female college student heights comes from a normal distributed population. We find the average height is 65 inches with a standard deviation of 3.6 inches. Find a 95% confidence interval on the variance.

9.13 A random sample of 16 female college student heights comes from a normal distributed population. We find the average height is 65 inches with a standard deviation of 3.6 inches. Find a 99% confidence interval on the standard deviation.

9.14 A random sample of 22 male college student heights comes from a normal distributed population. We find the average height is 70 inches with a standard deviation of 5 inches. Find a 99% confidence interval on the standard deviation.

9.15 A random sample of 22 female college student heights comes from a normal distributed population. We find the average height is 65 inches with a standard deviation of 3.6 inches. Find a 99% confidence interval on the variance.

9.16 A random sample of 15 female college student heights comes from a normal distributed population. We find the average height is 65 inches with a standard deviation of 3.6 inches. Find a 99% confidence interval on the standard deviation.

9.17 Find a 95% confidence interval for the true population mean, given the following data from a normal distribution: 20.093448 13.934039 29.179828 9.948202 10.434492 21.144037 16.182559 14.360259 6.681212 19.543834.

9.18 A study with 2,160 patients was conducted on a new vaccine for disease prevention. The study found 621 had a substantial side effect. Find a) whether or not the proportion follows an approximate normal distribution; and if so b) a 99% confidence interval for the true proportion with this side effect; c) determine the margin of error, E; d) if the CDC wants E to be 0.02, then what size sample is required?

9.19 You want to rent an unfurnished one-bedroom apartment in Durham, North Carolina, next year. The mean monthly rent for a random sample of 60 apartments advertised on Craigslist (a website that lists apartments for rent) is $1,000. Assume a population standard deviation of $200. Construct a 95% confidence interval.

9.20 How large a sample of one-bedroom apartments above would be needed to estimate the population mean within plus or minus $50 with 90% confidence?

9.21 Duncan Jones kept careful records of the fuel efficiency of his car. After the first 100 times he filled up the tank, he found the mean was 23.4 miles per gallon (mpg) with a population standard deviation of 0.9 mpg. Compute the 95% confidence interval for his mpg.

9.22 A sample of size $n = 100$, Alzheimer's patients are tested to assess the amount of time in stage IV sleep. It has been hypothesized that individuals displaying symptoms from Alzheimer's disease may spend less time per night in the deeper stages of sleep. Number of minutes spent in Stage IV sleep is recorded for 61 patients. The sample produced a mean of 48 minutes (S = 14 minutes) of stage IV sleep over a 24-hour period of time. Compute a 95% confidence interval for this data. What does this information tell you about a particular individual's (an Alzheimer's patient) stage IV sleep?

9.23 A university wants to know more about the knowledge of students regarding international events. The university is concerned that their students are uninformed in regards to news from other countries. A standardized test is used to assess students' knowledge of world events (national reported mean = 65, S = 5). A sample of 36 students are tested (sample mean = 58, standard error = 3.2). Compute a 99% confidence interval based on this sample's data. How do these students compare to the national sample?

9.24 A sample of students from an introductory statistics class was polled regarding the number of hours they spent studying for the last exam. All students anonymously submitted the number of hours on a three-by-five card. There were 24 individuals in the one section of the course polled. The data was used to make inferences regarding the other students taking the course. There data are below:

40.5	22	7	14.5	9	9	30.5	8	11	70.5	18	20
70.5	9	10.5	15	19	20.5	5	9	80.5	14	20	8

If we assume this data is approximately normally distributed, then compute a 95% confidence interval. What does this tell us?

9.25 Announcements for 84 upcoming engineering conferences were randomly picked from a stack of *IEEE Spectrum* magazines. The mean length of the conferences was 3.94 days, with a standard deviation of 1.28 days. Assume the underlying population is normal, construct and interpret a 95% confidence interval for the population mean for the conferences.

9.26 Suppose that an accounting firm does a study to determine the time needed to complete one person's tax forms. It randomly surveys 100 people. The sample mean is 23.6 hours and sample deviation is 7.0 hours. The population distribution is assumed to be normal. Construct a 95% confidence interval for the population mean time to complete the tax forms.

9.27 If the firm in Exercise 9.26 wished to increase its level of confidence and keep the error bound the same by taking another survey, what changes should it make? If the firm did another survey, kept the error bound the same, and only surveyed 49 people, what would happen to the level of confidence? Why? Suppose that the firm decided that it needed to be at least 96% confident of the population mean length of time to within one hour. How would the number of people the firm surveys change? Why?

9.28 A sample of 16 small bags of the same brand of candies was selected. Assume that the population distribution of bag weights is normal. The weight of each bag was then recorded. The mean weight was 2 ounces with a standard deviation of 0.12 ounces. The population standard deviation is known to be 0.1 ounce. a) Construct a 95% confidence interval; b) if s is not known and s is 0.1 ounce, compute the 95% confidence interval; c) why are the intervals different?

9.29 A camp director is interested in the mean number of letters each child sends during his or her camp session. The population is assumed to be normally distributed. A survey of 20 campers is taken. The mean from the sample is 7.9 with a sample standard deviation of 2.8. Compute a 95% and 99% confidence interval.

9.30 What is meant by the term "95% confident" when constructing a confidence interval for a mean?

9.31 The American Community Survey (ACS), part of the U.S. Census Bureau, conducts a yearly census similar to the one taken every 10 years, but with a smaller percentage of participants. The most recent survey estimates with 95% confidence that the mean household income in the United States falls between $69,720 and $69,922. Find the point estimate for mean U.S. household income and the error bound for mean U.S. household income. Assume you use Z.

9.32 The average height of young adult males has a normal distribution with standard deviation of 2.5 inches. You want to estimate the mean height of students at your college or university to within 1 inch with 95% confidence. How many male students must you measure?

9.33 In six packages of "M&Ms," there were 12 blue pieces. The total number of pieces in the six bags was 68. We wish to calculate a 95% confidence interval for the population proportion of blue pieces.

9.34 A random survey of enrollment at 35 community colleges across the United States yielded the following figures: 6,414; 1,550; 2,109; 9,350; 21,828; 4,300; 5,944; 5,722; 2,825; 2,044; 5,481; 5,200; 5,853; 2,750; 10,012; 6,357; 27,000; 9,414; 7,681; 3,200; 17,500; 9,200; 7,380; 18,314; 6,557; 13,713; 17,768; 7,493; 2,771; 2,861; 1,263; 7,285; 28,165; 5,080; 11,622. Assume the underlying population is normal. Construct a 95% confidence interval.

9.35 Suppose that a committee is studying whether or not there is waste of time in our judicial system. It is interested in the mean amount of time individuals spend at the courthouse waiting to be called for jury duty. The committee randomly surveyed 81 people who recently served as jurors. The sample mean wait time was eight hours with a sample standard deviation of four hours. Construct a 95% confidence interval.

9.36 Suppose that a committee is studying whether or not there is waste of time in our judicial system. It is interested in the mean amount of time individuals spend at the courthouse waiting to be called for jury duty. The committee randomly surveyed 81 people who recently served as jurors. The sample mean wait time was eight hours with a sample standard deviation of four hours. Construct a 99% confidence interval.

9.37 A pharmaceutical company makes tranquilizers. It is assumed that the distribution for the length of time they last is approximately normal. Researchers in a hospital used the drug on a random sample of nine patients. The effective period of the tranquilizer for each patient (in hours) was as follows: 2.7; 2.8; 3.0; 2.3; 2.3; 2.2; 2.8; 2.1; and 2.4. Construct a 95% confidence interval for the true population mean.

9.38 Suppose that 14 children, who were learning to ride two-wheel bikes, were surveyed to determine how long they had to use training wheels. It was revealed that they used them an average of six months with a sample standard deviation of three months. Assume that the underlying population distribution is normal. Construct a 99% confidence interval.

10

One Sample Hypothesis Testing

Objectives

1. **Know what a hypothesis test is.**
2. **Be able to set up a hypothesis test.**
3. **Be able to conduct the hypothesis test and reach a conclusion.**
4. **Be able to use and interpret a p-value.**

10.1 Introduction

You work for a researcher at a college and have collected data on the temperatures of students. The question you are trying to answer is whether or not the average temperature of an adult human being is still 98.6°F as was believed to be the case for many years. You have data on 1,000 students. The average temperature in the sample is 98.4°F, and the standard deviation is 0.6°F.

Is there enough evidence to support the claim that the average human temperature has decreased? Proving something is hard to do, but hypothesis testing is the closest we get to proving or disproving a claim.

A technique for inferring information about a parameter is a hypothesis test. A statistical hypothesis test is a claim about a single population characteristic or about values of several population characteristics. There is a null hypothesis (which is the claim initially favored or believed to be true, the "status quo") and is denoted by H_0. The other hypothesis, the alternate hypothesis, which is often what the researcher hopes to show, is denoted as H_a. We will always keep equality with the null hypothesis. The objective is to decide, based upon sample information, which of the two claims is more plausible. Typical hypothesis tests can be categorized by three cases. Using the parameter of interest as the mean:

Case 1: H_0: $\mu = \mu_0$ versus H_a: $\mu \neq \mu_0$

Case 2: H_0: $\mu \leq \mu_0$ versus H_a: $\mu > \mu_0$

Case 3: H_0: $\mu \geq \mu_0$ versus H_a: $\mu < \mu_0$

DOI: 10.1201/9781003317906-10

There are two types of errors that can be made in hypothesis testing, type I errors called α errors and type II errors called β errors. It is important to understand these. Consider the information provided in the table below.

State of Nature

		H_0 True	H_a True
Test Conclusion	Fail to Reject H_0	$1-\alpha$	β
	Reject H_0	α	$1-\beta$

We see the two errors are based on which is actually true (the "State of Nature"). If the null hypothesis is true, the error would be to say it is not true (to "Reject H_0"). This is a type I error, α. If, on the other hand, the null hypothesis is not true, so the alternative is true" the error would be to not reject the null hypothesis. This is the type II error, β.

Some important facts about both α and β:

(1) $\alpha = P(\text{reject } H_0 \mid \text{ is true}) = P(\text{type I error})$
(2) $\beta = P(\text{fail to reject } H_0 \mid \text{ is false}) = P(\text{type II error})$
(3) α is known as the level of significance of the test
(4) $1-\beta$ is known as the power of the test

Thus, referring to the table, we would like α to be small, since it is the probability that we reject H_0 when H_0 is true. We would also want $1-\beta$ to be large since it represents the probability that we reject H_0 when H_0 is false. Part of the modeling process is to determine which of these errors is the most costly and work to control that error as your primary error of interest. Unfortunately, we cannot simultaneously lower both types of errors. There is a trade-off: reducing one type of error will increase the other.

If you think of a trial where the person on trial is either guilty or innocent and the jury finds the person either guilty or not guilty. We can think of hypothesis testing in the same context as a criminal trial in the United States. A criminal trial in the United States is a familiar situation in which a choice between two contradictory claims must be made.

1. The accused of the crime must be judged either guilty or not guilty.
2. Under the U.S. system of justice, the individual on trial is initially presumed not guilty.
3. Only STRONG EVIDENCE to the contrary causes the not guilty claim to be rejected in favor of a guilty verdict.
4. The phrase "beyond a reasonable doubt" is often used to set the cutoff value for when enough evidence has been given to convict.

Theoretically, we should never say "The person is innocent," but instead, "There is not sufficient evidence to show that the person is guilty."

Consider the trade-off in errors in this context. A type I error would be rejecting the null hypothesis of "not guilty" when the person is, in fact, innocent. If we decide we never want a type I error, we could accomplish this with a system that never convicted. However, imagine what this does to the type II error: we fail to convict all the guilty people. The flip side, though, is if we wanted no type II error. We could find everyone guilty. We wouldn't let anyone who is guilty go unpunished, but we would have a huge type I error of convicting all the innocent people.

Now let's compare that to how we look at a hypothesis test.

1. The decision about the population parameter(s) must be judged to follow one of two hypotheses.
2. We initially assume that H_0 is true.

The null hypothesis H_0 will be rejected (in favor of H_a) only if the sample evidence strongly suggests that H_0 is false. If the sample does not provide such evidence, H_0 will not be rejected.

The analogy to "beyond a reasonable doubt" in hypothesis testing is what is known as the **significance level**. This will be set before conducting the hypothesis test and is denoted as α. We say we would rather let a guilty person go free than put an innocent person in jail, so this error is generally set to a fairly low probability. Common values for α are 0.1, 0.01, and 0.05; 0.05 is by far the most common.

The following is provided as a framework for hypothesis testing:

Step 0: Identify the parameter of interest and determine the value of α.

Step 1: Determine the null hypothesis, H_0, and the alternative hypothesis, H_a.

Step 2: Calculate a test statistic based on sample data.

Step 3: Determine the *p*-value or the critical region where your test statistics lie (rejection region or fail to reject region). We will focus on *p*-values.

Step 4: Make your statistical conclusion. Your choices are to either reject the null hypothesis or fail to reject the null hypothesis. Ensure the conclusion is scenario oriented.

We will illustrate each of these steps more in the examples throughout the remainder of the chapter.

10.2 Hypothesis Tests for a Population Mean; Large Samples ($n \geq 30$) with σ Known

We begin with tests for a mean when we have either a large sample, or we know the sample is from a population with a normal distribution, and assuming we know the variance in the population. This situation is similar to our first confidence intervals in the previous chapter.

We will be interested in testing the hypothesis:

$$H_0: \mu = \mu_0$$
$$H_a: \mu \neq \mu_0 \text{ or } \mu < \mu_0 \text{ or } \mu > \mu_0$$

The choice of the alternative hypothesis depends on the problem and is an important decision.

With the assumptions of either normality or large sample, we can assume:

$$\bar{X} \sim N\left(\mu, \frac{\sigma^2}{N}\right).$$

This allows us to form what is known as a test statistic. This is a value we compute using sample data and known parameters for which the distribution is known. For this situation, the test statistic is:

$$z = \frac{\bar{X} - \mu_0}{\sqrt{\sigma^2 / n}}.$$

If the null hypothesis is true and $\mu = \mu_0$, the distribution of the test statistic will be standard normal:

$$z \sim N(0,1) \text{ if } H_0 \text{ is true.}$$

We can thus compute the probability of obtaining various Z values and use this to determine how likely the Z value computed with our sample is to occur. If our sample produces a very unlikely Z, we would take this as evidence that the null hypothesis is not true. The standard normal distribution for the test statistic leads to referring to this test as a "Z test."

Hypothesis testing is be learned by considering examples. Along the way, we will use tools like a plot of the normal distribution to assist with understanding. We recommend always using these tools.

Example 1: Battery lifetimes.

A manufacturer of AA batteries claims that the mean lifetime of their batteries is 800 hours. Extensive quality control provides us with a known standard deviation of 22 hours. We randomly select 49 batteries and find the mean is 790 hours. Test the claim at a 5% level of significance.

Solution:

Step 0: Identify the parameter of interest, and determine the value of α.

The interest here is in the true population mean lifetime or:

μ = the population (true) mean lifetime of AA batteries produced by manufacturer

The problem tells us to use $\alpha = 0.05$. This value should always be determined prior to the hypothesis test.

Step 1: Determine the null hypothesis, H_0 and the alternative hypothesis, H_a:

The null hypothesis is the "status quo"—in this case, the claim from the manufacturer that the mean is 800 hours. It is generally the "=" value. The alternative could be not equal, less than, or greater than. There is nothing in the problem to suggest a "direction," so "not equal" is the proper choice here. If, for example, the problem involved a claim from a customer that the average lifetime is lower than claimed, we would use a "less than" alternative. Our hypotheses are written:

$$H_0: \mu = 800$$
$$H_a: \mu \neq 800$$

Step 2: Calculate a test statistic based on sample data:

$$z = \frac{\bar{X} - \mu_0}{\sqrt{\sigma^2 / n}} = \frac{790 - 800}{\sqrt{22^2 / 49}} = -3.1818.$$

Step 3: Determine the p-value or the critical region where your test statistics lies (rejection region or fail to reject region). We will focus on p-values.

There are two approaches to testing the null hypothesis. The first is the "critical region." In this approach, we draw the picture shown in Figure 10.1 of the distribution of the test statistic Z if the null hypothesis is true. We then identify a critical of unusual values of Z based on the type I error chosen, in this case $\alpha = 0.05$.

Since the alternative hypothesis is "not equal," unusual values of Z could be both positive or negative, so we divide the error in the two tails as shown in the figure. This should remind you of the confidence intervals in the previous chapter. The cut points are found in the exact same manner.

With $\alpha = 0.05$, the probabilities in each tail are 0.025, and there is probability of 0.95 in between the critical points. You should write these values on your sketch.

With $\alpha = 0.05$, we know the critical values are –1.96 and 1.96 from our confidence interval examples. As a quick review, we can obtain the positive value in R by finding the quantile

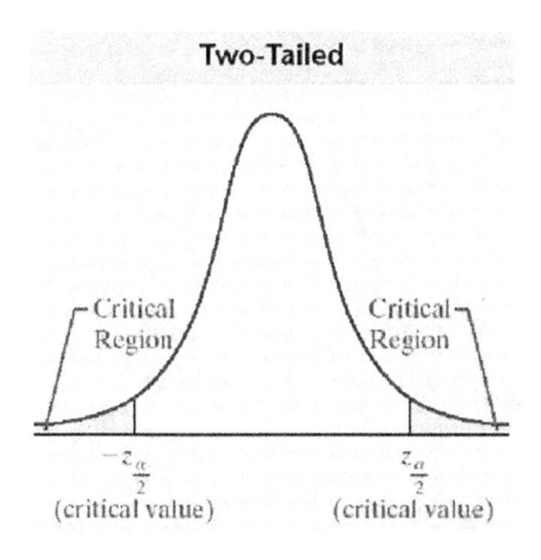

FIGURE 10.1
Critical region drawing for two-sided alternative.

("qnorm") with 0.975 probability below (0.025 in the left tail plus 0.95 in the middle) or the negative using 0.025 below:

```
> qnorm(0.975)
[1] 1.959964
> qnorm(0.025)
[1] -1.959964
```

The result means that any sample producing a Z larger, in absolute value, than 1.96 is evidence that the null hypothesis is not true. The probability of obtaining such a value is less than 0.05.

The second approach, which is much more prevalent in practice, is to compute what is known as a "p-value." This is the probability of obtaining a Z as unusual as the one observed with the sample if the null hypothesis is true. With a two-sided alternative hypothesis, we consider all values as large in absolute value as the observed Z and compute:

$$P(|Z| > z),$$

which is the same as $P(Z > z) + P(Z < -z) = 2P(Z > z)$. The situation is shown in the sketch in Figure 10.2 for this example (with $z = -3.2$).

The computation in R in two different ways (there are more!)

```
> pnorm(-3.1818) + pnorm(3.1818, lower.tail = FALSE)
[1] 0.001463629
> 2*pnorm(-3.1818)
[1] 0.001463629
```

The value is the probability of obtaining a z larger, in absolute value, than the value of -3.1818 from the sample. The p-value is quite small, suggesting such an occurrence is not likely if the true mean is 800.

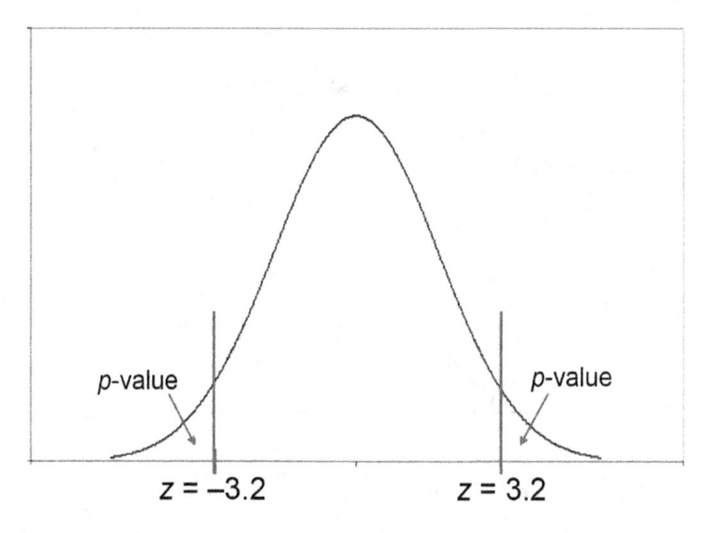

FIGURE 10.2
p-Value area for battery example.

STEP 4: Make your statistical conclusion. Your choices are to either reject the null hypothesis or fail to reject the null hypothesis. Ensure the conclusion is scenario oriented.

For the critical region approach, we reject the null hypothesis if the observed value is in the critical region. In our example, –3.1818 does fall in that region, as it is less than –1.96. For the *p*-value approach, we reject the null hypothesis is less than the stated α of 0.05 in this case. Since p = 0.0015 is less than 0.05, we reject the null hypothesis. The two approaches lead to the same conclusion and always should if computed properly.

We put this in context. Based on a sample of 49 batteries with a sample mean of 790, we have evidence that the true average lifetime is not equal to 800 as claimed (p = 0.0015 or p < 0.05 for the critical region approach).

Example 2: Worker pay.

In the midst of **labor-management negotiations**, the president of a company argues that the company's blue-collar workers, who are paid an average of $30,000 per year, are well paid, because the mean annual income of all blue-collar workers in the country is less than $30,000. That figure is disputed by the union, which does not believe that the mean blue-collar income is less than $30,000. To test the company president's belief, an arbitrator draws a random sample of 350 blue-collar workers from across the country and asks each to report his or her annual income. The mean is 29,500. If the arbitrator assumes that the blue-collar incomes are distributed with a population standard deviation of $8,000, can it be inferred at the 5% significance level that the company president is correct?

Solution:

Step 0: Identify the parameter of interest, and determine the value of α.

μ = the population (true) mean salary in dollars of workers

$\alpha = 0.05$

Step 1: Determine the null hypothesis, H_0 and the alternative hypothesis, H_a:

In this problem the union challenges the null hypothesis claim by saying the salary is "less than" $30,000 on average. This leads to a different alternative hypothesis than in our first example:

H_0: $\mu = 30{,}000$

H_a: $\mu < 30{,}000$

Step 2: Calculate a test statistic based on sample data:

$$z = \frac{\bar{X} - \mu_0}{\sqrt{\sigma^2/n}} = \frac{29500 - 30000}{\sqrt{8000^2/350}} = -1.169.$$

Step 3: Determine the *p*-value or the critical region where your test statistics lies (rejection region or fail to reject region). We will focus on *p*-values.

The "less than" alternative changes the pictures in Figures 10.1 and 10.2. We now only have the left side (negative) values and areas. For the critical region, the picture is shown in Figure 10.3. We leave the sketch for the *p*-value to you; it looks similar but with the value of $z = -1.169$ on the axis.

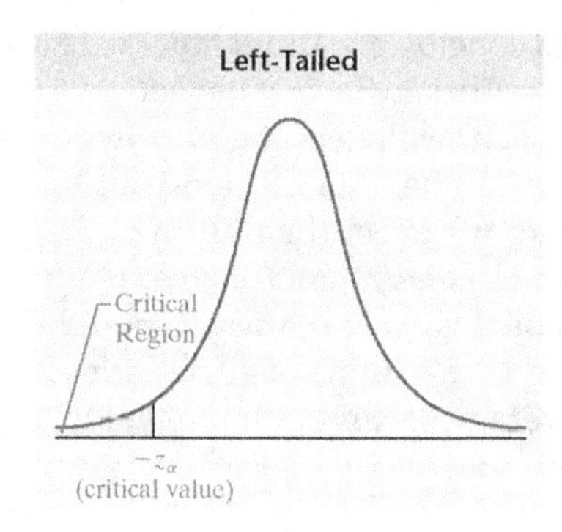

FIGURE 10.3
Critical region drawing for less than alternative.

The critical value is computed using "qnorm" and puts all the 0.05 probability in the left tail:

```
> qnorm(0.05)
[1] -1.644854
```

The p-value is computed using "pnorm":

```
> pnorm(-1.169)
[1] 0.1212018
```

Step 4: Make your statistical conclusion. Your choices are to either reject the null hypothesis or fail to reject the null hypothesis. Ensure the conclusion is scenario oriented.

For the critical region approach, we fail to reject the null hypothesis, as observed value of –1.169 does not fall in the critical region, and it is not below –1.645. For the p-value approach, we fail to reject the null hypothesis, as the p-value of 0.12 is not less than the stated α of 0.05 in this case.

We put this in context. Based on a sample of 350 workers with a sample mean of $29,500, we do not have evidence that the true average salary is less than $30,000 (p = 0.12 or p > 0.05 for the critical region approach).

10.3 Hypothesis Tests for a Population Mean; σ Unknown

As discussed in the chapter on confidence intervals, the true standard deviation is rarely known. As in the confidence interval case, we use the sample estimate, s, instead. The test statistic then becomes:

$$t = \frac{\bar{X} - \mu_0}{\sqrt{s^2/n}}.$$

As with the confidence intervals we developed in this case, we need either a large enough sample ($n > 30$) or samples from a normal population to then use the t distribution. If the assumptions hold, the distribution of the test statistic, if the null hypothesis is true, is:

$$T \sim t(n-1) \text{ if } H_0 \text{ is true}$$

The $df = n - 1$ is familiar from confidence intervals. The good news is that once we make these slight modifications, the hypothesis test is conducted in exactly the same manner as illustrated for the Z test. Not surprisingly, the test for unknown population standard deviation is known as a T test.

Example 3: Battery lifetimes revisited.

We return to the situation of Example 1: a manufacturer of AA batteries claims that the mean lifetime of their batteries is 800 hours. However, in this case the standard deviation in the population is not known. We randomly select 49 batteries and find their mean is 790 hours with a standard deviation in the sample of 22 hours. Test the claim at both a 5% level of significance and compare the results to those of Example 1.

Solution:

Step 0/1: No change from Example 1: μ = the population (true) mean lifetime of AA batteries produced by manufacturer, and $\alpha = 0.05$. The hypotheses are:

$H_0: \mu = 800$

$H_a: \mu \neq 800$

Step 2: The test statistic is defined differently since we do not know the population standard deviation and use the sample value. Since the sample standard deviation is the same as the population was in the previous example (an unlikely occurrence in practice), the value is the same. We use a T:

$$T = \frac{\bar{X} - \mu_0}{\sqrt{s^2/n}} = \frac{790 - 800}{\sqrt{22^2/49}} = -3.1818$$

Step 3: The critical value and p-value are computed from a t distribution instead of Z. With a sample of $n = 49$, df = 48:

```
> qt(0.975, df = 48)
[1] 2.010635
> 2*pt(-3.1818, df = 48)
[1] 0.002566815
```

Step 4:

We again reject the null hypothesis. The value of $T = -3.1818$ is greater in absolute value than our cut point of 2.01. Using the p-value, $p = 0.0026$ is well below 0.05.

The conclusion is the same as it was in Example 1. Note though that the cut point using a Z was smaller at 1.96, and the p-value was likewise smaller. Using the t makes it less likely we will reject the null hypothesis although in this case not enough to change the conclusion. Note that df = 48 is fairly large, so the t and Z do not differ by much.

Example 4: Newspaper reading.

The national average, μ, of time spent reading newspapers online is 8.6 minutes. Do people in leadership positions spend more time than the national average time per day reading newspapers? We sampled 100 leaders and found that they spend 8.66 minutes reading the paper (or from the web) with a standard deviation of 0.1 minutes.

Solution:

Step 0/1: μ = the population (true) mean reading time for leaders in minutes per day, and $\alpha = 0.05$. Note that a significance level is not given in the problem statement. We generally use 0.05 in such instances. The hypotheses are:

H_0: $\mu = 8.6$

H_a: $\mu > 8.6$

Step 2: Test statistic since we use a sample standard deviation, is a T statistic:

$$T = \frac{\bar{X} - \mu_0}{\sqrt{s^2/n}} = \frac{8.66 - 8}{\sqrt{0.1^2/100}} = 6.$$

Step 3: This is our first example with a "greater than" alternative hypothesis. The picture is thus as shown in Figure 10.4. The figure is for a generic Z but is the same with the t. In this example df = $n - 1 = 99$, and we will only compute the p-value:

```
> pt(6, df = 99, lower.tail = FALSE)
[1] 1.622958e-08
```

Step 4:

We reject the null hypothesis. Be careful in looking at the R output. The p-value is not 1.62. In fact, a p-value is a probability, so it cannot be greater than one. There is

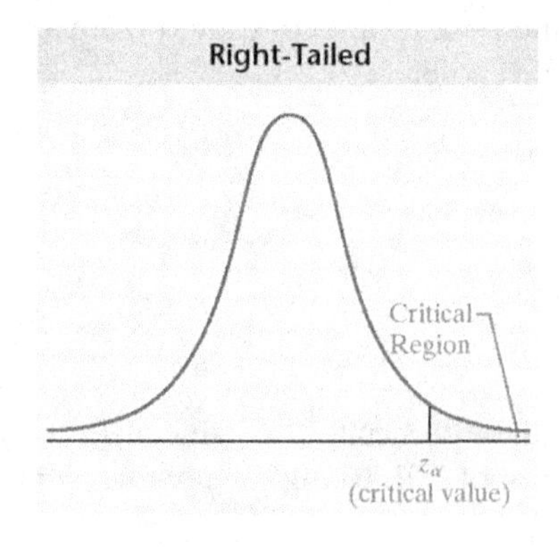

FIGURE 10.4
Critical region drawing for greater than alternative.

an "e−08" at the end of the value. This is scientific notation and means the decimal place moves 8 places to the left: 0.0000000162. In other words, a very small value. It is common to express this as $p < 0.0001$. Thus, with a sample of 100, we have evidence to suggest they read the news longer than the national average of 8 (sample average of 8.66, $p < 0.0001$).

In all our examples, the sample values were provided. Typically, however, we have a data set. Obviously, one option is to compute the sample mean and standard deviation from the data and then proceed to perform the test. R and most statistical software can essentially do the "work" for us with built-in commands for various hypothesis tests.

We will illustrate the "*t*.test" command in R in the next example. There is a Z test as well, which requires loading the BSDA package; the command is similarly "z.test." Since the population standard deviation is rarely known, the default t test command is almost always used for testing a mean.

Example 5: t Test in R with sample data.

We are given the following sample assumed to come from a normal population: 3, 7, 11, 0, 7, 0, 4, 5, 6, 2. We would like to test the claim that the population mean is greater than 3 at an α level of 0.05.

Solution:

Step 0/1: We still must set up the hypothesis test. Define μ = the population (true) mean, and $\alpha = 0.05$. The hypotheses are:

H_0: $\mu = 3$

H_a: $\mu > 3$

Step 2/3: R will do the computational steps for us. We create a variable for the sample data and then perform the test:

```
> x = c(3, 7, 11, 0, 7, 0, 4, 5, 6, 2)
> t.test(x, mu = 3, alternative = "greater")

    One Sample t-test

data: x
t = 1.3789, df = 9, p-value = 0.1006
alternative hypothesis: true mean is greater than 3
95 percent confidence interval:
  2.505919 Inf
sample estimates:
mean of x
    4.5
```

Examining the command, notice we define the mean of the null using the "mu" option and the alternative with the "alternative" option. The default if this option is not included is "not equal." The "less" option does the less than alternative.

The output provides us with the estimated sample mean, here 4.5. The key quantities for the t test are given: $t = 1.3789$ is the test statistic. We see there are 9 df, so the sample must be of $n = 10$. Finally, a p-value is provided.

Note that the 95% confidence interval is also given. Thus, the command allows us to do both the hypothesis test of this chapter as well as the computations for the last chapter.

Step 4: We conclude that there is not enough evidence to support a claim that the mean is greater than three. Our sample of 10 observations did have a higher mean at 4.5, but the result was not statistically significant (p = 0.1006).

10.4 Hypothesis Tests for a Population Proportion

Hypothesis testing for a proportion, *p*, involves similar changes to that of a mean that we saw for confidence intervals. We again need a reasonably normal distribution. A common check for this is:

$$min\{np, n(1-p)\} \geq 5.$$

The variance for a proportion is given by p(1-p), so we can modify our test statistic accordingly by using the hypothesis value for p. The resulting test statistic is then a standard normal, Z. The changes are essentially to the first two steps of our process.

Step 0: Identify the parameter of interest, and determine the value of α.

The parameter of interest changes to the true population or:

p = the population (true) proportion

Step 1: Determine the null hypothesis, H_0, and the alternative hypothesis, H_a.

These are now based on the proportion with hypothesized proportion (status quo) value denoted p_0:

H_0: $p = p_0$

H_a: $p \neq p_0$ or $p < p_0$ or $p > p_0$

Step 2: The test statistic based on our sample proportion $\hat{p}$ becomes:

$$z = \frac{\hat{p} - p_0}{\sqrt{p_0(1-p_0)/n}}$$

The standard normal distribution is then used to compute *p*-values or critical regions as we have done for the tests of the mean.

Example 6: Toothpaste.

In a television commercial, the manufacturer of a toothpaste claims that more than four out of five dentists recommend the ingredients in his product. To test that claim, a consumer protection group randomly samples 400 dentists and asks each one whether she or he would recommend a toothpaste that contained certain ingredients. The responses are 0 = No and 1 = Yes. There were 71 No answers and 329 Yes answers. At the 5% significance level, can the consumer group infer that the claim is true or not true?

Solution:

Step 0: Identify the parameter of interest, and determine the value of α.

p = the population (true) proportion of dentists who recommend the product

$\alpha = 0.05$

Step 1: Determine the null hypothesis, H_0, and the alternative hypothesis, H_a.

The claim, as is often the case, becomes the alternative hypothesis. The idea is to see if the sample provides sufficient evidence to reject the null hypothesis and thus support the claim. Here it is "greater than," since the claim is "more than four out of five":

H_0: $p = 4/5$

H_a: $p > 4/5$

Step 2: Test statistic.

We need to determine the sample proportion $\hat{p}$ first. We see that 329 dentists recommended the ingredients out of 400, so 329/400 is the sample proportion. The test statistic is then:

$$z = \frac{\frac{329}{400} - \frac{4}{5}}{\sqrt{0.8(1-0.8)/400}} = 1.125.$$

Step 3: Before we compute a p-value or critical region, we should check the assumption. We have:

$$min\{np, n(1-p)\} = \min(320, 80) = 80 \geq 5$$

We are safe to proceed. With the greater than alternative (you hopefully have the picture in mind by now), the critical value and the p-value computed in R are:

```
> qnorm(0.95)
[1] 1.644854
> pnorm(1.125, lower.tail = FALSE)
[1] 0.1302945
```

Step 4: Our test statistic of 1.125 is not in the critical region, as it is not greater than 1.645. Similarly, our p-value is not less than 0.05. We fail to reject the null hypothesis. With a sample of 400 dentists 82.25% recommended, but this is not enough evidence to conclude that the true proportion recommending the produce exceeds 4/5 (p = 0.13).

Example 7: Testing proportions in R.

We will use the previous example to illustrate how R can help with testing a proportion. The command is "prop.test." The command and its options from the R help are:

```
prop.test(x, n, p = NULL,
  alternative = c("two.sided", "less", "greater"),
  conf.level = 0.95, correct = TRUE)
```

The sample proportion is input using x = the number of successes and n = the number of trials. The null hypothesized value is "p." Options for the alternative are similar to the *t*.test command. The option "correct = TRUE" is a "continuity correction" for using a continuous distribution for discrete data. Setting this to "FALSE" will give the results we produced, but we generally use the default here.

We see that the results are similar:

```
> prop.test(329, 400, p = 4/5, alternative = "greater")

  1-sample proportions test with continuity correction

data: 329 out of 400, null probability 4/5
X-squared = 1.1289, df = 1, p-value = 0.144
alternative hypothesis: true p is greater than 0.8
95 percent confidence interval:
  0.787613 1.000000
sample estimates:
    p
0.8225
```

10.5 Hypothesis Tests and Inferences about a Population Variance

Less commonly, there is a need to test the variance. Methods for confidence intervals and hypothesis tests are similar conceptually for doing so. However, because the variance is always positive, the CLT or sum of normal random variables assumptions do not lead to a test statistic that is normal. The result is intervals and tests that look a little different.

If we assume our sample is from a normal distribution, there is theory to show that:

$$\frac{(n-1)s^2}{\sigma^2} \sim \chi^2(n-1).$$

The χ^2 is notation for a distribution known as chi-square (pronounced kigh). The parameter, as with the t distribution is known as the degrees of freedom (df), $n-1$ for the statistic above. We can use this fact to obtain confidence intervals on σ^2 as well as do hypothesis tests about σ^2.

The chi-square distribution has properties that differ from a normal distribution. The distribution is not symmetric, and is always positive. It is therefore a good model for the variance. A chi-square with 1 df is a "squared Z." This is one way to think of the distribution. Squaring the standard normal means all the negative values "flip" to the positive side.

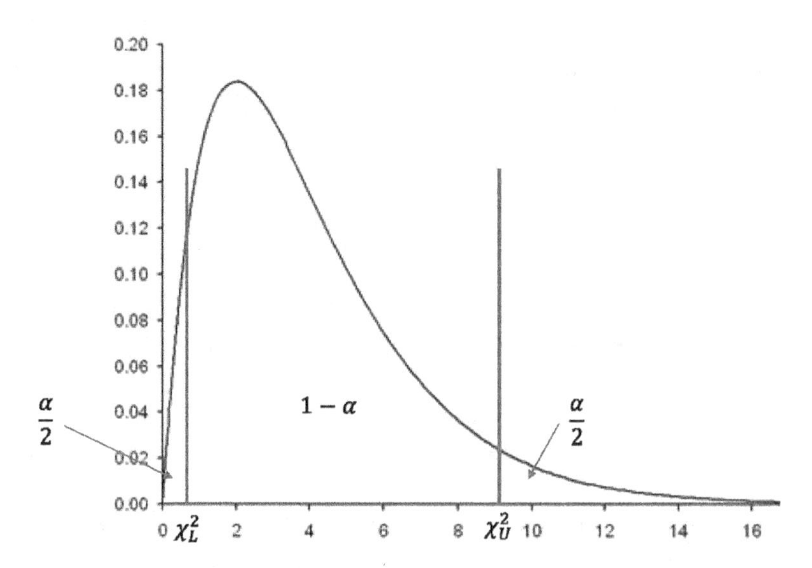

FIGURE 10.5
Critical region drawing for chi-square distribution.

The properties of the distribution lead to differences in the critical values. We recommend drawing a picture as in Figure 10.5. The key thing to note is that the lower and upper values will not be the same in absolute value as they were for the normal distribution. Thus, both must be computed for confidence intervals or two-sided hypothesis tests. The other noteworthy feature is that the lack of symmetry in the distribution leads to a lack of symmetry in interval estimates.

The 100(1 − α)% confidence interval for σ^2 from a normal population is given by the expression:

$$\frac{(n-1)s^2}{\chi_U^2}, \frac{(n-1)s^2}{\chi_L^2}.$$

The critical values in the denominators correspond to those in Figure 10.5. $U = 1 - \alpha/2$ and $L = \alpha/2$ in terms of the percentiles. The two may seem to be reversed in the confidence interval, but that is because they are in the denominator. The larger value, U, will lead to a smaller value of the interval.

The interval for the standard deviation is formed by taking the square root of the end points of the interval for the variance.

Example 8: Steel rods.

In quality assurance, the variability of items manufactured is important. The length of a steel rod is only acceptable if the standard deviation is at most 0.6 mm. A sample of 16 rods has a sample standard deviation is 0.75 mm. Form a 95% confidence interval for the true standard deviation. What does the interval suggest about quality control?

Solution:

We first obtain the critical values after drawing the sketch of Figure 10.5. We want 0.025 to the left of the lower critical value and 0.975 to the left of the upper. In R, "qchisq" is the command:

```
> qchisq(0.025, df = 15)
[1] 6.262138
> qchisq(0.975, df = 15)
[1] 27.48839
```

We substitute into our confidence interval formulas to obtain the interval for the variance:

$$\frac{(16-1)0.75^2}{27.48839}, \frac{(16-1)0.75^2}{6.262138}.$$

The interval is: [0.307, 1.347]. Taking square roots, the interval for the standard deviation is: [0.554, 1.161]. We are 95% confident this interval includes the true standard deviation. The desire is no greater than 0.6 for the standard deviation. This value is in the interval, so we cannot conclude statistically the value is greater. We do note that much of the interval is larger, however.

Hypothesis testing for variances follows the same steps as for means. The hypotheses we test are:

$$H_0: \sigma^2 = \sigma_0^2$$
$$H_a: \sigma^2 \neq \sigma_0^2 \text{ or. } \sigma^2 < \sigma_0^2 \text{ or. } \sigma^2 > \sigma_0^2$$

The test statistic is:

$$X^2 = \frac{(n-1)s^2}{\sigma_0^2}.$$

The distribution of the test statistic is chi-square with $n - 1$ degrees of freedom if the null hypothesis is true.

If the alternative is not equal, care must be used in computing p-values. Since the distribution of the test statistic is not symmetric, both the probability of being higher or lower than the computed value must be obtained, and then the p-value is two times the minimum of these values:

$$p = 2 \cdot min\left\{P\left(X^2 < \chi^2\right), P\left(X^2 > \chi^2\right)\right\}.$$

One-sided alternatives do not require such an adjustment.

Example 9: Steel rods revisited.

Form an appropriate test to determine if the variability in the length of a steel rods is not acceptable based on the previous example: acceptable standard deviation is at most 0.6 mm. Use a 0.05 level of significance for the sample of 16 rods with sample standard deviation of 0.75 mm.

Solution:

We write the hypothesis in terms of the variance, and the null is that the rods are acceptable:

$$H_0: \sigma^2 \leq 0.6^2 = 0.36$$
$$H_a: \sigma^2 > 0.36$$

The test statistic is then computed:

$$X^2 = \frac{(n-1)s^2}{\sigma_0^2} = \frac{(16-1)0.75^2}{0.36} = 23.44.$$

We can compute the rejection region by finding a critical value. Since the alternative is "greater than," we want the value such that 0.05 probability is to the right and 0.95 to the left (draw a figure!). In R:

```
> qchisq(0.95, df = 15)
[1] 24.99579
> qchisq(0.05, df = 15, lower.tail = FALSE)
[1] 24.99579
```

We do not reject the null hypothesis since 23.44 is not greater than this critical value. We can, alternatively, compute a p-value. We want the probability of a value 23.44 or larger:

```
> pchisq(23.44, df = 15, lower.tail = FALSE)
[1] 0.07523417
```

We again cannot reject the null hypothesis. We do not have statistical evidence from our sample to conclude the variability exceeds the desired limit ($p = 0.075$).

10.6 Chapter 10 Exercises

10.1 In a certain urban area, driving practices are observed: 48% of drivers did not stop at stop signs on county roads. Two months and a serious information campaign later: out of 800 drivers, 360 did not stop. Has the proportion of drivers who do not stop changed?

10.2 An intelligence agency claims that the proportion of the population that has access to computers in Afghanistan is at least 30%. A sample of 500 people is selected and 125 of these said they had access to a computer. Test the claim at a 5% level of significance.

10.3 As a business manager, you are asked to test a new item in the field. This item is claimed to be 95% reliable. You issue 250 of these items to your employees, and of these, 15 did not work properly. Perform a hypothesis test at a 5% level of significance.

10.4 You need steel cables for an upcoming mission that are at least 2.2 cm in diameter. You procure 35 of the cables and find through measurement that the mean diameter is only 2.05 cm. The standard deviation in the sample is 0.3 cm. Perform a hypothesis test of the cables at a 5% level.

10.5 For safety reasons, it is important that the mean concentration of a chemical used to make a volatile substance not exceed 8 mg/l. A random sample of 34 containers has a sample mean of 8.25 mg/l with a standard deviation of 0.9 mg/l. Do you conform to the safety requirements?

10.6 In a survey in 2003, adult Americans were asked which invention they hated the most but could not do without: 30% chose the cell phone. In a more recent survey, 363 of 1,000 adult Americans surveyed stated that the cell phone was the invention they hated the most and could not do without. Test at the 5% level of significance if the proportion of adult Americans who hate and have a cell phone is the same as it was in 2003.

10.7 The population mean earnings per share for financial services corporations including American Express, E*Trade Group, Goldman Sachs, and Merrill Lynch was $3 (*Business Week*, August 14, 2000). In 2001, a sample of 10 financial services corporations provided the following earnings per share:

1.92	2.16	3.63	3.16	4.02	3.14	2.20	2.34	3.05	2.38

Determine whether the population mean earnings per share in 2001 differ from $3 reported in 2000 using $\alpha = 0.05$.

10.8 Given a small sample size of $n = 9$ students from a normal distribution of pulse rates, we find the sample mean is 74.4 and $s = 10$. We want to test the claim that students' pulse rate is greater than 60 beats per minute.

10.9 The length of a steel rod is only acceptable if the standard deviation is at most 0.8 mm. Use a 0.05 level of significance to test the suitability of the steel rods. Our sample has 16 rods, and the sample standard deviation is 0.85 mm.

10.10 A random sample of male college student heights comes from a normally distributed population. We find the average height in the sample is 70 inches with a standard deviation of 5 inches. The claim is that the variance is no more than 5 inches. Test the claim at 0.05 level of significance.

11

Inferences Based on Two Samples

Objectives

1. **Inference on two population means, σ known.**
2. **Inference on two population means, σ unknown large sample:**
   a. **dependent.**
   b. **independent.**
3. **Inference on two population means, σ unknown small sample.**
4. **Inference on two population proportions.**
5. **Inference on two population variances.**

We have data for homicides in two neighboring cities. We might want to know if the average number of homicides in one city is the same as the other city. Perhaps we have two neighboring school districts and on the last statewide testing got average scores of 83.5 and 90.2. Are these schools different or the same in terms of the average scores? In this chapter, we will address how to answer this question.

Although we are interested, if $\mu_1 = \mu_2$ we will test the difference, $\mu_1 - \mu_2 = 0$.

If we are interested, if $\mu_1 - \mu_2 =$ "constant," we can test that as well.

11.1 Large Independent Samples (Both Sample Sizes m, $n \geq 30$)

To test the hypothesis about the difference of the means from two large independent samples, we will follow the procedure as follows.

Tests comparing two sample means from sample X and sample Y

H_0: $\mu_1 = \mu_2$, we write this as $\mu_1 - \mu_2 = 0$

H_a: This can be any of the following as required:

$\Delta\mu \neq 0$, $\Delta\mu < 0$ $\Delta\mu > 0$

Test Statistic: $Z = \dfrac{(\bar{x} - \bar{y})}{\sqrt{\frac{s_1^2}{m} + \frac{s_2^2}{n}}}$

Decision: Reject the claim, H_0 if for

$\Delta\mu \neq 0$ Either $Z \geq z_{\alpha/2}$ or $Z \leq -z_{\alpha/2}$

DOI: 10.1201/9781003317906-11

$\Delta\mu < 0 \; Z \leq -z_\alpha$

$\Delta\mu > 0 \; Z \geq z_\alpha$

Calculate the p-value

Two-tail test, P = 2 * $P(Z > |Z|)$

One-tail test, $p = P(Z > |z|)$

Example 1: Our pool manager thinks there has been a reduction in chorine levels in the community pool over the past two months. The pool manager collects samples three to four times a day, seven days a week. In June, he collected 85 samples and found a sample mean of 18.3 and a sample standard deviation of 1.2. In July, he took 110 samples and found a sample mean of 17.8 and a standard deviation of 1.8. Have the levels changed? Test with $\alpha = 0.05$.

$$H_0: \mu_1 - \mu_2 = \Delta\mu = 0$$
$$H_a: \Delta\mu \neq 0$$

Z critical at $\alpha = 0.5$ is $|z| = 1.645$. We will reject if our test $|Z| > |1.645|$

$$\text{Test Statistic: } Z = \frac{(\bar{x} - \bar{y})}{\sqrt{\frac{s_1^2}{m} + \frac{s_2^2}{n}}}$$

$$Z = \frac{(18.3 - 17.8)}{\sqrt{\frac{(1.2)^2}{85} + \frac{(1.8)^2}{110}}} = 2.32$$

Decision: Since 2.32 > 1.645, we reject the null hypothesis.

Two-tail test, P = 2 * $P(Z > |Z|)$ = 2 * $P(Z > 2.32)$ = 2 * 0.0102 = 0.0204, which is less than α, so we reject H_0.

11.2 Large Independent Samples (Only One Sample Size Is ≥30)

Consider a random sample of 15 women with low-income status delivering babies whose mean weight was 110 oz. with a sample standard deviation of 24 oz. Another sample of 76 mothers with medium-income status delivering babies whose average weight was 120 oz. with a standard deviation of 22 oz. Assume birth weights are normally distributed; determine if the lower-income is lower than the medium-income sample.

The procedure here is different, as now we use a t distribution.

$$H_0: \mu_1 = \mu_2, \text{ we write this as } \mu_1 - \mu_2 = 0$$
H_a: This can be any of the following as required:

$\Delta\mu \neq 0$, $\Delta\mu < 0$ $\Delta\mu > 0$

$$\text{Test Statistic: } t = \frac{(\bar{x} - \bar{y}) - (\mu_1 - \mu_2)}{S_p\sqrt{\frac{1}{m} + \frac{1}{n}}}$$

where $S_p^2 = \frac{(m-1)s_1^2 + (n-1)s_2^2}{m+n-2}$

$$\text{and } S_p = \sqrt{S_p^2}$$

Decision: Reject the claim, H_0 if for

$$\begin{array}{ll} \Delta\mu \neq 0 & \text{either } t \geq t_{\alpha/2} \text{ or } t \leq -t_{\alpha/2} \\ \Delta\mu < 0 & t \leq -t_{\alpha} \\ \Delta\mu > 0 & t \geq t_{\alpha} \end{array}$$

and for df use $m + n - 2$

Example 2: Woman births described above.

H_0: $\mu_1 = \mu_2$, we write this as $\mu_1 - \mu_2 = 0$

H_a: $\Delta\mu<0$

$$\text{Test Statistic: } t = \frac{(\bar{x}-\bar{y})-(\mu_1-\mu_2)}{S_p\sqrt{\frac{1}{m}+\frac{1}{n}}} = (110 - 120) \ / \ (22.32 * 0.2825) = -1.5859$$

where $S_p^2 = \frac{(m-1)s_1^2+(n-1)s_2^2}{m+n-2} = 498.5$

$$\text{and } S_p = \sqrt{S_p^2} = 22.32$$

Decision: Reject the claim, H_0 if for

$$\Delta\mu<0 \; t \leq -t_{\alpha 89} = -1.6621$$

$$P(T > |t|) = P(T > |-1.5859|) = 0.05815$$

We fail to reject H_0 since the p-value is > than α.

Example 3: Testing of 16 students in class 1 showed a mean of 107 and a standard deviation of 10, while the testing of class 2 of 14 students showed a mean of 112 and a standard deviation of 8. Assuming the testing is normally distributed, is there a significant difference between the testing of these two classes at $\alpha = 0.1$?

$$\text{We know } n = 16, \bar{x} = 107, s_1 = 10, n = 14, \overline{y = 112, s_2 = 8}$$

Since the two standard deviation are close in value, we assume equal variance.

H_0: $\mu_1 - \mu_2 = 0$

H_a:

$\Delta\mu \neq 0$,

$$\text{Test Statistic: } t = \frac{(\bar{x}-\bar{y})-(\mu_1-\mu_2)}{S_p\sqrt{\frac{1}{m}+\frac{1}{n}}} = (107 - 112) \ / \ (9.12 * 0.3659) = -1.498$$

where $S_p^2 = \frac{(m-1)s_1^2+(n-1)s_2^2}{m+n-2} = 83.3$

$$\text{and } S_p = \sqrt{S_p^2} = 9.12$$

Decision: Reject the claim, H_0 if for

$$\Delta\mu\neq 0 \text{ Either } t \geq t_{\alpha/2} \text{ or } t \leq -t_{\alpha/2} \text{ so } |t| = 2.7632$$

and for df use $m + n - 2$

$$p\text{-value} = 2 * P(t > |-1.498|) = 2 * 0.0726 = 0.1453$$

Since 0.1453 > 0.01, we fail to reject the null hypothesis.

11.3 Paired Data

Often when the lengths of the data are the same $m = n$, then we might use paired data.

A sampling method is **independent** when the individuals selected for one sample does not dictate which individuals are to be in a second sample. A sampling method is **dependent** when the individuals selected to be in one sample are used to determine the individuals to be in the second sample. Dependent samples are often referred to as **matched pairs** samples. An underlying assumption is that the samples must be approximately normally distributed.

Step 1. A claim is made regarding the mean of the difference, d. We let X_1 represent the first sample and X_2 represent the second sample; then $d = X_1 - X_2$. The claim is used to determine the null hypothesis and alternative hypothesis as before.

Two-Tail	Left-Tail	Right-Tail
$H_0: \mu_d = 0$	$H_0: \mu_d \geq 0$	$H_0: \mu_d \leq 0$
$H_a: \mu_d \neq 0$	$H_a: \mu_d < 0$	$H_a: \mu_d > 0$

Step 2. Select a level of significance based upon the serious of making a type I error. The level of significance is usually 0.05 or 0.01. To obtain the critical value we use

$|t_{a/2,}\ df|$ for two-tail test or $|t_{a,}\ df|$ for a one-tail test.

Step 3. Compute the test statistic, t.

$$t = \frac{\bar{d} - 0}{\frac{s_d}{\sqrt{n}}}$$

which approximately follows a Student's t distribution–with $n - 1$ degrees of freedom. The values of $\bar{d}$ and s_d are the mean and standard deviation of the difference, $d = X_1 - X_2$

Step 4. Compare the test statistic to the t critical value.

Step 5. Make decision.

Step 6. Compute the p-value.

Two-tail test $2*P(T > |t|)$

One-tail test $P(T > |t|)$

Example 4: Rental cars.

We want to know if Thrift is cheaper than Hertz in the 10 cities for which we collected data. We assume that the cost for rental cars follows a normal distribution. Test the claim at a level of significance of 0.01.

City	Thrifty	Hertz	d
1	21.81	18.99	**2.82**
2	29.89	37.99	**−8.1**
3	17.9	19.99	**−2.09**
4	27.98	35.99	**−8.01**
5	24.61	25.6	**−0.99**
6	21.96	22.99	**−1.03**
7	20.9	19.99	**0.91**
8	37.75	36.99	**0.76**
9	33.81	26.99	**6.82**
10	33.49	30.99	**−.5**

We compute the mean and standard deviation of d as −0.641 and 4.645, respectively. Since there are 10 data pairs, $n = 10$ and df = 10 − 1 = 9.

$$H_0: \mu_d \geq 0$$
$$H_a: \mu_d < 0$$
$$\text{t-critical} = t_{0.05,9} = -1.833$$
$$\text{Test statistics } t = -0.641/\ (4.645/\text{sqrt}\ (10)) = -0.436$$

Since −0.436 is not < −1.833, we fail to reject the null hypothesis.

p-Value is $P(T < -0.436) = 0.3364$. Since 0.3364 is not less than α we fail to reject H_0.

Example 5: It is stated that using a study module improves your scores. We have collected information on 20 students who took the test twice, pre-module and post-module. We assume testing follow an approximate normal distribution.

Student	Pre-Module Score	Post-Module Score	d
1	18	22	4
2	21	25	4
3	16	17	1
4	22	24	2
5	19	16	3
6	24	29	5
7	17	20	3
8	21	23	2
9	23	19	−4
10	18	20	2
11	14	15	1
12	16	15	−1
13	16	18	2
14	19	26	7
15	18	18	0
16	20	24	4
17	12	8	6
18	22	25	3
19	15	19	4
20	17	16	−1

Calculating the mean and standard deviation gives $\bar{d} = 2.35$ and $s_d = 2.5808$ and sd/sqrt $(n) = 2.5808/\text{sqrt}(20) = 0.57708$

$$H_0: \mu_d \leq 0$$
$$H_a: \mu_d > 0$$
$$\text{t-critical} = t_{0.05,19} = 2.093$$
$$\text{Test statistics } t = 2.35/0.57708 = 4.072$$

Since $4.072 > 2.093$ we reject H_0.

p-Value is $P(T > 4.072) = 0.000325$ since 0.000325 is less than α we reject H_0.

There is evidence to suggest that using the module helps improve the score.

Here we provide an example using R.

Matched Pairs

There are two methods, and both yield the same value.

Method 1 (Note *x*, *y* must be of the same length) with raw data

```
t.test(x, y, paired = TRUE, alternative = "two.sided")
```

`Enter data for x and y.`

> d1<-c(11.5, 12.3, 21, 19, 18.3, 9.9, 11, 12.5, 13)

> d2<-c(10.9, 11.5, 20, 21, 19.5, 10, 10, 12.8, 13)

In this example, compute d = d1 – d2, call it d3.

> d3<-d1 – d2

> d3

[1] 0.6 0.8 1.0 –2.0 –1.2 –0.1 1.0 –0.3 0.0

> *t*.test(d1, d2, paired = TRUE, alternative = "two.sided")

Paired *t*-test

data: d1 and d2

t = –0.064541, df = 8, *p*-value = 0.9501

alternative hypothesis: true difference in means is not equal to 0

95% confidence interval:

–0.8162025 0.7717581

sample estimates:

mean of the differences

–0.02222222

Conclusion, we fail to reject(FTR) the null hypothesis, H_0, that the differences = 0.

Method 2: Find differences first

> *t*.test(d3,mu = 0)

One-sample *t*-test

data: d3
$t = -0.064541$, df = 8, *p*-value = 0.9501
alternative hypothesis: true mean is not equal to 0
95% confidence interval:
−0.8162025 0.7717581
sample estimates:
mean of *x*
−0.02222222

Conclusion, we fail to reject (FTR) H_0 that the differences = 0.
Note that the conclusions and results are identical.

11.4 Two Sample Tests on Proportions

These days in 2020, we might be concerned with the testing for COVID-19 vaccines. Most tests have a placebo for one sample and the "real" drug for the other samples. We will want lots of people in each sample for statistical relevance. The claim will be that number of protected people in the "real" drug sample is higher than the placebo sample. So, how will we test this? Table 11.1 provides a framework.

TABLE 11.1

Test about $p_1 - p_2$ for Two Independent Samples

	Two-Tail Test	Right-Tail Test	Left-Tail Test
Step 1.	H_0: $p_1 - p_2 = 0$ Ha= : $p_1 - p_2 \neq 0$	H_0: $p_1 - p_2 = 0$ Ha= : $p_1 - p_2 > 0$	H_0: $p_1 - p_2 = 0$ Ha= : $p_1 - p_2 < 0$
Step 2.	Choose α =	Choose α =	Choose α =
Step 3.	Test statistic $z = \frac{p_1 - p_2}{\sqrt{\hat{p}\hat{q}\left(\frac{1}{m} + \frac{1}{n}\right)}}$	Test statistic $z = \frac{p_1 - p_2}{\sqrt{\hat{p}\hat{q}\left(\frac{1}{m} + \frac{1}{n}\right)}}$	Test statistic $z = \frac{p_1 - p_2}{\sqrt{\hat{p}\hat{q}\left(\frac{1}{m} + \frac{1}{n}\right)}}$
Step 4.	*Determine rejection regions* $\lvert z \rvert > z_{a/2}$	Determine rejection regions. $z \geq z_a$	*Determine rejection regions.* $z \leq z_a$
Step 5.	Compare test statistic to rejection region, make decision.	Compare test statistic to rejection region, make decision.	Compare test statistic to rejection region, make decision.
Step 6.	Compute *p*-value and make decision $P = 2 * \text{minimum } (P(Z \geq \lvert z \rvert))$	Compute *p*-value and make decision $P = P(Z \geq z)$	Compute *p*-value and make decision $P = P(Z \leq z)$

To test if our test will be valid, we must see if

$p_1 * n$ and $(1 - p_1)n > 5$ and $p_2 * m$ and $(1 - p_2)m$ are each greater than or equal to 5; that ensures the normality of our proportions.

Example 6: Monterey County has proposed widening Highway 156 from the 1 to the 101. They are concerned about the constituents in the county. A sample of 3,000 voters in the city of Monterey revealed 930 favored the widening, while a sample of 1,800 voters outside the city yielded 500 who support widening the road. Does this information indicate that sentiment for widening the road is different in the two groups? Use a level of significance of 0.05.

Check if we can do this test first and meet the normality requirement, and we find we can do the test: (0.31) * 3000, 0.69 * 3000, 0.278 * 1800, 0.722 * 1800, and all > 5.

1. p_i = the proportion of people in i who support widening the road.
2. H_0: $p_1 - p_2 = 0$
3. H_a: $p_1 - p_2 \neq 0$
4. Information $p_1 = 930/3000 = 0.31$, $m = 3000$, $p_2 = 500/1800 = 0.278$, $n = 1800$
5. *Finding p-hat*

   P-hat = (930 + 500) / (3000 + 1800) = 0.2979, *q-hat* = 1 − *p-hat* = 0.7021
6. Test-statistic:

$$z = \frac{0.31 - 0.278}{\sqrt{(0.2979)(0.7021)\left[\frac{1}{3000} + \frac{1}{1800}\right]}} = 2.34$$

7. We will reject if and only if $|z| \geq z_\alpha$. *Reject if* $|z| \geq z_{0.05/2} \geq 1.96$.
8. Since 2.34 is in the rejection region, we reject the null hypothesis at 0.05. The two sample proportions are different.

Example 7: Consider our drug trials mentioned at the beginning of this section. Let's assume that 2,000 were in the real drug sample, and 1,750 were protected. In the placebo trial sample, there were 1,800 people, and 1,200 were protected. We want to test if sample 1 is better than sample 2.

$$H_0: p_1 - p_2 = 0$$
$$H_a: p_1 - p_2 > 0$$
$$p_1 = 1750/2000 = 0.875 \quad p_2 = 1200/1800 = 0.666$$

We check that we can do the test and find all calculations are greater than 5.

(0.8759) (2000), (0.125) (2000) and (0.666) (1800), (0.334) (1800) are all > 5

The test statistic is found using: $z = \dfrac{p_1 - p_2}{\sqrt{\hat{p}\hat{q}\left(\frac{1}{m} + \frac{1}{n}\right)}}$

First we find $\hat{p} = \dfrac{x + y}{n + m}$

$$\hat{p} = \frac{1750 + 1200}{2000 + 1800} = \frac{2950}{3800} = 0.7763$$

If we assume an α level of 0.05 then our rejection region is $z \geq 1.645$.

We find

$$z = \frac{0.875 - 0.666}{\sqrt{(0.7763)(0.2237)\left[\frac{1}{2000} + \frac{1}{1800}\right]}} = 15.387$$

We find 15.387 > 1.645. We also find the p-value.

The p-value is $P(Z > 15.387) \approx 0$.

Our decision is that we reject H_0.

Example 8: Revisit Example 7 using R.

First, we must make sure that we may use the test,

$$p_1 = 1750/2000 = 0.875 \; p_2 = 1200/1800 = 0.666$$

(0.8759 ((2000), (0.125) (2000) and (0.666)(1800), (0.334) (1800) = 400.84 are all > 5.

```
> res<-prop.test(x = c(1750,1200), n = c(2000,1800), alternative = "greater")
> res
   2-sample test for equality of proportions with continuity
   correction
data: c(1750, 1200) out of c(2000, 1800)
X-squared = 235.59, df = 1, p-value < 2.2e–16
alternative hypothesis: greater
95% confidence interval:
 0.18585161.0000000
sample estimates:
prop 1    prop 2
0.8750000 0.6666667
```

We reject the null hypothesis because the p-value (2.2e–16) is smaller than α.

11.5 Two Sample Tests on Variances

If independent sample of size m and n are taken from normal populations having equal variances then we use an F distribution to characterize

$F = s_1^2/s_2^2$, the ratio of the two sample variances has an F distribution with $m - 1$ and $n - 1$ degrees of freedom.

We follow Table 11.2 for the hypothesis test.

Example 9: Data has been randomly collected from two pizza delivery chains. There was 25 customers in each random sample from Chain A and Chain B. The data provided standard deviations of 20 minutes and 11 minutes, respectively. Does this data support the claim that one delivery chain is more consistent than the other? Let's test using a level of significance of 0.01.

$$H_0: \sigma 2 = \sigma 02$$
$$H_a: \sigma 2 \neq \sigma 02$$
$$\alpha = 0.01$$

TABLE 11.2

Hypothesis Test for Variances

	Two-Tail Test	Right-Tail Test	Left-Tail Test
Step 1.	H_0: $\sigma^2=\sigma_0^2$; Ha= : $\sigma^2 \neq \sigma_0^2$;	H_0: $\sigma^2=\sigma_0^2$; Ha= : $\sigma^2 > \sigma_0^2$;	H_0: $\sigma^2=\sigma_0^2$; Ha= : $\sigma^2 < \sigma_0^2$;
Step 2.	Choose α =	Choose α =	Choose α =
Step 3.	Test statistic $F = s_1^2/s_2^2$	Test statistic $F = s_1^2/s_2^2$	Test statistic $F = s_1^2/s_2^2$
Step 4.	*Determine rejection Regions* $f \geq F_{\alpha/2,\, m-1,n-1}$ or $f \leq 1/F_{\alpha/2,m-1,n-1}$	Determine rejection regions. $f \geq F_{\alpha,\, m-1,n-1}$	*Determine rejection regions.* $f \leq 1/F_{\alpha,m-1,n-1}$
Step 5.	Compare test statistic to rejection region, make decision.	Compare test statistic to rejection region, make decision.	Compare test statistic to rejection region, make decision.
Step 6.	Compute p-value and make decision $P = 2 *$ minimum $(P(F \geq f), P(F \leq f)$ where F is $F_{m-1,\, n-1}$	Compute p-value and make decision $P = (P(F \geq f)$ where F is $F_{m-1,\, n-1}$	Compute p-value and make decision $P = P(F \leq f)$ where F is $F_{m-1,\, n-1}$

The test statistic $F = 20^2/11^2 = 3.3057$

Rejection region

From R, we obtain

2.967 and 0.337

Since 3.3057 > 2.967, we reject H_0.

p-Value is found using R.

11.6 Chapter 11 Exercises

11.1 In order to investigate the relationship between mean job tenure in years among workers who have a bachelor's degree or higher and those who do not, random samples of each type of worker were taken, with the following results.

	N	Mean	s
Bachelor's degree or higher	155	5.5	1.3
No degree	210	5	1.5

Test, at the 1% level of significance the claim that mean job tenure among those with higher education is greater than among those without, against the default that there is no difference in the means. Compute the observed p-value of the test.

11.2 Records of 40 used passenger cars and 40 used pickup trucks (none used commercially) were randomly selected to investigate whether there was any difference in the mean time in years that they were kept by the original owner before being sold. For cars, the mean was 6.3 years with standard deviation

2.8 years. For pickup trucks, the mean was 7.3 years with standard deviation 3.0 years. Test the hypothesis that there is a difference in the means against the null hypothesis that there is no difference. Compute the observed p-value of the test.

11.3 We believe the average number of patients arriving per hour in an emergency room on a weekend is the same as in the evening of weekdays. Two samples are compared with statistics collected. Test whether the sample means are the same or different. Compute and interpret the Pp-value.

	n	**Mean**	S
Weekdays	144	12.6	2.6
Weekends	121	13.8	3.3

11.4 In previous years, the average number of patients per hour at a hospital emergency room on weekends exceeded the average on weekdays by 6.3 visits per hour. A hospital administrator believes that the current weekend mean exceeds the weekday mean by fewer than 6.3 hours.

	N	**Mean**	s
Weekdays	350	8.6	2.65
Weekends	310	13.8	3.1

Determine whether the current weekend mean exceeds the weekday mean by fewer than 6.3 patients per hour. Compute the p-value of the test.

11.5. A sociologist surveys 50 randomly selected citizens in each of two countries to compare the mean number of hours of volunteer work done by adults in each. Among the 50 inhabitants of Country A, the mean hours of volunteer work per year was 52, with standard deviation 11.8. Among the 50 inhabitants of Country B, the mean number of hours of volunteer work per year was 37, with standard deviation 7.2. Test the claim that the mean number of hours volunteered by all residents of Country A is more than 10 hours greater than the mean number of hours volunteered by all residents of Country B. Compute the p-value of the test.

11.6. A university administrator asserted that upper classmen spend more time studying than under classmen. A survey is completed and the following data is collected:

	N	**Mean hours**	S
Under classmen	150	2.34	4.1
Upper classmen	100	5.7	2.9

Test this claim against the default that the average number of hours of study per week by the two groups is the same, using the following information based on random samples from each group of students. Compute and interpret the p-value?

11.7. A small college administrator asserted that upper classmen spend more time studying than under classmen. A survey is completed and the following data is collected:

	N	**Mean hours**	s
Under classmen	25	2.34	4.1
Upper classmen	28	5.7	2.9

Test this claim against the default that the average number of hours of study per week by the two groups is the same, using the following information based on random samples from each group of students. Compute the p-value.

11.8 A kinesiologist claims that the resting heart rate of men aged 18–25 who exercise regularly is more than five beats per minute less than that of men who do not exercise regularly. Men in each category were selected at random, and their resting heart rates were measured. A university administrator asserted that upper classmen who exercise regularly have better resting heart rate than those who do not exercise regularly. The following data is collected:

	n	Mean hours	S
Regular exercise	42	62	1.5
No regular exercise	32	70	1.2

Perform the relevant test of hypotheses; compute and interpret the p-value.

11.9 Children in two elementary school classrooms were given two versions of the same test, but with the order of questions arranged from easier to more difficult in Version A and in reverse order in Version B. Randomly selected students from each class were given Version A, and the rest were given Version B. The results are shown in the table:

	n	Mean hours	s
Version A	26	84	4.5
Version B	28	79	5.0

(a) Construct the 95% confidence interval for the difference in the means of the populations of all children taking Version A of such a test and of all children taking Version B of such a test.

(b) Test at the 1% level of significance the hypothesis that the A version of the test is easier than the B version (even though the questions are the same).

(c) Compute and interpret the p-value.

11.10 What assumptions must hold before we use a two sample proportion test?

11.11 What assumptions are required for a two sample t test with small sample sizes ($n < 30$)?

11.12 We have conducted an educational study on two classrooms of 36 students using two different teaching methods. The first method had 50% of students pass a standardized test, and the classroom using the second teaching method had 60% of the students pass. Perform a test to determine if the rates were the same or different.

11.13 Two mathematics professors grade the same 49 tests. The result of the grading showed that professor one had an average of 82.3 with a standard deviation of 16.4 while the second professor had an average of 79.9 and a standard deviation of 22.5. Perform a test on the variances to determine if the variances are the same or different. Assume the grades follow a normal distribution.

11.14 Two mathematics professors grade the same 49 tests. The result of the grading showed that professor one had an average of 82.3 with a standard deviation of 16.4 while the second professor had an average of 79.9 and a standard deviation of 22.5. Perform a test on the standard deviations to determine if the standard deviations are the same or different. Assume the grades follow a normal distribution.

11.15 We are comparing two random samples to determine if the mean of sample 1 is greater than the mean of sample 2. The statistics were: sample 1: $m = 64$, mean of 72.4, standard deviation of 25.5 and sample 2: $n = 25$, mean of 69.95, and a standard deviation of 24.4. Test the claim, and provide the p-value.

11.16 Suppose the Acme Drug Company develops a new drug, designed to prevent colds. The company states that the drug is equally effective for men and women. To test this claim, they choose a simple random sample of 250 women and 280 men from a population of about 100,000 volunteers. At the end of the study, 38% of the women caught a cold, and 50% of the men caught a cold. Based on these findings, can we reject the company's claim that the drug is equally effective for men and women? Find and interpret the p-value (assume initially a 0.05 level of significance.).

11.17 Suppose the Acme Drug Company develops a new drug, designed to prevent colds. The company states that the drug is more effective for women than for men. To test this claim, they choose a simple random sample of 100 women and 200 men from a population of 100,000 volunteers. At the end of the study, 38% of the women caught a cold, and 51% of the men caught a cold. Based on these findings, can we conclude that the drug is more effective for women than for men?

11.18 Two college instructors are interested in whether or not there is any variation in the way they grade math exams. They each grade the same set of 30 exams. The first instructor's grades have a variance of 61.1. The second instructor's grades have a variance of 73.3. Test the claim that the first instructor's variance is smaller. (In most colleges, it is desirable for the variances of exam grades to be nearly the same among instructors.) Assume the grades are normally distributed. What can we conclude about the variance?

11.19 Energizer and Duracell batteries manufacturers are at it again. Energizer says it lasts longer than Duracell. We test Energizer and find after we randomly select 40 batteries their mean is 790 hours with a standard deviation of 22 hours. Duracell is randomly tested with 49 batteries, and we find a mean of 782 hours and a standard deviation of 25 hours. Perform a test of Energizer's claim.

11.20 Energizer and Duracell are at it again. Energizer says it lasts longer than Duracell. We test Energizer and find after we randomly select 49 batteries their mean is 790 hours with a standard deviation of 22 hours. Duracell is randomly tested with 25 batteries, and we find a mean of 782 hours and a standard deviation of 25 hours. Assume both samples come from a normal distribution. Perform a test of Energizer's claim.

11.21 A study was conducted to investigate the effectiveness of hypnotism in reducing pain. Results for randomly selected subjects are shown in the figure below. A lower score indicates less pain. The "before" value is matched to an "after" value, and the differences are calculated. The differences have a normal distribution. Are the sensory measurements, on average, lower after hypnotism? Find and interpret the p-value for the test.

Reported Pain Data								
Subject:	A	B	C	D	E	F	G	H
Before	6.6	6.5	9.0	10.3	11.3	8.1	6.3	11.6
After	6.8	2.4	7.4	8.5	8.1	6.1	3.4	2.0

11.22 A study was conducted to investigate how effective a new diet was in lowering cholesterol. Results for the randomly selected subjects are shown in the table. The differences have a normal distribution. Are the subjects' cholesterol levels lower on average after the diet? Test at the 5% level. Find and interpret the p-value.

Cholesterol Levels									
Subject	A	B	C	D	E	F	G	H	I
Before	209	210	205	198	216	217	238	240	222
After	199	207	189	209	217	202	211	223	201

12

Reliability Modeling (Modified and Adapted from Military Reliability Modeling by Fox and Horton)

Objectives

1. **Know and understand the concept of reliability.**
2. **Know and understand the concept of mean time to failure.**
3. **Know and apply series and parallel components.**
4. **Know application of redundant and standby redundant systems.**
5. **Be able to solve problems involving large systems.**

12.1 Introduction

You are a New York City policeperson. You are on a stakeout and must occupy a position at least the next 24 hours. Hourly situation reports must be made by radio. You must carry all necessary food, equipment, and supplies for the 24-hour period. The stakeout is ineffective unless other members of the force can communicate with you in a timely manner. Therefore, radio communications must be reliable. The radio has several components that affect its reliability, an essential one being the battery.

Batteries have a useful life that is not deterministic (we do not know exactly how long a battery will last when we install it). Its lifetime is a variable that may depend on previous use, manufacturing defects, weather, and so on. The battery that is installed in the radio prior to leaving for the stakeout could last only a few minutes or for the entire 24 hours.

DOI: 10.1201/9781003317906-12

Since communications are so important to this mission, we are interested in modeling and analyzing the reliability of the battery.

We will use the following definition for reliability:

> If T is the time to failure of a component of a system, and $f(t)$ is the probability distribution function of T, then the components' *reliability* at time t is
>
> $R(t) = P(T > t) = 1 - F(t)$
>
> $R(t)$ is called the reliability function, and $F(t)$ is the cumulative distribution function of $f(t)$.

A measure of this reliability is the probability that a given battery will last more than 24 hours. If we know the probability distribution for the battery life, we can use our knowledge of probability theory to determine the reliability. If the battery reliability is below acceptable standards, one solution is to have the policepersons carry spares. Clearly, the more spares they carry, the less likely there is to be a failure in communications due to batteries. Of course, the battery is only one component of the radio. Others include the antenna, handset, and so on. Failure of any one of the essential components causes the system to fail. This is a relatively simple example of one of many military applications of reliability.

This chapter will show we can use elementary probability to generate models that can be used to determine the reliability of equipment and systems of equipment.

12.2 Modeling Component Reliability

In this section, we will discuss how to model component reliability. The reliability function, $R(t)$, is defined as:

$$R(t) = P(T > t) = P(\text{component fails after time } t)$$

This can also be stated, using T as the component failure time, as:

$$R(t) = P(T > t) = 1 - P(T \le t) = 1 - \int_{-\infty}^{t} f(x)\,dx = 1 - F(t).$$

Thus, if we know the probability density function $f(t)$ of the time to failure T, we can use probability theory to determine the reliability function $R(t)$. We normally think of these functions as being time dependent; however, this is not always the case. The function might be discrete such as the lifetime of a cannon tube. It is dependent on the number of rounds fired through it (a discrete random variable).

A useful probability distribution in reliability is the exponential distribution. Recall that its density function is given by:

$$f(t) = \begin{cases} \lambda e^{-\lambda t} & t \ge 0 \\ 0 & else \end{cases}$$

where the parameter λ is a rate.

We know for the exponential distribution the mean is $\mu = \frac{1}{\lambda}$. If T denotes the time to failure of a piece of equipment or a system, this is the mean time to failure expressed in units of time. For applications of reliability, we will use the parameter λ, the average number of failures per unit time, or failure rate, to define the distribution. For example, if a light bulb has a time to failure that follows an exponential distribution with mean time to failure of 50 hours, then its failure rate is 1 light bulb per 50 hours so $\lambda = 1/50 = 0.02$. Note the at the mean of *T*, the time to failure of the component, is then 50.

Example 1: Batteries.

Let's consider the example presented in the introduction. Let the random variable *T* be defined as follows:

***T* = time until a randomly selected battery fails**

Suppose radio batteries have a time to failure that is exponentially distributed with a mean of 30 hours. In this case, we could write:

$$T \sim exp\left(\lambda = \frac{1}{30}\right).$$

Since $\lambda = 1/30$, we know the density function is $f(t) = 1/30\ e^{-t/30}$ for $t \geq 0$ and $F(t)$ is as follows:

$$F(t) = \int_0^t \frac{1}{30} e^{-x/30} dx$$

F(*t*), the CDF of the exponential distribution, can be integrated to obtain:

$$F(t) = 1 - e^{-t/30}, \quad t > 0.$$

Now we can compute the reliability function for a battery:

$$R(t) = 1 - F(t) = 1 - \left(1 - e^{-t/30}\right) = e^{-t/30}.$$

Recall that in the earlier example, the police must occupy the stakeout for 24 hours. The reliability of the battery for 24 hours is:

$$R(t) = e^{-24/30} = 0.44933.$$

In R, we can compute values of *F*(*t*) for the exponential using the "pexp" command, so the reliability calculations is then:

```
> 1-pexp(24,1/30)
[1] 0.449329
```

The probability that the battery lasts more than 24 hours is 0.44933.

Example 2: New nickel cadmium battery.

We have the option to purchase a new nickel cadmium battery for our stakeout. Testing has shown that the distribution of the time to failure can be modeled using a parabolic function:

$$f(x)=\begin{cases}\dfrac{x}{384}\left(1-\dfrac{x}{48}\right) & 0\le x\le 48\\ 0 & \text{otherwise}\end{cases}$$

Let the random variable T be defined as follows:

T = time until a randomly selected battery fails.

The CDF is computed as:

$$F(t)=\int_0^t \frac{x}{384}\left(1-\frac{x}{48}\right)dx$$

Recall that in the earlier example the persons must man the stakeout for 24 hours. The reliability of the battery for 24 hours is therefore:

$$R(24)=1-F(24)=1-\int_0^{24}\frac{x}{384}\left(1-\frac{x}{48}\right)dx=0.500.$$

This represents an improvement over the batteries from Example 1. Therefore, we should prefer the new battery.

12.3 Modeling Series and Parallel Components

12.3.1 Modeling Series Systems

Now we consider a system with n components $C_1, C_2, \ldots, C_n$, where each of the individual components must work in order for the system to function. A model of this type of system is shown in Figure 12.1.

If we assume these components are mutually independent, the reliability of this type of system is easy to compute. We denote the reliability of component i at time t by $R_i(t)$. In other words, $R_i(t)$ is simply the probability that component i will function continuously

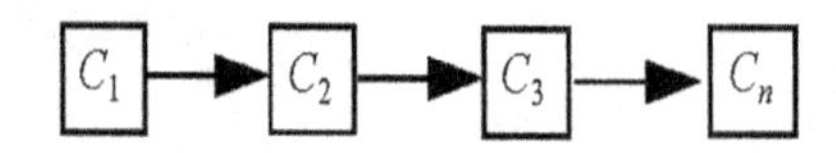

FIGURE 12.1
Series system.

from time 0 through until time t. We are interested in the reliability of the entire system of n components, but since these components are mutually independent, the system reliability is

$$R(t) = R_1(t) \cdot R_2(t) \cdot \cdots \cdot R_n(t)$$

Example 3: Radio with components.

Our radio has several components. Let us assume that there are four major components: they are (in order) the handset, the battery, the receiver-transmitter, and the antenna. Since they all must function properly for the radio to operate, we can model the radio with the diagram shown in Figure 12.2.

Suppose we know that the probability that the handset will work for at least 24 hours is 0.6703, and the reliabilities for the other components are 0.4493, 0.7261, and 0.9531, respectively. If we assume that the components work *independently* of each other, then the probability that the entire system works for 24 hours is:

$$R(24) = R1(24) \cdot R2\,(24) \cdot R3\,(24) \cdot R4\,(24) = (.6703)(.4493)(.7261)(.9531) = 0.2084$$

Recall that two events A and B are *independent* if $P(A \mid B) = P(A)$.

12.3.2 Modeling Parallel Systems (Two Components)

Now we consider a system with two components where only one of the components must work for the system to function. A system of this type is depicted in Figure 12.3.

Notice that in this situation the two components are *both* put in operation at time 0; they are both subject to failure throughout the period of interest. Only when *both* components fail before time t does the system fail. Again, we also assume that the components are independent. The reliability of this type of system can be found using the following well-known addition model presented in Chapter 4:

$$P(A \cup B) = P(A) + P(B) - P(A \cap B)$$

Since the components are independent, this reduces to

$$P(A \cup B) = P(A) + P(B) - P(A) * P(B)$$

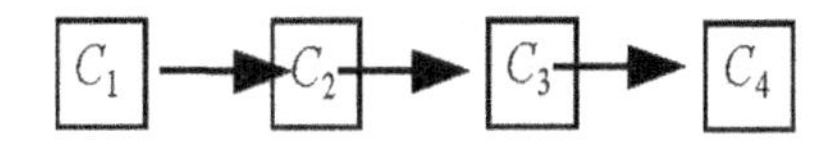

FIGURE 12.2
Radio system.

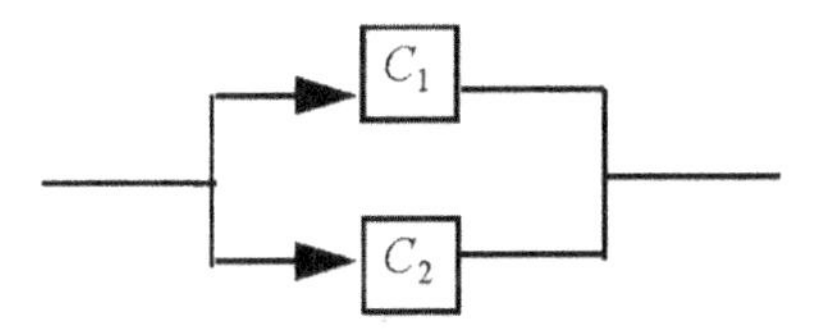

FIGURE 12.3
Parallel system of two components.

In this case, A is the event that the first component functions for longer than some time, t, and B is the event that the second component functions longer than the same time, t. Since reliabilities *are* probabilities, we can translate the above formula into the following:

$$R(_t) = R_1(t) + R_2(t) - R_1(t)R_2(t)$$

Example 4: Bridges.

Suppose we have two bridges in the area for a Boy Scout hike. It will take three hours to complete the crossing for all the hikers. The crossing will be successful as long as at least one bridge remains operational during the entire crossing period. You estimate that the bridges are in bad shape and that bridge 1 has a one-third chance of being destroyed and there is a one-fourth chance of bridge 2 being destroyed in the next three hours. Assume the destruction of the bridges are independent. What is the probability–that our boy scouts can complete the crossing?

Solution:

First, we compute the individual reliabilities:

$$R_1\ (3) = 1 - 1/3 = 2/3$$

and

$$R_2\ (3) = 1 - 1/4 = 3/4$$

Now it is easy to compute the system reliability:

$$R_s(3) = R_1(3) + R_2(3) - R_1(3)R_2(3)$$
$$= 2/3 + 3/4 - (2/3)(3/4) = 0.9166667$$

12.4 Modeling Active Redundant Systems

Consider the situation in which a system has n components, all of which begin operating (are active) at time $t = 0$. The system continues to function properly as long as at least k of the components do not fail. In other words, if $n - k + 1$ components fail, the system fails. This type of component system is called an active redundant system. The active redundant system can be modeled as a parallel system of components as shown in Figure 12.4.

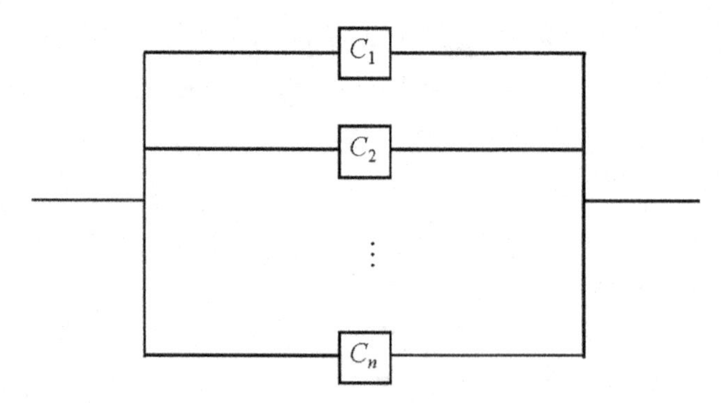

FIGURE 12.4
Active redundant system.

We assume that all n components are identical and will fail independently.

If we let Ti be the time to failure of the ith component, then the Ti terms are independent and identically distributed for $i = 1, 2, 3, \ldots, n$. Thus $Ri\,(t\,)$, the reliability at time t for component i, is identical for all components.

Recall in active redundant systems that our system operates if at least k components function properly. Now we define the random variables X and T as follows:

$$X = \text{number of components functioning at time } t$$
$$T = \text{time to failure of the entire system}$$

Then we have

$$R(t) = P(T > t) = P(X \geq k) = 1 - P(X \leq (k - 1)$$

It is easy to see that we now have n identical and independent components with the same probability of failure by time t. This situation corresponds to a binomial experiment, and we can solve for the system reliability using the binomial distribution with parameters n and $p = p(s) = R_i(t)$.

Example 5: Machines.

A manufacturing company has 15 different machines to make item A. They estimate if at least 12 are operating that they will be able to make all the items necessary to meet demand. Machines are assumed to be in parallel (active redundant)—that is, they fail independently. If we know that each machine has a 0.6065 probability of operating properly for at least 24 hours, we can compute the reliability of the entire machine system for 24 hours.

Define the random variable: X = number of machines working after 24 hours.

Clearly, the random variable X is binomially distributed with $n = 15$ and $p = 0.6065$. In the language of mathematics, we write this sentence as

$$X \sim b(15, 0.6065) \text{ or } X \sim \text{BINOMIAL}(15, 0.6065).$$

We know that the reliability of the machine system for 24 hours is (using R):

$$R(24) = P(X \geq 12) = P(12 \leq X \leq 15) = 0.0990$$

```
> pbinom(11, 15, 0.6065, lower.tail = FALSE)
[1] 0.09903638
```

Thus, the reliability of the system for 24 hours is only 0.0990.

Example 6: Exponential system components.

We have nine independent components in an active redundant system where at least six must work. We have the components with an exponential time to failure with parameter 0.01 per day. We want to find the reliability for 20 days.

Solution:

We observed earlier that the reliability for one component using the exponential distribution is:

$$R(t) = 1 - F(t) = 1 - \left(1 - e^{-t/100}\right) = e^{-t/100}.$$

Thus, we get the reliability of a single component for 20 days as:

$$R(20) = e^{-(0.01)(20)} = 0.818$$

In R, we compute this as:

```
> 1-pexp(20, rate = 0.01)
[1] 0.8187308
```

$n = 9$, $p(s) = 0.818$ and for the system, X, the number of components working after 20 days, is then modeled as a binomial random variable, $X \sim \text{Binomial}(n = 9, p = 0.818)$. We can then find $P(X \geq 6)$ using R:

```
> pbinom(5, 9, 0.818, lower.tail = FALSE)
[1] 0.9362637
```

The probability the system works 20 days is 0.936.

12.5 Modeling Standby Redundant Systems

Active redundant systems can sometimes be inefficient. These systems require only k of the n components to be operational, but all n components are initially in operation and thus subject to failure. An alternative is the use of spare components. Such systems have only k components initially in operation, exactly what we need for the whole system to be operational. When a component fails, we have a spare "standing by," which is immediately put in to operation. For this reason, we call these *Standby Redundant Systems.*

Suppose our system requires k operational components and we initially have $n - k$ spares available. When a component in operation fails, a decision switch causes a spare or standby component to activate (becoming an operational component). The system will continue to function until there are less than k operational components remaining. In other words, the system works until $n - k + 1$ components have failed. We will consider only the case where one operational component is required (the special case where $k = 1$) and there are $n - 1$ standby (spare) components available. We will assume that a decision switch (DS) controls the activation of the standby components instantaneously and 100% reliably. We use the model shown in Figure 12.5 to represent this situation.

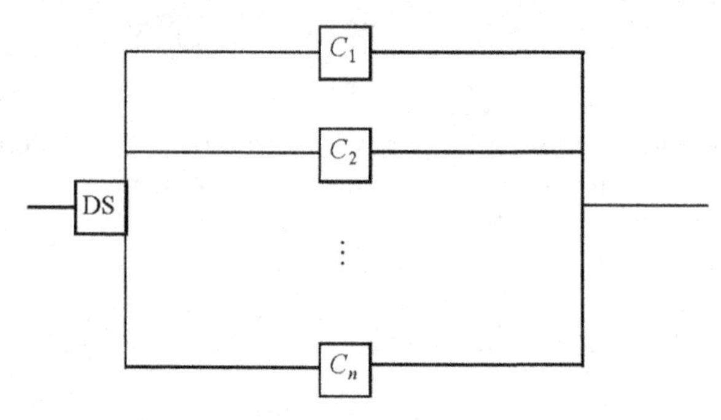

FIGURE 12.5
Standby redundant system.

If we let Ti be the time to failure of the ith component, then the Ti's are independent and identically distributed for $i = 1, 2, 3, \ldots, n$. Thus, $R_i(t)$ is identical for all components. Let T = time to failure of the entire system. Since the system fails only when all n components have failed, and component $i + 1$ is put into operation only when component i fails, it is easy to see that

$$T = T_1 + T_2 + \ldots + Tn.$$

In other words, we can compute the system failure time easily if we know the failure times of the individual components.

Finally, we can define a random variable

X = number of components that *fail* before time t in a standby redundant system

Now the reliability of the system is simply equal to the probability that less than n components fail during the time interval $(0, t)$. In other words,

$$R(t) = P(X < n)$$

It can be shown that X follows a Poisson distribution (review the Section 5.4 on this distribution to see why) with parameter $\lambda = \alpha t$, where α is the failure rate, so we write:

$$X \sim \text{POISSON}(\lambda)$$

For example, if time is measured in seconds, then <I>α</I> is the number of failures per second. The reliability for some specific time t then becomes:

$$R(t) = P(X < n) = P(0 \leq X \leq n - 1)$$

Example 7: Radio battery revisited.

Consider the reliability of a radio battery. We determined previously that one battery has a reliability for 24 hours of 0.4493. In light of the importance of communications, you decide that this reliability is not satisfactory. Suppose we carry two spare batteries. The addition of the spares should increase the battery system reliability. We do not want to dwell on how to calculate the failure rate, α, for a battery given the reliability (0.4493 in this case). So we will provide this to you: $\alpha = 1/30$ per hour. We know that $n = 3$ total batteries.

Therefore:

$$X \sim Poisson\left(\lambda = \alpha t = \frac{24}{30} = 0.8\right)$$

and

$$\text{R}(24) = P(X < 3) = P(0 \leq X \leq 2) = 0.9526$$

The reliability of the system with two spare batteries for 24 hours is now 0.9526.

Example 8: Number of spares.

If the police stakeout must stay out for 48 hours without resupply, how many spare batteries must be taken to maintain a reliability of 0.95? We can use trial and error to solve this problem. We start by trying our current load of two spares. We have

$$X \sim Poisson\left(\lambda = \alpha t = \frac{48}{30} = 1.6\right)$$

and we can now compute the system reliability

$$R(48) = P(X < 3) = P(0 \leq X \leq 2) = 0.7834 < 0.95$$

which is not good enough. Therefore, we try another spare so $n = 4$ (three spares), and we compute:

$$R(48) = P(X < 4) = P(0 \leq X \leq 3) = 0.9212 < 0.95$$

which is still not quite good enough, but we are getting close! Finally, we try $n = 5$, which turns out to be sufficient:

$$R(48) = P(X < 5) = P(0 \leq X \leq 4) = 0.9763 \geq 0.95.$$

Therefore, we conclude that the stakeout should take out at least four spare batteries for a 48-hour mission.

12.6 Models of Large-Scale Systems

In our discussion of reliability up to this point, we have discussed series systems, active redundant systems, and standby redundant systems. Unfortunately, things are not always this simple. The types of systems listed above often appear as subsystems in larger arrangements of components that we shall call "large-scale systems." Fortunately, if you know how to deal with series systems, active redundant systems, and standby redundant systems, finding system reliabilities for large-scale systems is easy. Consider the following example.

The first and most important step in developing a model to analyze a large-scale system is to draw a picture. Consider the network that appears as Figure 12.6 below. Subsystem A is the standby redundant system of three components (each with failure rate five per year) with the decision switch on the left of the figure. Subsystem B_1 is the active redundant system of three components (each with failure rate three per year), where at least two of the three components must be working for the subsystem to work. Subsystem B_1 appears in the upper right portion of the figure. Subsystem B_2 is the two-component parallel system in the lower right portion of the figure. We define subsystem B as being subsystems B_1 and B_2 together. We assume all components have exponentially distributed times to failure with failure rates as shown in Figure 12.6.

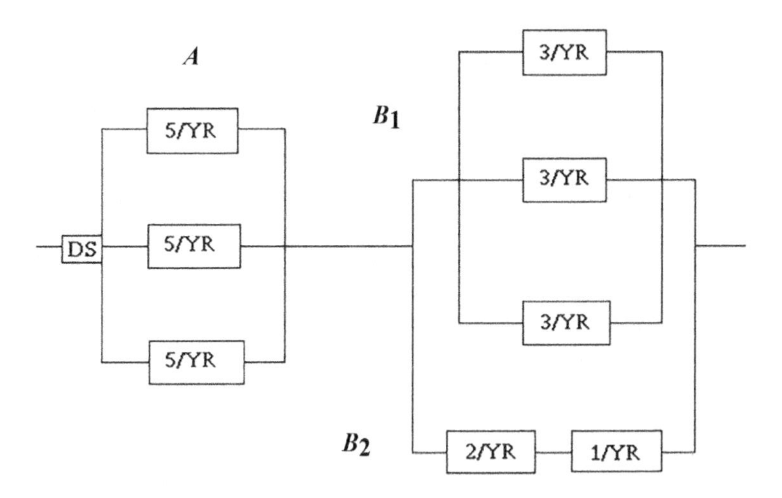

FIGURE 12.6
Network example.

Suppose we want to know the reliability of the whole system for six months. Observe that you already know how to compute reliabilities for the subsystems A, B_1, and B_2. Let's review these computations and then see how we can use them to simplify our problem.

Subsystem A is a standby redundant system, so we will use the Poisson model. We let

$$X = \text{the number of components which fail in one year}$$

Since six months is 0.5 years, we seek $R_A\,(0.5) = P(X < 3)$, where X follows a Poisson distribution with parameter λ <I>11 </I>= <I>αt </I>= (5)(0.5) = 2.5. Then

$$R_A\,(0.5) = P(X < 3) = P(0 \le X \le 2) = 0.5438$$

Now we consider subsystem B_1. In Section 12.2 we learned how to find individual component reliabilities when the time to failure followed an exponential distribution. For subsystem B_1, the failure rate is three per year, so our individual component reliability is

$$R(0.5) = 1 - F(0.5) = 1 - (1 - e^{-(3)(0.5)}) = e^{-(3)(0.5)} = 0.2231$$

Now recall that subsystem B_1 is an active redundant system where two components of the three must work for the subsystem to work. If we let

$$Y = \text{the number of components that function for six months}$$

and recognize that Y follows a binomial distribution with $n = 3$ and $p = 0.2231$, we can quickly compute the reliability of the subsystem B_1 as follows:

$$R_B\,(0.5) = P(Y \ge 2) = 1 - P(Y < 2) = 1 - P(Y \le 1) = 1 - 0.8729 = 0.1271$$

Finally, we can look at subsystem B_2. Again, we use the fact that the failure times follow an exponential distribution. The subsystem consists of two components; obviously, they both need to work for the subsystem to work. The first component's reliability is

$$R(0.5) = 1 - F(0.5) = 1 - (1 - e^{-(2)(0.5)}) = e^{-(2)(0.5)}$$
$$= 0.3679, \text{ and for the other component the reliability is}$$
$$R(0.5) = 1 - F(0.5) = 1 - (1 - e^{-(1)(0.5)}) = e^{-(1)(0.5)}$$
$$= 0.6065$$

Therefore, the reliability of the subsystem is

$$R_B\,(0.5) = (0.3679)(0.6065) = 0.2231$$

Our overall system can now be drawn as shown in Figure 12.7.

From here, we determine the reliability of subsystem B by treating it as a system of two independent components in parallel where only one component must work. Therefore,

$$R_B\,(0.5) = R_B\,(0.5) + R_B\,(0.5) - R_B\,(0.5) \cdot R_B\,(0.5)$$
$$= 0.1271 + 0.2231 - (0.1271)(0.2231) = 0.3218$$

Finally, since subsystems A and B are in series, we can find the overall system reliability for six months by taking the product of the two subsystem reliabilities:

$$R\text{system}\,(0.5) = R_A\,(0.5) \cdot R_B\,(0.5) = (0.5438)(0.3218) = 0.1750$$

We have used a network reduction approach to determine the reliability for a large-scale system for a given time period. Starting with those subsystems that consist of components independent of other subsystems, we reduced the size of our network by evaluating each subsystem reliability one at a time. This approach works for any large-scale network consisting of basic subsystems of the types we have studied (series, active redundant, and standby redundant).

We have seen how methods from elementary probability can be used to model military reliability problems. The modeling approach presented here is useful in helping students simultaneously improve their understanding of both the military problems addressed and the mathematics behind these problems. The models presented also motivate students to appreciate the power of mathematics and its relevance to today's world.

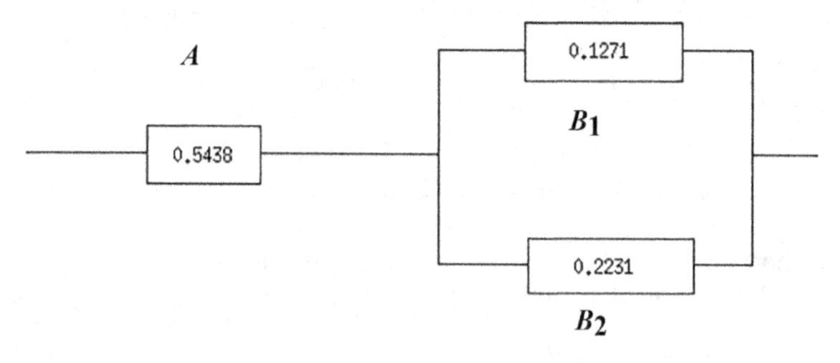

FIGURE 12.7
Simplified network example.

12.7 Summary of Reliability

Example Case 1: Series.

Suppose a system consists of four components arranged in series. The first two components have reliabilities of 0.9 at time $t = 1$ year and the other two components have reliabilities of 0.8 at $t = 1$ year. What is the overall reliability of the system at one year?

Solution:

The probability of the system operating successfully at the end of the first year is:

$$(0.9)^2(0.8)^2 = 0.5184$$

Example Case 2: Series.

A truck cab assembly line is under development. The line will utilize five welding robots.

a) If each robot has 95% reliability, what is the total robot system reliability for the line?
b) To have a robot system reliability of 95%, what must each individual robot's reliability be?

Solution:

(a) $0.95^5 = 0.7738$
(b) Find $0.95 = R^5$ so $R = 0.9897$

Example Case 3: Exponential case.

Five components in series are each distributed exponentially with a hazard rate of 0.2 failures per hour. What is the reliability and MTTF of the system?

Solution:

$$\because \lambda_s = 0.2 \quad i = 1, 2, 3, 4, 5$$

$$\because R_s = e^{-\lambda_s t}$$

$$\lambda_s = \sum_{i=1}^{5} \lambda_s = (0.2)^5 = 1$$

$$\therefore R_s = e^{-t}$$

$$MTTF_s = \frac{1}{\lambda_s} = 1$$

Example Case 4: Parallel.

Suppose a system consists of four components arranged in parallel. The first two components have reliabilities of 0.9 and time $t = 1$ year and the other two components have reliabilities of 0.8 at $t = 1$ year. What is the overall reliability of the system at one year?

$$R = 1 - [(1 - 0.9)^2(1 - 0.8)^2] = 0.996$$

12.8 References and Suggested Readings

Devore, J. L. (1995). *Probability and Statistics for Engineers and Scientists,* 4th ed., Duxbury Press, Belmont, CA.

Fox, W. (1990). *Program Director Notes,* Unpublished MA 206.

Fox, W. and S. Horton (1992). *Military Mathematical Modeling,* D. Arny, ed., USMA, West Point, NY.

Resnick, S. L. (1992). *Adventures in Stochastic Processes,* Birkhäuser, Boston, MA.

12.9 Chapter 12 Exercises

12.1 A continuous random variable Y, representing the time to failure of a 0.50 mm tube, has a probability density function given by

$$f(y)=\begin{cases} 1/3e^{-y/3} & y \geq 0 \\ 0 & otherwise \end{cases}$$

a. Find the reliability function for Y.

b. Find the reliability for 1.2 time periods, R (1.2).

12.2 The lifetime of a car engine (measured in time of operation) is exponentially distributed with a MTTF mean time to failure of 400 hours. You have received a mission that requires 12 hours of continuous operation. Your log book indicates that the car engine has been operating for 158 hours.

a. Find the reliability of your engine for this mission.

b. If your vehicle's engine had operated for 250 hours prior to the mission, find the reliability for the mission.

12.3 A criminal must be captured. You are on the police SWAT team, which decides to use helicopters to help capture the key criminal. The SWAT aviation battalion is tasked to send four helicopters. On their way to the target area, these helicopters must fly over foggy territory for approximately 15 minutes during which time they are vulnerable to accidents. The lifetime of a helicopter over this territory is estimated to be exponentially distributed with a mean of 18.8 minutes. It is further estimated that two or more helicopters are required to capture the criminal. Find the reliability of the helicopters in accomplishing their mission (assuming the only reason a helicopter fails to reach the target is the foggy weather).

12.4 For the mission in Exercise 12.3, the police commissioner determines that to justify risking the loss of helicopters, there must be at least an 80% chance of capturing the criminal. How many helicopters should the aviation battalion recommend be sent? Justify your answer.

12.5 Mines are dangerous obstacles. Most mines have three components—the firing device, the wire, and the mine itself (casing). If any of these components fail, the systems fails. These components of the mine start to "age" when they are unpacked from their sealed containers. All three components have MTTF mean time to failures (MTTF) that are exponentially distributed with means of 60 days, 300 days, and 35 days, respectively.

a. Find the reliability of the mine after 90 days.

b. What is the MTTF of the mine?

c. What assumptions, if any, did you make?

12.6 You are a project manager for a new system being developed in Huntsville, Alabama. A critical subsystem has two components arranged in a parallel configuration. You have told the contractor that you require this subsystem to be at least 0.995 reliable. One of the subsystems came from an older system and has a known reliability of 0.95. What is the minimum reliability of the other component so that we meet your specifications?

12.7 You are in charge of stage lighting for an outdoor concert. There is some concern about the reliability of the lighting system for the stage. The lights are powered by a 1.5 KW generator that has a MTTF of 7.5 hours.

a. Find the reliability of the generator for 10 hours if the generator's reliability is exponential.

b. Find the reliability of the power system if two other identical 1.5 KW generators are available. First consider as active redundant and then as standby redundant. Which would improve the reliability the most?

c. How many generators would be necessary to ensure a 99% reliability?

12.8 Consider the system below with the reliability for each component as indicated. Assume all components are independent and the radars are active redundant.

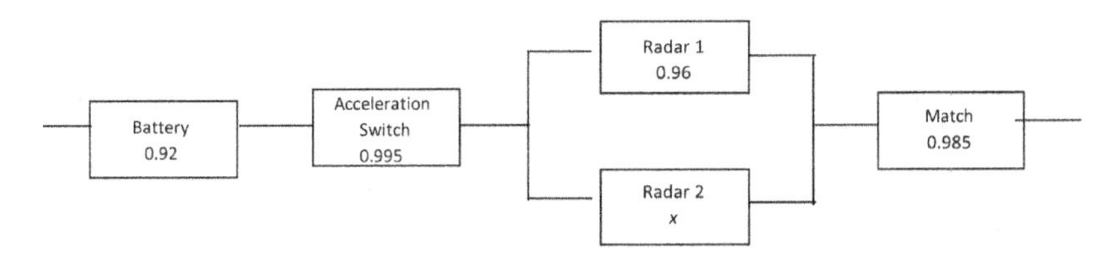

a. Find the system reliability for six months when $x = 0.96$.

b. Find the system reliability for six months when $x = 0.939$.

12.9 A major system has components as shown in diagram below. All components have exponential times to failure with mean times to failure shown. All components operate independently of each other. Find the reliability for this weapon system for two hours.

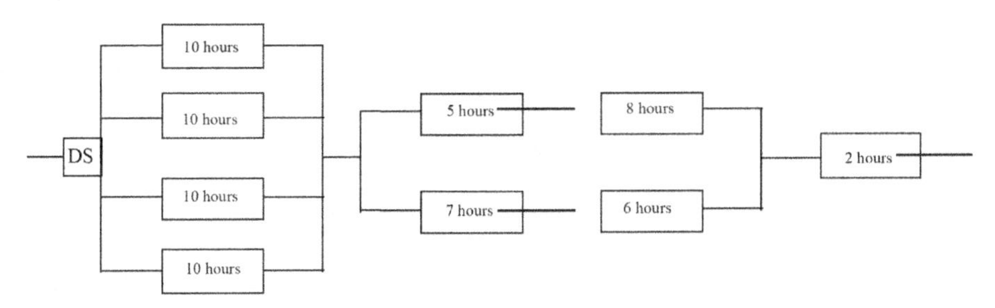

12.10 Write a short essay about how you can use elementary probability in reliability of modeling business problems and provide one example.

12.11 A system is composed of three components. These components have constant failure rates of 0.0004, 0.0005, 0.0003 failures per hour. The system will stop working if any one of its components fails. Calculate the following:

The reliability of the system at 2,500-hour running time?

12.12 Solve the following system reliabilities:

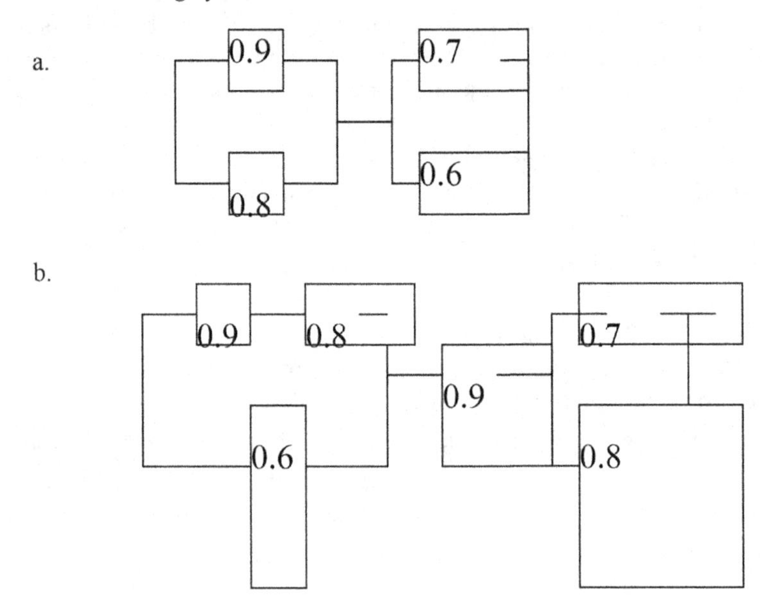

12.13 For a two-out-of-three system (a special case of k-out-of-n system), each component has reliability of 0.9. What is the system reliability?

12.14 A space shuttle requires three out of four of its main engines to achieve orbit. Each engine has a reliability of 0.97. What is the probability of achieving orbit?

13

Introduction to Regression Techniques

Objectives

1. **Know and understand the concept of correlation.**
2. **Know and understand the linear relationship and correlation.**
3. **Know how to obtain and interpret a linear regression model.**
4. **Know how to apply diagnostics to test model adequacy.**

We have data that relates time to demand for a specific item for the past two years. We would like to forecast next year's demand assuming the thing remains almost unchanged. Can we do this? If so, how do we proceed?

Why regression techniques? Often we have data, and we want or need to find a relationship between the independent and dependent variables. Notice that I have said relationship. Relationships are linear or nonlinear. The concept of correlation ONLY applies to linear relationships! Why do we obtain a model via regression? We might want to explain a behavior or predict a future or intermediate result. Several issues need to be examined besides just building or obtaining the model. These include: how good is the model we have built, do the results pass the "common sense" test, and did I use the appropriate regression technique? These sections help answer these important questions.

13.1 Correlation and Misconceptions

13.1.1 Correlation and Covariance

More students fail to understand correlation as a concept more than anything else we do.

When two variables X and Y are not independent, it is often of interest to measure how strongly they are related to one another. Covariance, in many statistics books, has several formulas that we might use, such as

$$COV(X, Y) = E[XY] - \mu_x\mu_y \text{ or}$$

$\Sigma x \Sigma y\, (x - \mu_x)(y - \mu_y) * p(x, y)$ both use probability information.

We will return to this later when we do regression.

DOI: 10.1201/9781003317906-13

13.1.2 Correlation: A Measure of Linear Relationship

Excel's definition and description are NOT correct as are many other definitions on the Internet.

"Returns the correlation coefficient of the array1 and array2 cell ranges. Use the correlation coefficient to determine the relationship between two properties. For example, you can examine the relationship between a location's average temperature and the use of air conditioners."

13.1.2.1 Correlation Is the Measure of Linear Relationship between Variables

Correlation is computed into what is known as the correlation coefficient, which ranges between −1 and +1. Perfect positive correlation (a correlation coefficient of +1) implies that as one variable moves, either up or down, the other variable will move in lockstep, in the same direction. Alternatively, perfect negative correlation means that if one variable moves in either direction, the variable that is perfectly negatively correlated will move in the opposite direction. If the correlation is 0, the movements of the variable are said to have no correlation; they are completely random, or their relationship is nonlinear.

In real life, perfectly correlated variables are extremely rare; rather you will find variables with some degree of correlation.

In the textbook by Jay Devore, *Probability and Statistics for Engineering and Sciences,* he says that for descriptive purposes, the correlation relationship, called ρ, that measures the linear relationship of the variables will be described as

Strong when $|\rho| \geq 0.8$,
Moderate if $0.5 < |\rho| < 0.8$,
and Weak if $|\rho| \leq 0.5$.

For social science data, the correlation measures scales are different.

Correlation level	Values
None	$0.0 \leq \lvert\rho\rvert \leq 0.1$
Small	$0.1 < \lvert\rho\rvert \leq 0.3$
Medium	$0.3 < \lvert\rho\rvert \leq 0.5$
Strong	$0.5 < \lvert\rho\rvert \leq 1$

Caution must be used when discussing correlation. We provide two examples both that show that there is perhaps a relationship between x and y but not always linear.

13.1.2.2 Correlation: A Measure of Linear Relationship

The correlation is one of the most common and most useful statistics. A correlation is a single number that describes the degree of linear relationship between two variables. Let's work through an example to show you how this statistic is computed.

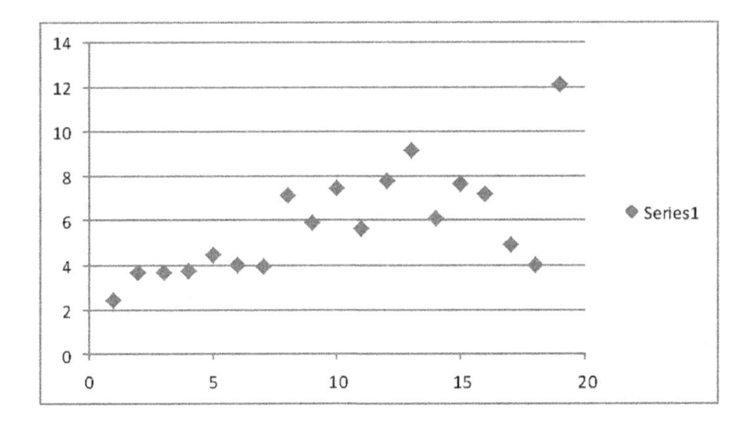

The correlation between x and y is 0.779. The plot appears "reasonably" linear.

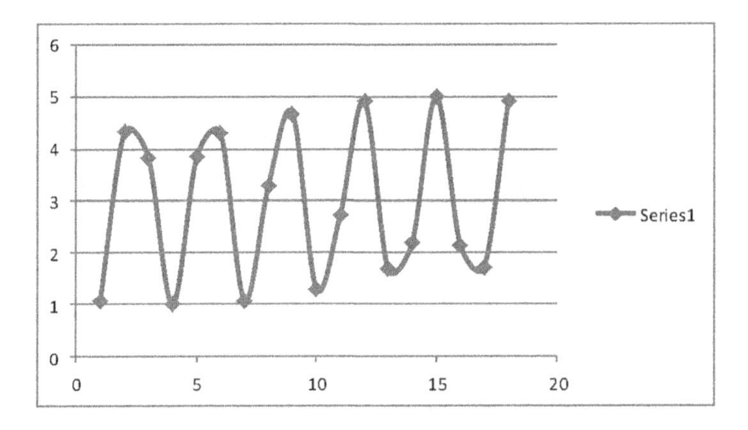

The correlation here is 0.108. The plot is not remotely linear yet clearly, we see the oscillating relationship between x and y. The relationship is nonlinear.

FIGURE 13.1
Illustrations of some plots and correlation.

13.1.3 Calculating the Correlation

Now we're ready to compute the correlation value. The formula for the correlation is:

$$\rho = \frac{n\sum xy - (\sum x)(\sum y)}{\sqrt{\left[n\sum x^2 - \left(\sum x\right)^2\right]\left[n\sum y^2 - \left(\sum y\right)^2\right]}}$$

We use the symbol **r or** ρ **(rho)** to stand for the correlation. Through the magic of mathematics, it turns out that $\boldsymbol{\rho}$ will always be between –1.0 and +1.0. If the correlation is negative, we have a negative relationship; if it's positive, the relationship is positive. You don't need to know how we came up with this formula unless you want to be a statistician. But you probably will need to know how the formula relates to real data and how you can use the formula to compute the correlation.

Let's assume we have six data pairs for grades in two classes; suppose x = class 1 and y = class 2.

x	70	92	80	74	65	83
y	74	84	63	87	78	90

Through data manipulation, we find $\Sigma x = 464$, $\Sigma x^2 = 36354$, $\Sigma y = 476$, $\Sigma y^2 = 38254$, and $\Sigma xy = 36926$. We substitute into our formula for ρ.

$$\rho = \frac{6(36926)-(464)(476)}{\sqrt{[6(36354)-(464)^2(6(38254)(476)^2}} = 0.2396$$

In R this is much easier. The single command is

```
> cor(x, y)
```

where x and y are data sets.

Example 1: Correlation of grades.

```
mg = c(70, 92, 80, 74, 65, 83)
> eg = c(74, 84, 63, 87, 78, 90)
> cor(mg, eg)
[1] 0.2396639
```

If this data were scientific data, our correlation would indicate a weak linear relationship, but if our data was social science data (nonscientific data), the correlation would indicate a small linear relationship.

13.1.4 Testing the Significance of a Correlation with Hypothesis Testing

Once you've computed a correlation, you can determine the probability that the observed correlation occurred by chance. That is, you can conduct a significance test. Most often you are interested in determining the probability that the correlation is a real one and not a chance occurrence. In this case, you are testing the mutually exclusive hypotheses:

Null hypothesis:	$\rho = 0$
Alternative hypothesis:	$\rho \neq 0$

The easiest way to test this hypothesis is to find a statistics book that has a table of critical values of ρ. Most introductory statistics texts would have a table like this, and since we are not using one, we include it here.

As in all hypothesis testing, you need to first determine the significance level. Here, we use the common significance level of alpha = 0.05. This means that we are conducting a test where the odds that the correlation is a chance occurrence are no more than

5 out of 100. Before we look up the critical value in a table, we also have to compute the degrees of freedom or *df*. The *df* is simply equal to $N - 2$ where N is the number of data pairs, or, in this example, *df* is $20 - 2 = 18$. Finally, we have to decide whether we are doing a one-tail or two-tail test. In this example, since we have no strong prior theory to suggest whether the relationship between height and self-esteem would be positive or negative, we will opt for the two tail test. With these three pieces of information—the significance level ($\alpha = 0.05$), degrees of freedom (df = 18), and type of test (two tailed)—we can now test the significance of the correlation we found. When we look up this value in the table at the back of my statistics book, we find that the critical value is 0.468. This means that if the correlation is greater than 0.468 or less than −0.468 (remember, this is a two-tail test), we can conclude that the odds are less than 5 out of 100 that this is a chance occurrence. Since our correlation of 0.73 is actually quite a bit higher, we conclude that it is not a chance finding and that the correlation is "statistically significant" (given the parameters of the test). We can reject the null hypothesis in favor of the alternative hypothesis.

13.2 Model Fitting and Least Squares

Consider a simple spring-mass system, such as the one in Figure 13.2.

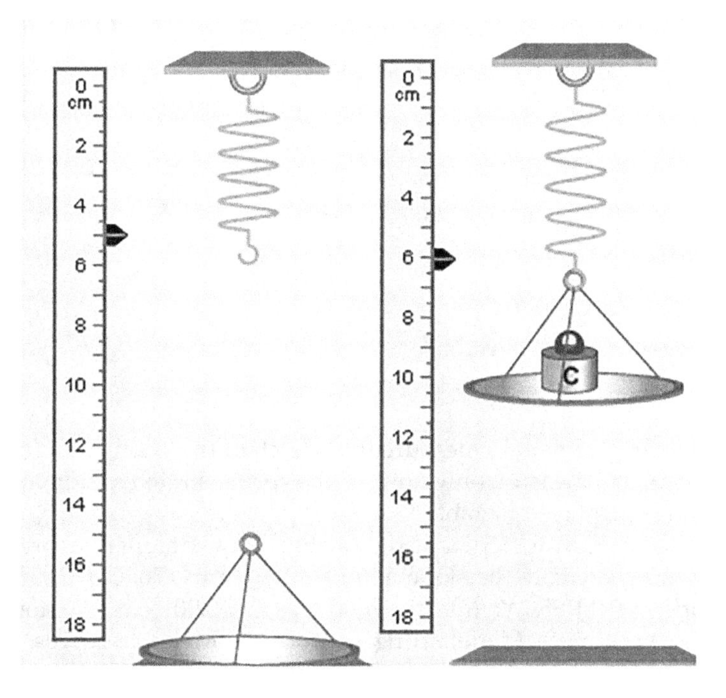

FIGURE 13.2
Spring-mass systems.

TABLE 13.1
Spring-Mass System

Mass (grams)	Stretch (m)
50	0.1
100	0.1875
150	0.275
200	0.325
250	0.4375
300	0.4875
350	0.5675
400	0.65
450	0.725
500	0.80
550	0.875

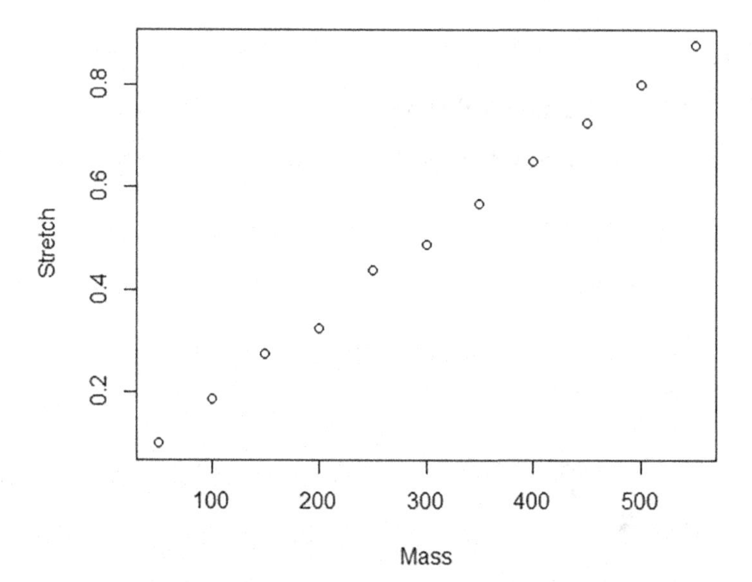

FIGURE 13.3
Plot of spring-mass data that appears linear.

We conducted an experiment to measure the stretch of the spring as a function of mass placed on the spring. We measure only how far the spring stretched from its original position, and this data is displayed in Table 13.1.

The data plot, seen in Figure 13.3, looks reasonably like a straight line through the origin. Our next step was to calculate the slope. We use the points (50, 0.1) and (550, 0.875) and compute the slope as 0.00155. We find the model as $F = 0.00155\ S$. We now want a more exact fit of our line to the data. Model fitting, especially with least squares, will be how we obtain a better fit.

This chapter focuses on the analytical methods to arrive at a model for a given data set using a prescribed criterion. Again, from the family $y = kx^2$, the parameter k can be determined analytically by using a curve-fitting criterion, such as least squares, Chebyshev's, or minimizing the sum of the absolute error and then solving the resulting optimization problem. We concentrate on the presentation of least squares in this chapter. We present the R commands that solve the least-squares optimization problem with analysis of the "goodness of the fit" of the resulting model. Remark: In R we can obtain a least-squares fit of our data.

R commands

Example 2: Spring-mass system.

```
Mass = c(50, 100, 150, 200, 250, 300, 350, 400, 450, 500, 550)
Stretch = c(0.1, 0.1875, 0.275, 0.325, 0.4375, 0.4875, 0.5675, 0.65, 0.725, 0.8, 0.875)

cor(Mass, Stretch)
[1] 0.9992718
> plot(Mass, Stretch)
```

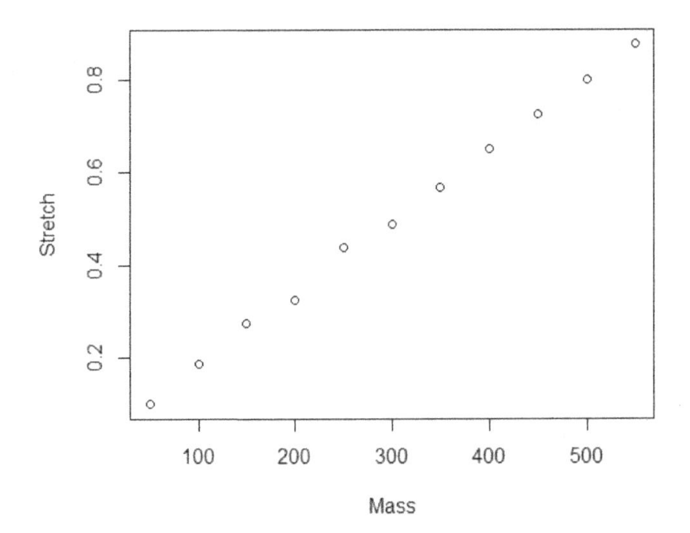


```
Model by R
> fit<-lm(Stretch~Mass)
> fit

Call:
lm(formula = Stretch ~ Mass)

Coefficients:
(Intercept) Mass
0.032455 0.00153
```

The model is Stretch = 0.02455 + 0.00153 Mass. We can use R to overlay the data and the least-squares line,

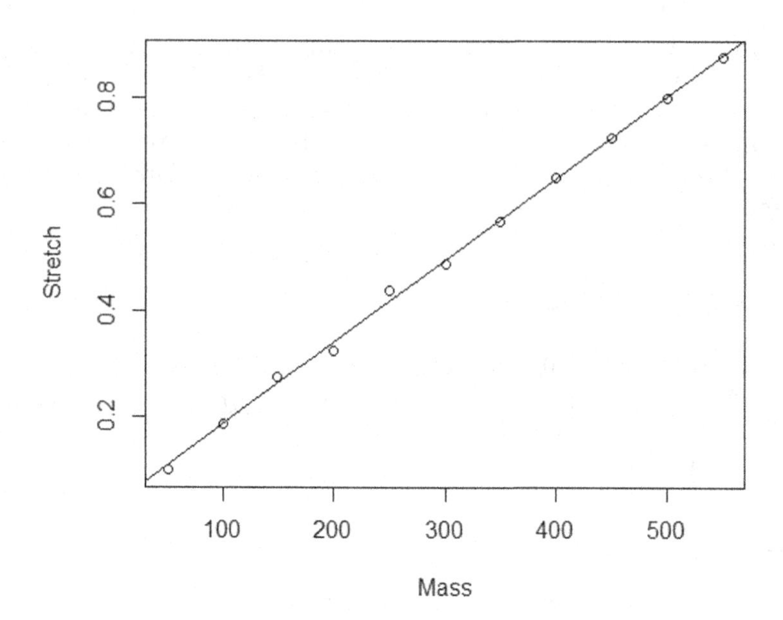

We go ahead, at this time, and get the summary output from R, and we will explain this output later in the chapter.

```
> summary(fit)
Call:
lm(formula = Stretch ~ Mass)
Residuals:
   Min        1Q       Median      3Q       Max
−0.014909 −0.004568 −0.001091 0.001977 0.020727
Coefficients:
Estimate                Std. Error  t value    Pr(>|t|)
(Intercept) 3.245e−02  6.635e−03   4.891    0.000858 ***
Mass 1.537e−03         1.957e−05   78.569   4.44e−14 ***
Signif. codes: 0 '***' 0.001 '**' 0.01 '*' 0.05 '.' 0.1 '' 1
Residual standard error: 0.01026 on 9 degrees of freedom
Multiple R-squared: 0.9985,    Adjusted R-squared: 0.9984
F-statistic: 6173 on 1 and 9 DF, p-value: 4.437e−14
```

Example 3: Car loans.

```
Data
year = c(0, 1, 2, 3, 4)
> rate = c(9.34, 8.5, 7.62, 6.93, 6.6)
> plot(year, rate,
+ main = "Commercial Banks Interest Rate for 4 Year Car Loan",
+ sub = "www.federalreserve.gov/releases/g19/20050805/")
```

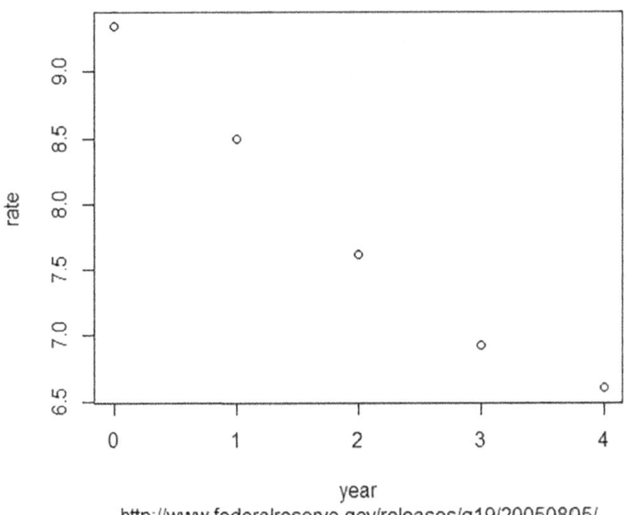

FIGURE 13.4
Plot of car loans in R showing a negative linear relationship.

The scatterplot, Figure 13.4, shows a negative trend.

```
cor(year, rate)
[1] -0.9880813
```

The least-squares fit is

```
fit<-lm(rate~year)
> fit<-lm(rate~year)
Call:
lm(formula = rate ~ year)
Coefficients:
(Intercept) year
9.208 -0.705
```

$$R^2 = (-0.98808)^2 = 0.976$$

```
> summary(fit)
Call:
lm(formula = rate ~ year)
Residuals:
  1      2      3      4     5
0.132 -0.003 -0.178 -0.163 0.212
Coefficients:
              Estimate  Std. Error  t value   Pr(>|t|)
(Intercept) 1419.20800  126.94957     11.18   0.00153 **
year          -0.70500    0.06341    -11.12   0.00156 **
—
```

Signif. codes: 0 '***' 0.001 '**' 0.01 '*' 0.05 '.' 0.1 '' 1

Residual standard error: 0.2005 on 3 degrees of freedom

Multiple R-Squared: 0.9763, Adjusted R-squared: 0.9684

F-statistic: 123.6 on 1 and 3 DF, p-value: 0.0

```
anova(fit)
Analysis of Variance Table
Response: rate
Df Sum Sq Mean Sq F value Pr(>F)
year 1 4.9702 4.9702 123.61 0.001559 **
Residuals 3 0.1206 0.0402
—
Signif. codes: 0 '***' 0.001 '**' 0.01 '*' 0.05 '.' 0.1 '' 1


>
```

Residual plot

We see a curved trend in Figure 13.5, so the model is not adequate.

Example 4: Age versus height of a tree over a five-year span

Data on the age of the tree and its most recent five-year height growth (in cm) are given below:

```
### CLEAR CUT EXAMPLE ###
age<-c(5,  9,  9, 10, 10, 11, 11, 12,
       13, 13, 14, 14, 15, 15, 18, 18)
five_year<-c(70, 150, 260, 230, 255, 165, 225, 340,
             305, 335, 290, 340, 225, 300, 380, 400)
```

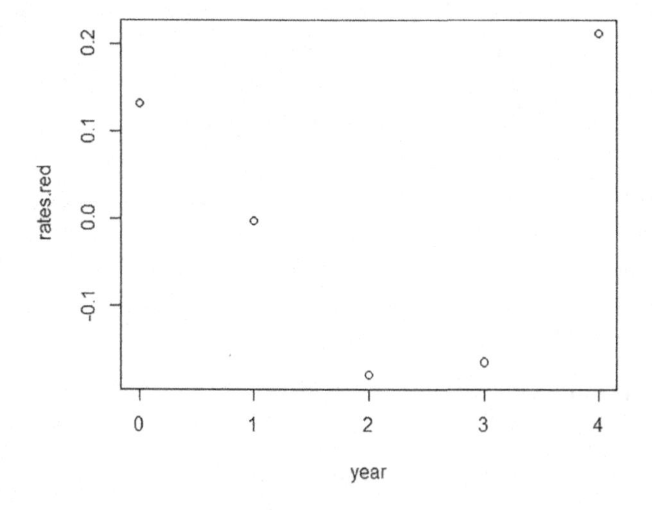

FIGURE 13.5
Residual plot.

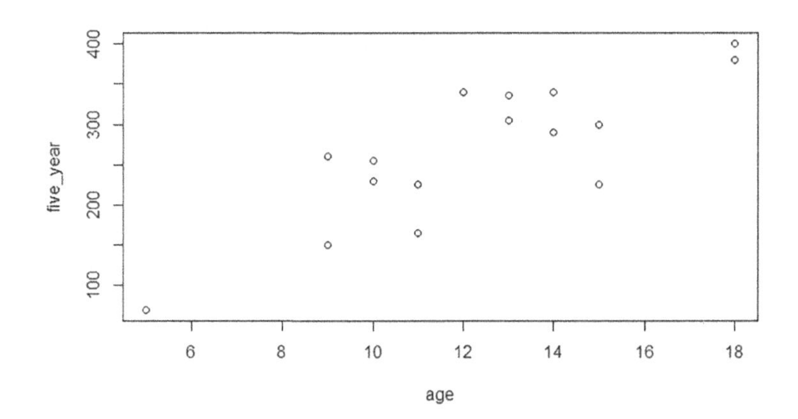

FIGURE 13.6
Residual plot with linear trend.

We describe the relationship between age and five-year growth by examining the plot:

The plot, Figure 13.6 appears linear with a positive trend. The correlation is found in R is:

```
> cor(age, five_year)
[1] 0.8263674
```

We think this is a strong enough linear relationship to continue.

```
model<-glm(five_year~age)
> summary(model)

Call:
glm(formula = five_year ~ age)

Deviance Residuals:
  Min      1Q    Median   3Q    Max
-99.177 -28.373  1.858  37.216 79.788

Coefficients:
            Estimate Std. Error t value Pr(>|t|)
(Intercept)  4.352    49.511    0.088   0.931
age         21.322     3.883    5.491   7.95e-05 ***
—

Signif. codes: 0 '***' 0.001 '**' 0.01 '*' 0.05 '.' 0.1 ' ' 1

(Dispersion parameter for Gaussian family taken to be 2645.52)

Null deviance: 116794 on 15 degrees of freedom
Residual deviance: 37037 on 14 degrees of freedom
```

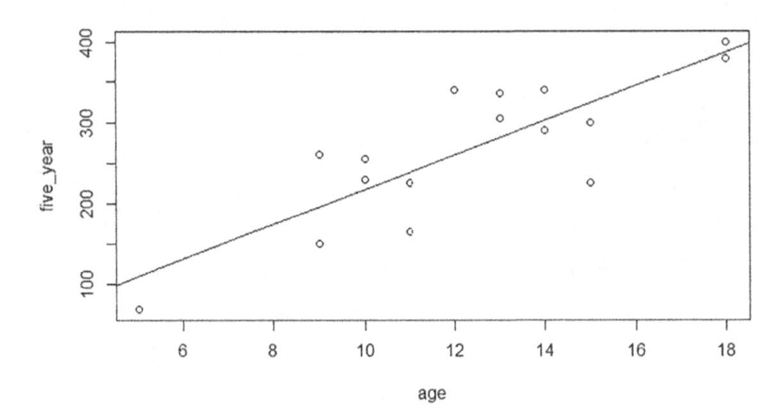

FIGURE 13.7
Overlay of data and line.

AIC: 175.36

The model is $five\_year\ growth = 4.352 + 21.322\ {}^*age$

```
### FIT LSRL ####
mod<-lm(five _ year~age)

plot(age, five _ year, pch = 16,
                    xlab = "Age (years)",
                    ylab = "Five Year Growth (cm)",
                    main = "Scatterplot of Age vs Five Year Growth")
abline(coefficients(mod), lwd = 2, lty = 2,
        col = "red")
```

Figure 13.7 shows an adequate fit.

13.2.1 Coefficient of Determination, R-square

In looking at the data and/or the scatterplot, not all the five-year growths are the same. Therefore, there is some variation in the response variable. The hope is that the least-squares regression line will fit between the data points in a manner that will "explain" quite a bit of that variation. The closer the data points are to the regression line, the higher proportion of the variation in the response variable that's explained by the regression line.

R-square is the percent (or proportion) of the variation in the most recent five-year tree growth is explained by the regression of tree growth on age. (This is called R-square, or R^2.)

```
### R-SQUARED ###
R2<-summary(mod)$r.squared
R2
## [1] 0.682883
```

Use the value of R-square to determine the correlation coefficient.

```
### CORRELATION ###
sqrt(R2)
## [1] 0.8263674
cor(five_year, age)
## [1] 0.8263674
```

13.3 Curve-Fitting Criterion, Least Squares

We will briefly our curve-fitting criterion: least squares or linear regression.

The method of least-squares curve-fitting, also known as **ordinary least squares** and **linear regression**, is simply the solution to a model that minimizes the sum of the squares of the deviations between the observations and predictions as shown in Equation 13.1. Least squares will find the parameters of the function, $f(x)$ that will

$$\text{Minimize } S = \sum_{j=1}^{m} \left[y_1 - f(x_j) \right]^2 \tag{13.1}$$

For example, to fit a proposed model $y = kx^2$ to a set of data, the least-squares criterion requires the minimization of Equation 13.2. Note in Equation 13.2, k is a slope.

$$\text{Minimize } S = \sum_{j=1}^{5} \left[y_i - kx_j^2 \right]^2 \tag{13.2}$$

Minimizing Equation 13.2 is achieved using the first derivative, setting it equal to zero, and solving for the unknown parameter, k.

$$\frac{ds}{dk} = -2\sum x_j^2 (y_j - kx_j^2) = 0. \text{ Solving for } k\text{: } k = \left(\sum x_j^2 y_j\right) / \left(\sum x_j^4\right) \tag{13.3}$$

Given the data set in Table 13.2, we will find the least-squares fit to the model, $y = kx^2$.

Solving for k: $k = \left(\sum x_j^2 y_j\right) / \left(\sum x_j^4\right) = (195.0)/(61.1875) = 3.1869$ and the model $y = kx^2$ becomes $y = 3.1869x^2$

Let's assume we prefer the model, $y = ax^2 + bx + c$.

$$xv := [.5, 1, 1.5, 2, 2.5];$$

$$yv := [.7, 3.4, 7.2, 12.4, 20.1];$$

TABLE 13.2

Least Squares Data Points

X	0.5	1.0	1.5	2.0	2.5
Y	0.7	3.4	7.2	12.4	20.1

The least-squares fit is

$$y = 3.2607x^2 - 0.22223x + 0.12630$$

Next, we illustrate the least-squares fit, Figure 13.8, applied to our model $y = kx^2$ for the data set for our example. As obtained previously, the least-squares model is $y = 3.1870x^2$ (rounded to three decimal places). The fit will not be as good nor appear as good, although this might not always be visible to the eye. Why?

Example 5: A least-squares fit for explosive data.

Assume we have developed a proportionality model, $V = kD^3$. *Let's demonstrate how to determine a constant of proportionality, using the least-squares criterion. V represents the volume of the crater from TNT and diameter³ is the estimate of the volume of the crater created. This example illustrates the fit command analytically fitting the model V = k diameter³* to the same data set. Using our data:

Diameter	14.5	12.5	17.25	14.5	12.625	17.75	14.125	12.625
Size, *V*	27	17	41	26	17	49	23	16

Our least-squares model is

$$y = 0.0084011x^3$$

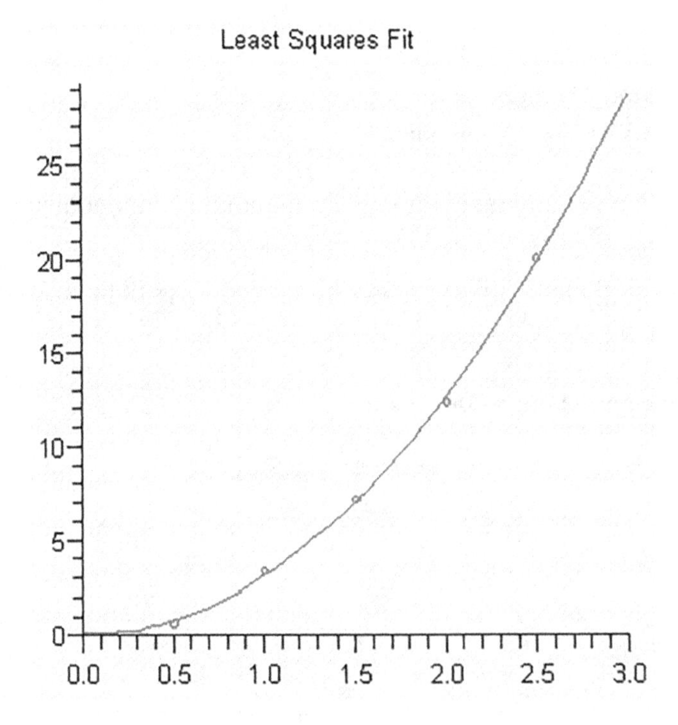

FIGURE 13.8
The least squares for the model $y = kx^2$ plotted with the data.

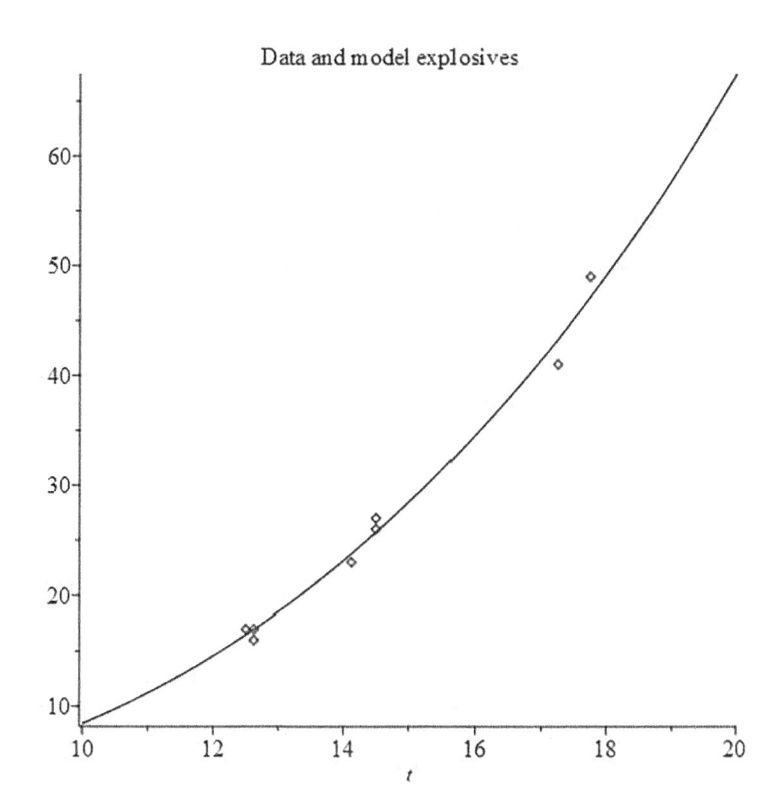

FIGURE 13.9
Plot of data and model for explosives.

The least-squares estimate of the proportionality constant in this model is $k = 0.0084365$. Thus, the model is $W = 0.0084365L^3$. The graph of the least-squares fit with the original data above shows that the model does capture the trend of the data; see Figure 13.9.

13.4 Chapter 13 Exercises

Find the correlation that fit the following data with the models using least squares:

13.1

X	1	2	3	4	5
Y	1	1	2	2	4

a) $y = b + ax$
   $y = ax^2$

13.2 Stretch of a spring data:

x (× 10^{-3})	5	10	20	30	40	50	60	70	80	90	100
y (× 10^5)	0	19	57	94	134	173	216	256	297	343	390

a) $y = ax$
b) $y = b + ax$
c) $y = ax^2$

13.3 Data for the ponderosa pine:

x	17	19	20	22	23	25	28	31	32	33	36	37	39	42
y	19	25	32	51	57	71	113	140	153	187	192	205	250	260

a) $y = ax + b$
b) $y = ax^2$
c) $y = ax^3$
d) $y = ax^3 + bx^2 + c$

13.5 Diagnostics and Their Interpretations

13.5.1 Coefficient of Determination: Statistical Term: R^2

In statistics, the **coefficient of determination**, R^2, is used in the context of statistical models whose main purpose is the prediction of future outcomes on the basis of other related information. It is the proportion of variability in a data set that is accounted for by the statistical model. It provides a measure of how well future outcomes are likely to be predicted by the model.

There are several different definitions of R^2 that are only sometimes equivalent. One class of such cases includes that of linear regression. In this case, R^2 is simply the square of the sample correlation coefficient between the outcomes and their predicted values, or in the case of simple linear regression, between the outcome and the values being used for prediction. In such cases, the values vary from 0 to 1. Important cases where the computational definition of R^2 can yield negative values, depending on the definition used, arise where the predictions that are being compared to the corresponding outcome have not derived from a model-fitting procedure using those data.

13.5.2 $R^2 = 1 - SSE/SST$

Values of R^2 outside the range 0–1 can occur where it is used to measure the agreement between observed and modeled values and where the "modeled" values are not obtained by linear regression and depending on which formulation of R^2 is used. If the first formula above is used, values can never be greater than 1.

13.5.3 Plotting the Residuals for a Least-Squares Fit

In the previous section, you learned how to obtain a least-squares fit of a model, and you plotted the model's predictions on the same graph as the observed data points in order to get a visual indication of how well the model matches the trend of the data. A powerful technique for quickly determining where the model breaks down or is adequate is to plot the actual deviations or residuals between the observed and predicted values as a function of the independent variable or the model. We plot the residuals in the y-axis and the model or the independent variable on the x-axis. The deviations should be randomly distributed and contained in a reasonably small band that is commensurate with the accuracy required by the model. Any excessively large residual warrants further investigation of the data point in question to find the cause of the large deviation. A pattern or trend in the residuals indicates that a predictable effect remains to be modeled, and the nature of the pattern gives clues on how to refine the model, if a refinement is called for. We illustrate the possible patterns for residuals in Figure 13.10 (a)–(e). Our intent is to provide the modeler with knowledge concerning the adequacy of the model they have found. We will leave further investigations into correcting the patterns to follow on courses in statistical regression.

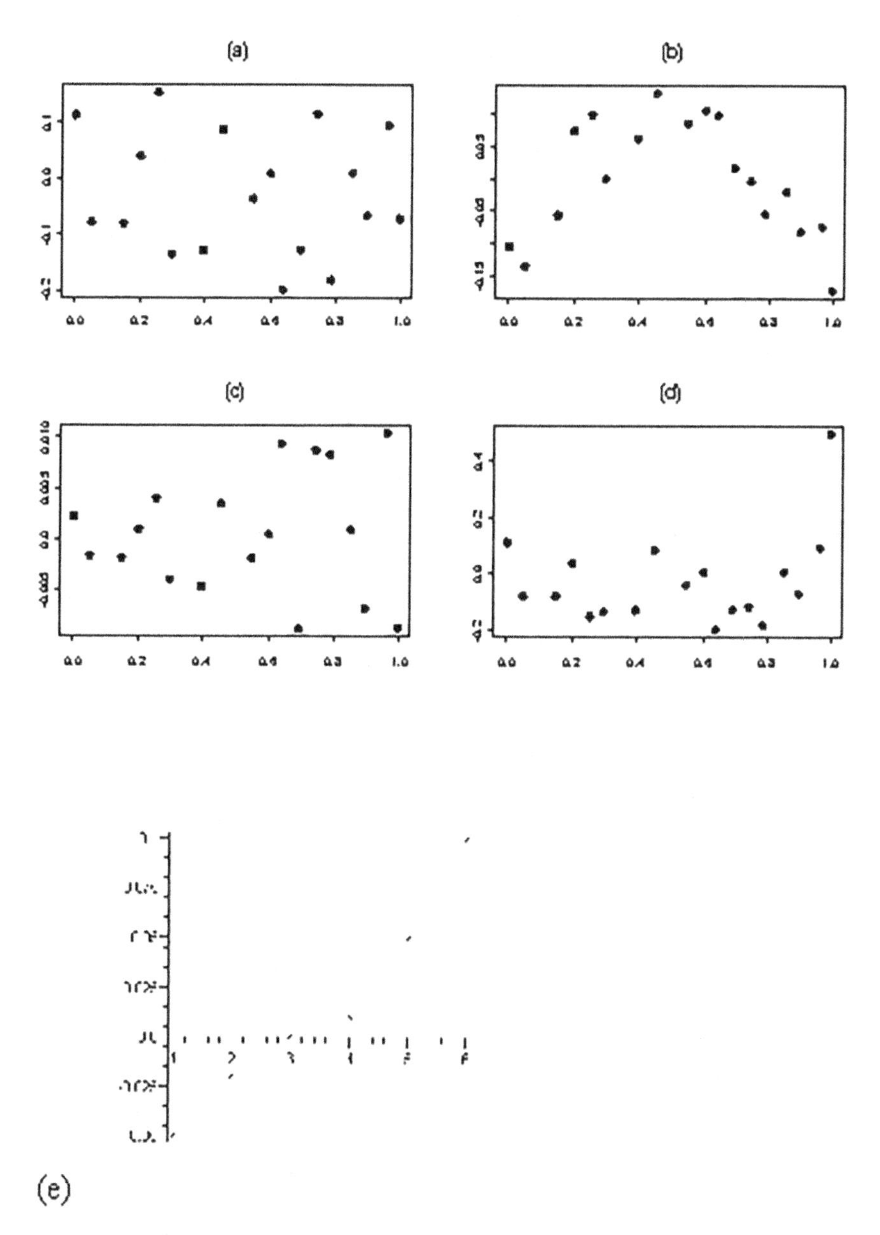

FIGURE 13.10
Patterns for residuals (a) No pattern, (b) Curved pattern, (c) Fanning Pattern, (d) Outliers, (e) Linear trend.

In R, we can always request a residual plot. After fitting a specified model, the difference between the observed and predicted values can be calculated by using the array manipulations.

Example 6: Explosives.

In this example, the relationship was modeled by the following expression:

$$V = 0.00854365\ diameter^3$$

Using our data:

Diameter	14.5	12.5	17.25	14.5	12.625	17.75	14.125	12.625
Size, V	27	17	41	26	17	49	23	16

The residuals are the differences between the predicted and observed values. Mathematically, these are called errors, deviations, or residuals and are found by

$$y_i - f(x_i)$$

In our example, the errors found using the model

$$V_i - 0.00854365^* D_i^3$$

are 1.388,0.592, −2.498, 0.388, 0.194, 1.619, −0.550, −0.806. We plot these errors versus the model.

Note that the residuals, Figure 13.11, are randomly distributed and contained in a relatively small band about zero. There are no outliers, or unusually large residuals, and there

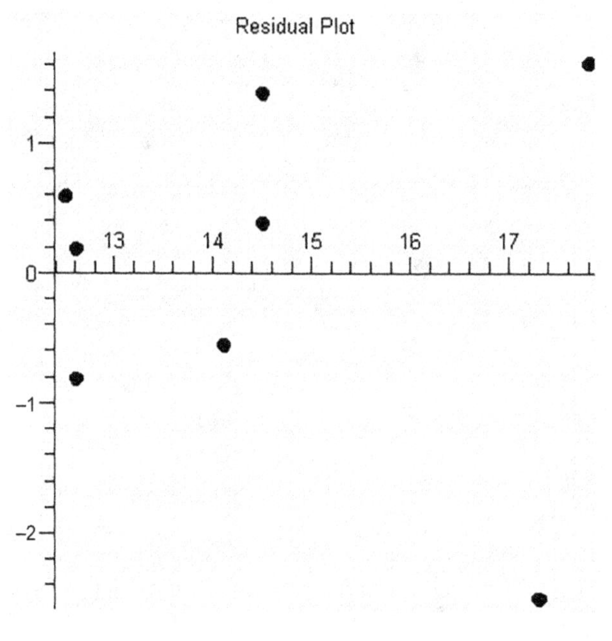

FIGURE 13.11
The explosive model's residuals.

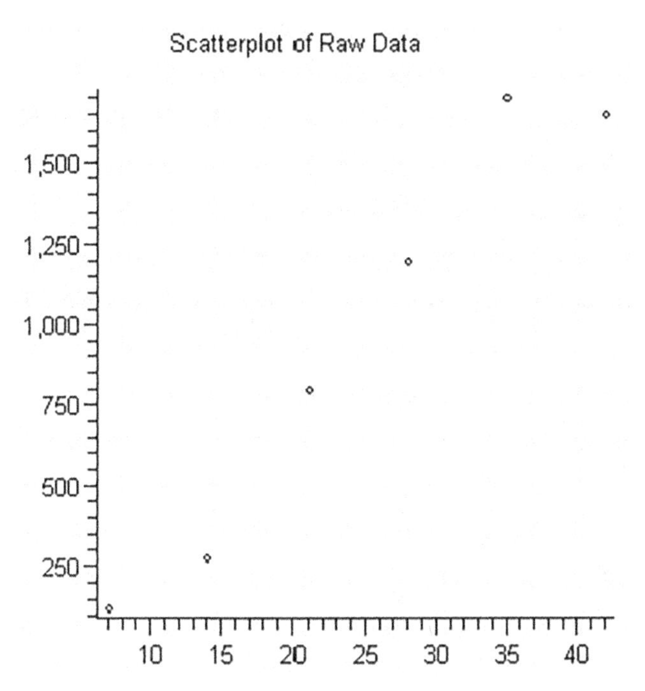

FIGURE 13.12
Scatterplot for Example 7.

appears to be no pattern in the residuals. Based on these aspects of the plot of the residuals, the model approximates the data, and we consider it adequate.

Example 7: We decide to fit a cubic equation to the following data:

T	7	14	21	28	35	42
P	125	275	800	1200	1700	1650

There appears to be a change in concavity in the data, so we decided to try a cubic equation; see Figure 13.12. The least-squares model is approximately

$$y = -0.1066x^3 + 7.436x^2 - 95.7814x + 466.6667$$

We use this model, and we find the residuals as 0.9920635, –15.6746, 52/7777. –74.206, 47.817, –11.706.

There does not appear to be any pattern to the residual plot (Figure 13.13), so we conclude the model is adequate.

13.5.4 Percent Relative Error

When using a model to predict information, we really want to know how well the model appears to work. We will use percent relative error (% Re*lERR*).

$$\% \operatorname{Re} lERR = 100\% \cdot \frac{|y_{actual} - y_{model}|}{y_{actual}}$$

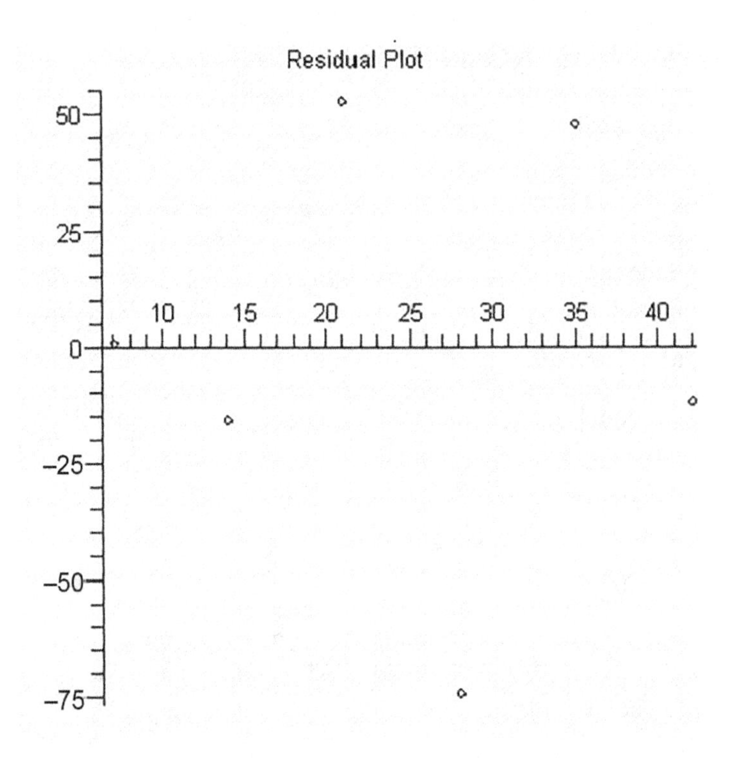

FIGURE 13.13
Residual plot.

We really want these percent relative errors to be small (less than 10%–20% on average).

Example 8: Revisit explosive example

We see that all our percent relative errors are less than 10%. The maximum percent relative error is just more than 6%. This is additional support for a good model.

RESIDUAL OUTPUT

Observation	Predicted Y	Residuals	Percent relative Errors
1	25.72051951	1.279480487	4.738816617
2	16.47804819	0.521951806	3.070304743
3	43.30549707	−2.305497073	5.623163592
4	25.72051951	0.279480487	1.074924948
5	16.97734953	0.022650468	0.133238047
6	47.18139766	1.818602343	3.711433354
7	23.7761263	−0.776126305	3.374462194
8	16.97734953	−0.977349532	6.108434576

Example 9: Revisit the cubic model. We note that although some of the residuals appear large (52.77 and 74.206), all the percent relative error are less than 7%—that is, there is support for a good model.

RESIDUAL OUTPUT			
Observation	Predicted Y	Residuals	Percent Relative Error
1	124.0079	0.992063492	0.793650794
2	290.6746	−15.67460317	5.6998557
3	747.2222	52.77777778	6.597222222
4	1274.206	−74.20634921	6.183862434
5	1652.183	47.81746032	2.812791783
6	1661.706	−11.70634921	0.709475709

13.6 Examples with Diagnostics and Inferential Statistics

Example 10: Revisit spring-mass system.

```
> summary(fit)
Call:
lm(formula = Stretch ~ Mass)
Residuals:
   Min        1Q      Median     3Q       Max
-0.014909 -0.004568 -0.001091 0.001977 0.020727
Coefficients:
             Estimate  Std. Error  t value   Pr(>|t|)
(Intercept) 3.245e-02  6.635e-03   4.891   0.000858 ***
Mass        1.537e-03  1.957e-05   78.569  4.44e-14 ***
—
Signif. codes: 0 '***' 0.001 '**' 0.01 '*' 0.05 '.' 0.1 '' 1
Residual standard error: 0.01026 on 9 degrees of freedom
```

Testing the coefficients of the model with hypothesis testing.

The hypotheses tests being run are

H_0: $B_i = 0$
H_a: $B_i \neq 0$

for the slope and the intercept. The p-values for the slope (4.44e−14) and the intercept (0.000858) are both less than 0.05, our alpha level, and we conclude both are nonzero and need to be part of the equation.

Example 11: Simple linear regression model with complete explanation summary in R.

```
> year = c(1, 2, 3, 4, 5, 6, 7, 8, 9, 10, 11, 121, 31, 41, 5, 16, 17, 18, 19, 20)
> quantity = c(50, 47, 51, 46, 45, 44, 46, 38, 39, 37, 36, 32, 30, 32, 30, 28, 26, 24, 25, 22)
> fit_1<-lm(quantity~year)
> fit_1
```

```
Call:
lm(formula = quantity ~ year)
Coefficients:
(Intercept)   year
52.537      - 1.537
> summary(fit_1)
Call:
lm(formula = quantity ~ year)
Residuals:
  Min      1Q    Median   3Q    Max
-2.5579 -0.9053 0.1000  0.5579 4.2211
Coefficients:
             Estimate  Std. Error t value   Pr(>|t|)
(Intercept)  52.53684   0.82043   64.03  < 2e-16 ***
year         -1.53684   0.06849  -22.44  1.31e-14 ***
—
Signif. codes: 0 '***' 0.001 '**' 0.01 '*' 0.05 '.' 0.1 '' 1
Residual standard error: 1.766 on 18 degrees of freedom
Multiple R-squared: 0.9655,   Adjusted R-squared: 0.9636
F-statistic: 503.5 on 1 and 18 DF, p-value: 1.309e-14
> plot(year, quantity)
> abline(fit_1)
```

The model is written as quantity = 52.53684 – 1.5684 year

We see in Figure 13.14 that this linear model appears to fit the data quite well.

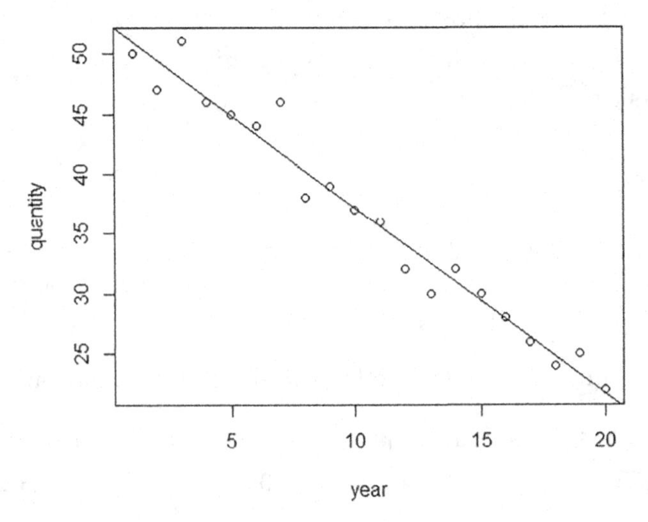

FIGURE 13.14
Line and data overlaid.

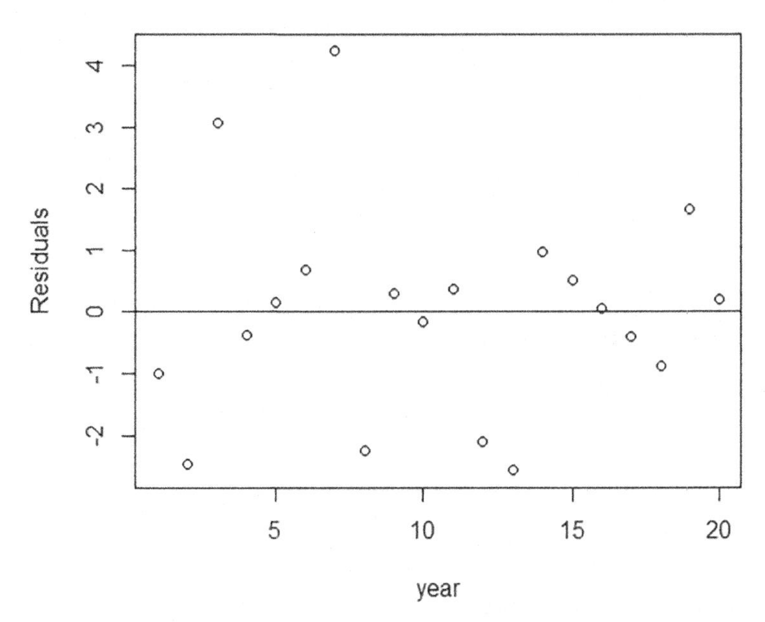

FIGURE 13.15
Residual plot for Example 11.

13.6.1 Diagnostics

(1) Residual plot; it appears pretty random, Figure 13.15.
Hypothesis tests results all show significance. That is good.

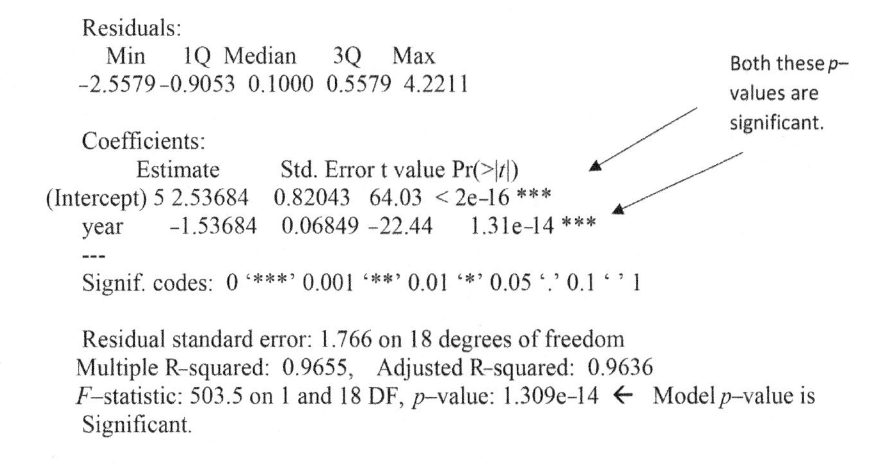

```
Residuals:
    Min      1Q  Median      3Q     Max
-2.5579 -0.9053  0.1000  0.5579  4.2211

Coefficients:
            Estimate    Std. Error t value Pr(>|t|)
(Intercept) 5 2.53684   0.82043   64.03  < 2e-16 ***
year        -1.53684    0.06849  -22.44   1.31e-14 ***
---
Signif. codes:  0 '***' 0.001 '**' 0.01 '*' 0.05 '.' 0.1 ' ' 1

Residual standard error: 1.766 on 18 degrees of freedom
Multiple R-squared:  0.9655,    Adjusted R-squared:  0.9636
F-statistic: 503.5 on 1 and 18 DF, p-value: 1.309e-14 ← Model p-value is
Significant.
```


(4) SSE and SSR

```
sse = sum((fitted(fit_1) – mean(quantity))^2)
> ssr = sum((fitted(fit_1) – quantity)^2)
> 1 – (ssr/(sse + ssr))
[1] 0.965486
>
```

```
> sse
[1] 1570.653
> ssr
[1] 56.14737
```

13.7 Polynomial Regression in R

The equation that minimizes $S = \Sigma(y - (b_0 + b_1x + b_2x^2)^2$. We take the partial derivatives with respect to our unknowns $\{b_0, b_1, b_2\}$ and obtain the normal equations:

$$\Sigma y = \Sigma b_0 + b_1\Sigma x + b_2\Sigma x^2$$
$$\Sigma xy = b_0\Sigma x + b_1\Sigma x^2 + b_2\Sigma x^3$$
$$\Sigma x^2y = b_0\Sigma x^2 + b_1\Sigma x^3 + b_2\Sigma x^4$$

We could solve this system of equations for our parameters or use R.

Example 12: Recovery level.

First, we see if we want to fit a polynomial like $y = b_0 + b_1x + b_2x^2$. We can set up and let x be time, t.

We have the following data for time and levels of results. We plot the data in Figure 13.16 and see a curved trend.

```
> Level = c(54, 50, 45, 37, 35, 25, 20, 16, 18, 13, 8, 11, 8, 4, 6)
> time = c(2, 5, 7, 10, 14, 19, 26, 31, 34, 38, 45, 52, 53, 60, 65)
> plot(time, Level)
```

The scatterplot is provided in Figure 13.16.

We obtain a correlation value in R.

```
cor(time, Level)
[1] -0.9410528
```

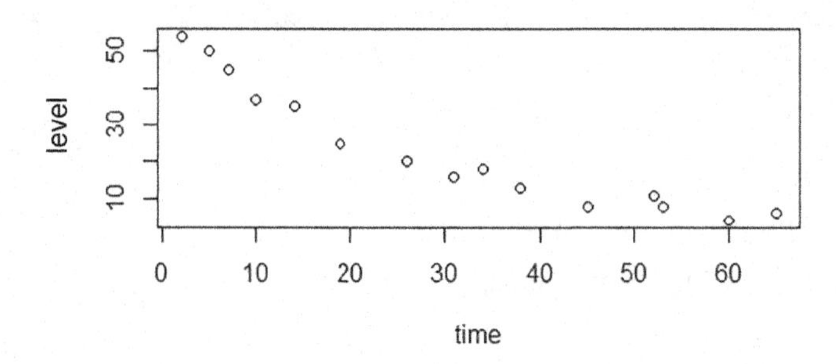

FIGURE 13.16
Scatterplot of data: time versus level.

This implies a strong linear relationship. However, due to the curve trend we see in the scatterplot, we decide to model with a quadratics function.

```
recovery_model2 <- lm(Level ~ time + I((time)^2)
+)
> summary(recovery_model2)
Call:
lm(formula = Level ~ time + I((time)^2))
Residuals:
  Min      1Q   Median  3Q     Max
-3.6724 -1.2896  0.2194  1.3841 4.0736
Coefficients:
              Estimate Std. Error t value  Pr(>|t|)
(Intercept) 55.822134  1.649202   33.848  2.81e-13 ***
time        -1.710263  0.124797  -13.704  1.09e-08 ***
I((time)^2)  0.014807  0.001868    7.927  4.13e-06 ***
—
Signif. codes: 0 '***' 0.001 '**' 0.01 '*' 0.05 '.' 0.1 '' 1
Residual standard error: 2.455 on 12 degrees of freedom
Multiple R-squared: 0.9817,    Adjusted R-squared: 0.9786
F-statistic: 321.1 on 2 and 12 DF, p-value: 3.812e-11
```

The model is recovery level = 55.822123 – 1.700263 time + 0.014807 time^2.

Residual plot, Figure 13.17.

The sum of squared error, sse, is:

```
[1] 3870.991
> ssr
[1] 56.14737
> 1 - (ssr/(sse + ssr))
[1] 0.9857027
```

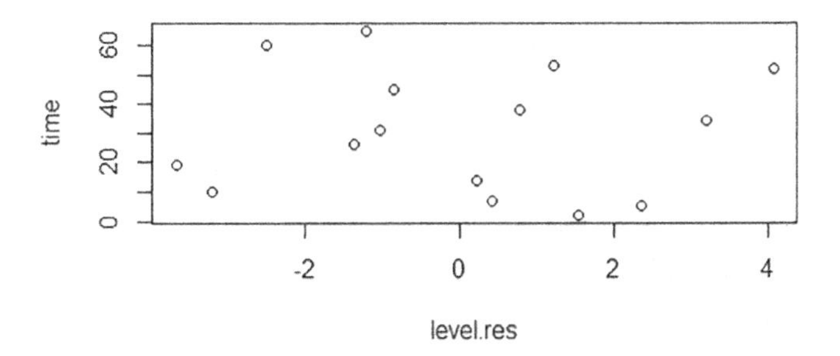

FIGURE 13.17
Residual plot for Example 12, recovery.

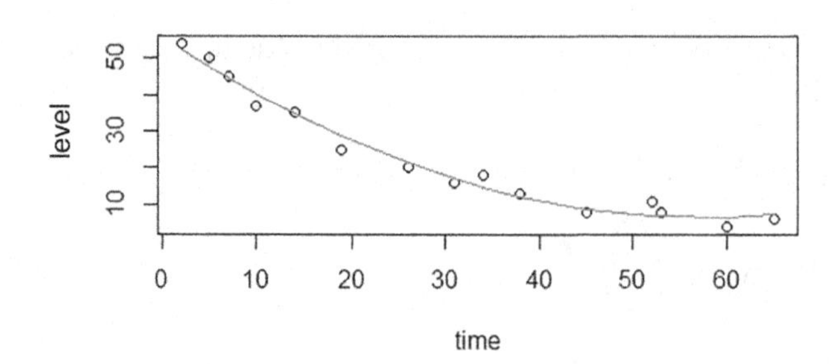

FIGURE 13.18
Plot of data and model for Example 12.

Percent relative error

Per_rel_err

1	2	3	4	5	6
−2.8503045	−4.7180144	−0.9425917	8.6491871	−0.6268164	14.6897820
7	8	9	10	11	12
6.8239618	6.4591042	−17.8332119	−6.0510012	10.5545422	−37.0323886
13	14	15			
−15.3637082	62.7838691	20.2396576			

All indicators are it appears to be a good model; see Figure 13.18. However, when we go to predict the levels after 100 days, we obtain an answer that does not pass a common sense test. We expect the value of levels to get closer to 0; however, f (100) = 33.82.

```
fm <- lm(Level ~ poly(time, 2), BOD)
> plot(Level ~time, BOD)
> lines(fitted(fm) ~ time, BOD, col = "red")
```

Example 13: Polynomial regression.

Let's assume we return to Example 2 of the last section but instead of quantity = function (time) as a linear model, we desire a quadratic polynomial such as $quantity = b_0+b_1\ time+ b_2\ year^2$.

We will use all the same diagnostics as before.

```
model = lm(quantity~poly(year,2))
> summary(model)
Call:
lm(formula = quantity ~ poly(year, 2))
Residuals:
   Min      1Q   Median    3Q    Max
-2.5271 -0.9411  0.0956  0.5727 4.2450
Coefficients:
              Estimate  Std. Error  t value  Pr(>|t|)
(Intercept)   36.4000    0.4063     89.591   < 2e-16 ***
```

poly(year, 2) 1 −39.6315 1.8170 −21.812 7.22e−14 ***
poly(year, 2) 2 0.1509 1.8170 0.083 0.935

—

Signif. codes: 0 '***' 0.001 '**' 0.01 '*' 0.05 '.' 0.1 '' 1
Residual standard error: 1.817 on 17 degrees of freedom
Multiple R-squared: 0.9655, Adjusted R-squared: 0.9614
F-statistic: 237.9 on 2 and 17 DF, *p*-value: 3.728e−13

Interpretation of model: In the model, we find that a p-value of 0.935 is not significant, meaning we can leave that variable out of the model and recompute.

13.8 Chapter 13 Exercises

13.4 a) Find the least-squares regression line for the following set of data $\{(-1, 0), (0, 2), (1, 4), (2, 5)\}$

b) Plot the given points and the regression line in the same rectangular system of axes.

13.5 The values of x and the corresponding values of y are shown in the table below.

x	0	1	2	3	4
y	2	3	5	4	6

a) Find the least-squares regression line $y = ax + b$.

b) Estimate the value of y when $x = 10$.

13.6 The sales of a company (in million dollars) for each year are shown in the table below.

x (year)	2005	2006	2007	2008	2009
y (sales)	12	19	29	37	45

a) Find the least square regression line $y = ax + \mathrm{b}$.

b) Use the least-squares regression line as a model to estimate the sales of the company in 2012.

13.7 For the data on student and books.

Semester	Students	Books
1	36	31
2	28	29
3	35	34
4	39	35
5	30	29
6	30	30
7	31	30
8	38	38

Semester	Students	Books
9	36	34
10	38	33
11	29	29
12	26	26

a) Obtain a scatterplot of the number of books sold versus the number of registered students.

b) Find the correlation coefficient, and interpret it in terms of this problem.

c) Give the regression equation, and interpret the coefficients in terms of this problem.

d) If appropriate, predict the number of books that would be sold in a semester when 30 students have registered.

13.8 Consider the data on e-commerce sales and advertising dollars.

Online Store	Monthly E-commerce Sales (in 1000s)	Online Advertising Dollars (in 1000s)
1	368	1.7
2	340	1.5
3	665	2.8
4	954	5
5	331	1.3
6	556	2.2
7	376	1.3

a) Determine the correlation coefficient.

b) Build a simple linear model and interpret its diagnostics.

c) Build a quadratic model and interpret its diagnostics.

13.9 You have to examine the relationship between the age and price for used cars sold in the last year by a car dealership company.

Here is the table of the data:

Car Age (in years)	Price (in dollars)
4	6,300
4	5,800
5	5,700
5	4,500
7	4,500
7	4,200
8	4,100
9	3,100
10	2,100
11	2,500
12	2,200

a) Determine the correlation coefficient.

b) Build a simple linear model, and interpret its diagnostics.

c) Build a quadratic model, and interpret its diagnostics

13.10 The time x in years that an employee spent at a company and the employee's hourly pay, y, for five employees are listed in the table below.

a) Calculate and interpret the correlation coefficient. Include a plot of the data in your discussion.

b) Build a simple linear model.

c) Interpret the results.

x	y
5	25
3	20
4	21
10	35
15	38

13.11 This data set of size $n = 15$ contains measurements of yield from an experiment done at five different temperature levels. The variables are y = yield and x = temperature in degrees Fahrenheit. The table below gives the data used for this analysis.

i	**Temperature**	**Yield**
1	50	3.3
2	50	2.8
3	50	2.9
4	70	2.3
5	70	2.6
6	70	2.1
7	80	2.5
8	80	2.9
9	80	2.4
10	90	3.0
11	90	3.1
12	90	2.8
13	100	3.3
14	100	3.5
15	100	3.0

a) Determine the correlation coefficient.

b) Build a simple linear model, and interpret its diagnostics.

c) Build a quadratic model, and interpret its diagnostics.

13.9 Multiple Regression in R

Sometimes we have more than one independent variable. One such example would be modeling the cost of a new house as a function of squares footage, number of bedrooms, and number of bathrooms.

13.9.1 Examples of Multiple Regression

Example 14: A researcher has collected data on three psychological variables, four academic variables (standardized test scores), and the type of educational program the student is in for 600 high school students. She is interested in how the set of psychological variables is related to the academic variables and the type of program the student is in.

Example 15: A doctor has collected data on cholesterol, blood pressure, and weight. She also collected data on the eating habits of the subjects (e.g., how many ounces of red meat, fish, dairy products, and chocolate consumed per week). She wants to investigate the relationship between the three measures of health and eating habits.

Example 16: A researcher is interested in determining what factors influence the health of African violet plants. She collects data on the average leaf diameter, the mass of the root ball, and the average diameter of the blooms, as well as how long the plant has been in its current container. For predictor variables, she measures several elements in the soil, as well as the amount of light and water each plant receives.

Assume we have two independent variables (x_1, x_2) and one dependent variable (y)

Four useful models to try are:

(1) $y = b_0 + b_1x_1 + b_2x_2 + e$

(2) $y = b_0 + b_1x_1 + b_2x_2 + +b_3x_1^2 + b_4x_2^2 + e$

(3) $y = b_0 + b_1x_1 + b_2x_2 + b_3x_1x_2 + e$

(4) $y = b_0 + b_1x_1 + b_2x_2 + b_3x_1^2 + b_4x_2^2 + b_5x_1x_2 + e$

Discussion of approach:

(1) As you fix x_1 or x_2 to a constant (any constant), we still get a straight line in two dimensions with constant slopes, line are parallel.

(2) As you fix x_1 or x_2 to a constant (any constant), we still get quadratic parallel curves in two dimensions.

(3) As you fix x_1 or x_2 to a constant (any constant), we still get a straight line in two dimensions with nonconstant slopes; the lines will intersect.

(4) As you fix x_1 or x_2 to a constant (any constant), we still get quadratic curves in two dimensions that intersect.

Example 17: We want to see how y varies with x_1 and x_2.

x1	*x2*	*y*
8.5	*2*	*30.9*
8.9	*3*	*32.7*
10.6	*3*	*36.7*
10.2	*20*	*41.9*
9.8	*22*	*40.9*
10.8	*20*	*42.9*
11.6	*31*	*46.3*
12	*32*	*47.6*
12.5	*31*	*47.2*
10.9	*28*	*44*
12.2	*36*	*47.7*
11.9	*28*	*43.9*
11.3	*30*	*46.8*
13	*27*	*46.2*
12.9	*24*	*47*
12	*25*	*46.8*
12.9	*28*	*45.9*
13.1	*28*	*48.8*
11.4	*32*	*46.2*
13.2	*28*	*47.8*
11.6	*35*	*49.2*
12.1	*34*	*48.3*
11.3	*35*	*48.6*
11.1	*40*	*50.2*
11.5	*45*	*49.6*
11.6	*50*	*53.2*
11.7	*55*	*54.3*
11.7	*57*	*55.8*

```
> model <- lm(y ~ x1 + x2)
> summary(model)
```

```
> x1<-data6[,c("x1")]
> x2<-data6[,c("x2")]
> y<-data6[,c("y")]
> model <- lm(y ~ x1 + x2)
> summary(model)
```

```
Call:
lm(formula = y ~ x1 + x2)

Residuals:
   Min       1Q     Median    3Q      Max
-2.10062 -0.60544 -0.03045 1.00419 1.66205

Coefficients:
            Estimate  Std. Error  t value   Pr(>|t|)
(Intercept) 19.43976   2.18829     8.884   3.30e-09 ***
x1           1.44228   0.20764     6.946   2.79e-07 ***
x2           0.33563   0.01814    18.507   4.18e-16 ***
—
Signif. codes: 0 '***' 0.001 '**' 0.01 '*' 0.05 '.' 0.1 ' ' 1
Residual standard error: 1.094 on 25 degrees of freedom
Multiple R-squared: 0.9645, Adjusted R-squared: 0.9616
F-statistic: 339.3 on 2 and 25 DF, p-value: < 2.2e-16
res<-c((19.4398+1.4423* x1+0.3356*x2) - y)

> res
[1] 1.47055 0.58307 -0.96502 -1.03674 0.05754 -1.17136 0.27408
[8] -0.11340 0.67215 0.55767 1.41746 2.09997 -0.99421 1.05090
[15] -0.90013 -1.66260 1.54227 -1.06927 0.42122 0.07496 -1.28352
[22] 0.00203 -1.11621 -1.32667 1.52825 -0.24952 0.47271 -0.35609
> plot(res, y)
>
```

The residual plot, Figure 13.19, shows no apparent trends, so we deem the model adequate. **Interpretation**: The model as stated is adequate.

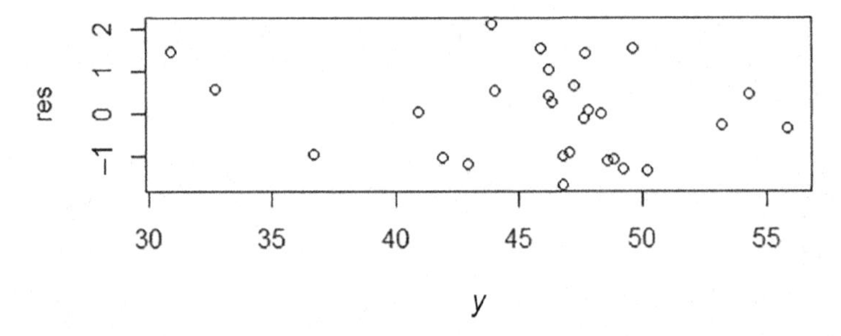

FIGURE 13.19
Residual plot for Example 17.

Example 18: Sales.

We have data for sales (y_i), size of target population (x_{1i}), and per capita discretionary income (x_{2i}) in dollars as shown in Table 13.3.

```
Predictor       Coef        StDev       T      P
Constant       3.612        2.465    1.47  0.169
tarP        0.495445     0.006205   79.84  0.000
Dollars    0.0092033    0.0009888    9.31  0.000

S = 2.217 R-Sq = 99.9% R-Sq(adj) = 99.9%

Analysis of Variance

Source          DF     SS     MS        F      P
Regression       2  53843  26921  5475.70  0.000
Residual Error  12     59      5
Total           14  53902

Source    DF   Seq SS
tarP       1    53417
Dollars    1      426
```

Interpretation: Only the coefficient for the target population and discretionary dollars are significant and not the intercept.

TABLE 13.3

Data for Sales

Region, i	Sales in Quantities of Gross Jars, y_i	Target Population, x_{1i}	Per Capita Discretionary Income, x_{2i}
1	162	274	2,450
2	120	180	3,254
3	223	375	3,802
4	131	205	2,838
5	67	86	2,347
6	169	265	3,782
7	81	98	3,008
8	192	330	2,450
9	116	195	2,137
10	55	53	2,500
11	252	430	4,020
12	232	372	4,427
13	144	236	2,660
14	103	157	2,088
15	212	370	2,605

Example 19: Earning per month as a function of entrance score and age.

Person	Score	Age	Earnings/Month
1	89	21	2,625
2	93	24	2,700
3	91	21	3,100
4	122	23	3,150
5	115	27	3,175
6	100	18	3,100
7	98	19	2,700
8	105	16	2,475
9	112	23	3,625
10	109	28	3,525
11	130	20	3,225
12	104	25	3,450
13	104	20	2,425
14	111	26	3,025
15	97	28	3,625
16	115	29	2,750
17	113	25	3,150
18	88	23	2,600
19	108	19	2,525
20	101	16	2,650

```
model1 <- lm(earnings ~ score + age)
> summary(model1)

Call:
lm(formula = earnings ~ score + age)

Residuals:
  Min      1Q  Median  3Q    Max
-606.17 -244.81 -26.14  269.52 573.96

Coefficients:
             Estimate  Std. Error  t value  Pr(>|t|)
(Intercept) 1059.902   789.619    1.342    0.1972
score          8.174     7.092    1.153    0.2650
age           46.768    19.292    2.424    0.0268 *
—

Signif. codes: 0 '***' 0.001 '**' 0.01 '*' 0.05 '.' 0.1 '' 1

Residual standard error: 335.3 on 17 degrees of freedom
Multiple R-squared: 0.3376, Adjusted R-squared: 0.2597
F-statistic: 4.333 on 2 and 17 DF, p-value: 0.03015
```

Interpretation: We find that only the variable age is significant.
The residual plot, Figure 13.20, appears to show a linear trend.

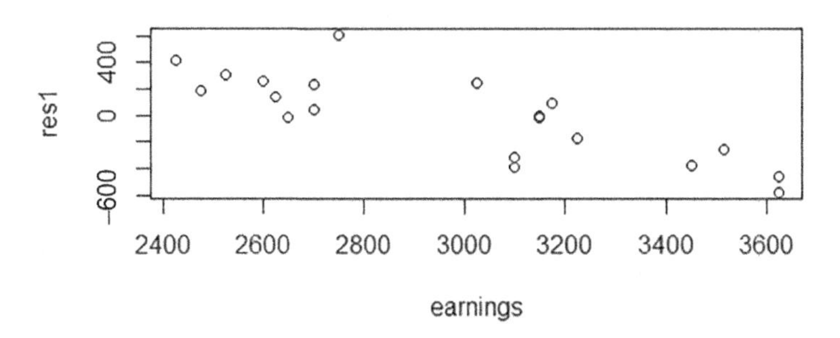

FIGURE 13.20
Residual plot for Example 19.

Example 20: Predicting fish weight from length and girth

Length, l (in.)	14.5	12.5	17.25	12.6	17.75	14.2	14.5	12.7
Girth, g (in.)	9.75	8.4	11	8.5	12.25	9	9.5	8.5
Weight, W (oz.)	27	17	41	17	49	23	26	16

```
model2<-lm(wght ~ xlen + girth)
> summary(model2)

Call:
lm(formula = wght ~ xlen + girth)

Residuals:
      1       2       3       4       5       6        7        8
-0.76656 1.12562 0.26152 0.34982 0.20133 0.06956 -0.37282 -0.86848

Coefficients:
             Estimate Std. Error t value  Pr(>|t|)
(Intercept) -58.2426    2.0969  -27.775  1.13e-06 ***
xlen          2.1830    0.6389    3.417   0.01890 *
girth         5.5750    0.9468    5.888   0.00201 **
---
Signif. codes: 0 '***' 0.001 '**' 0.01 '*' 0.05 '.' 0.1 '' 1

Residual standard error: 0.7725 on 5 degrees of freedom
Multiple R-squared: 0.9971, Adjusted R-squared: 0.9959
F-statistic: 850.4 on 2 and 5 DF, p-value: 4.652e-07

>
```

Residual plot, Figure 13.21, appears to have a curved pattern, indicating the model is not adequate.

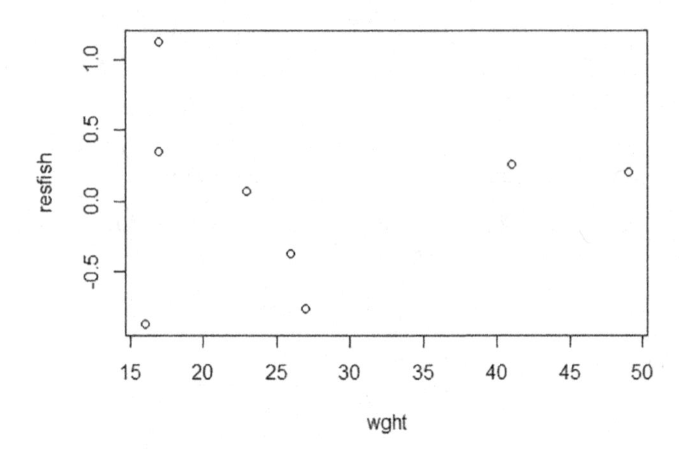

FIGURE 13.21
Residual plot for Example 20.

Interpretation: Although all the coefficients appear significant, the residual plot indicates the model may not be adequate.

13.10 Chapter 13 Exercises

13.12 Heart disease.

An article in *Newsweek* (January 22, 1996) stated that "Compared with real wine-drinking countries, the United States is practically dry. That may be a reason, scientists say, that our rate of heart disease is higher." Dr. Charles Fuchs is also quoted saying that drinking beer versus wine may produce extraordinary differences in life expectancy. Questions: 1) Does wine, beer, or liquor consumption provide an explanation of death rates from heart disease? 2) Does wine consumption significantly lower death rates from heart disease? The following data on average country death rates and average country alcoholic beverage consumption, in liters per capita, were provided in the *Newsweek* article. "Heart disease" is defined as the death rate per 100,000.

	Death Rate from Heart Disease	Wine*	Beer*	Liquor*
France	61.1	63.5	40.1	2.5
Italy	94.1	58.0	25.1	0.9
Switzerland	106.4	46.0	65.0	1.7
Australia	173.0	15.7	102.1	1.2
Britain	199.7	12.2	100.0	1.5
USA	176.0	8.9	87.8	2.0
Russia	373.6	2.7	17.1	3.8
Czech Republic	283.7	1.7	140.0	1.0
Japan	34.7	1.0	55.0	2.1
Mexico	36.4	0.2	50.4	0.8

Note: *Consumption per capita.

13.13 Body weight: Build a multiple regression model for weight as a function of height and waist size.

Weight (lb.)	Height (in.)	Waist size (in.)
132	60	24
136	61	24.5
141	62	25
145	63	26
150	64	27
160	65	27.5
165	66	28
170	67	28.5
175	68	29
180	69	29.5
185	70	30
190	71	30.5
195	72	32
200	73	34

14

*Advanced Regression Models: Nonlinear, Sinusoidal, and Binary Logistics Regression Using R**

Objectives

1. **Know the advanced regression methods for nonlinear, logistic, and sine regression.**
2. **Be able to apply them in R.**

14.1 Introduction

Analysis in data science and digital business requires analysis of the data and, in many cases, the use of regression techniques. This chapter discusses some simple regression and advanced regression techniques that have been used often in the analysis of data for business, industry, and government. Regression is not a one-method-fits-all approach. Regression takes good approaches and common sense to complement the mathematical and statistical approaches used. We also discuss methods to check for model adequacy after the regression model is found. We also believe because of the popularity of R that we would illustrate this chapter using R, and at the end of this chapter, we provide the commands that were used in our examples.

Often we might want to model the data in order to make predictions or explain within the domain of the data. Besides the models, we provide insights into the adequacy of the model through various approaches including regression ANOVA output, residual plots, and percent relative error.

In general, we suggest using the following steps in regression analysis.

Step 1. Enter the data (x, y), obtain a scatterplot of the data, and note the trends.

Step 2. If necessary, transform the data into "y" and "x" components.

Step 3. Build or compute the regression equation. Obtain all the output. Interpret the ANOVA output for R^2, *F-test*, and *p-values* for coefficients.

Step 4. Plot the regression function and the data to obtain a visual fit.

Step 5. Compute the predictions, the residuals, and percent relative error as described later.

Step 6. Ensure the predictive results pass the common sense test.

Step 7. Plot the residual versus prediction to determine model adequacy.

* Modified from previously published work by Fox and Hammond in References

DOI: 10.1201/9781003317906-14

We present several methods to check for model adequacy. First, we suggest your predictions pass the "common sense" test. If not, return to your regression model as we show with our exponential decay model in Section 14.4.1. The residual plot is also very revealing. Figure 14.1 shows possible residual plot results where only random patterns indicate model adequacy

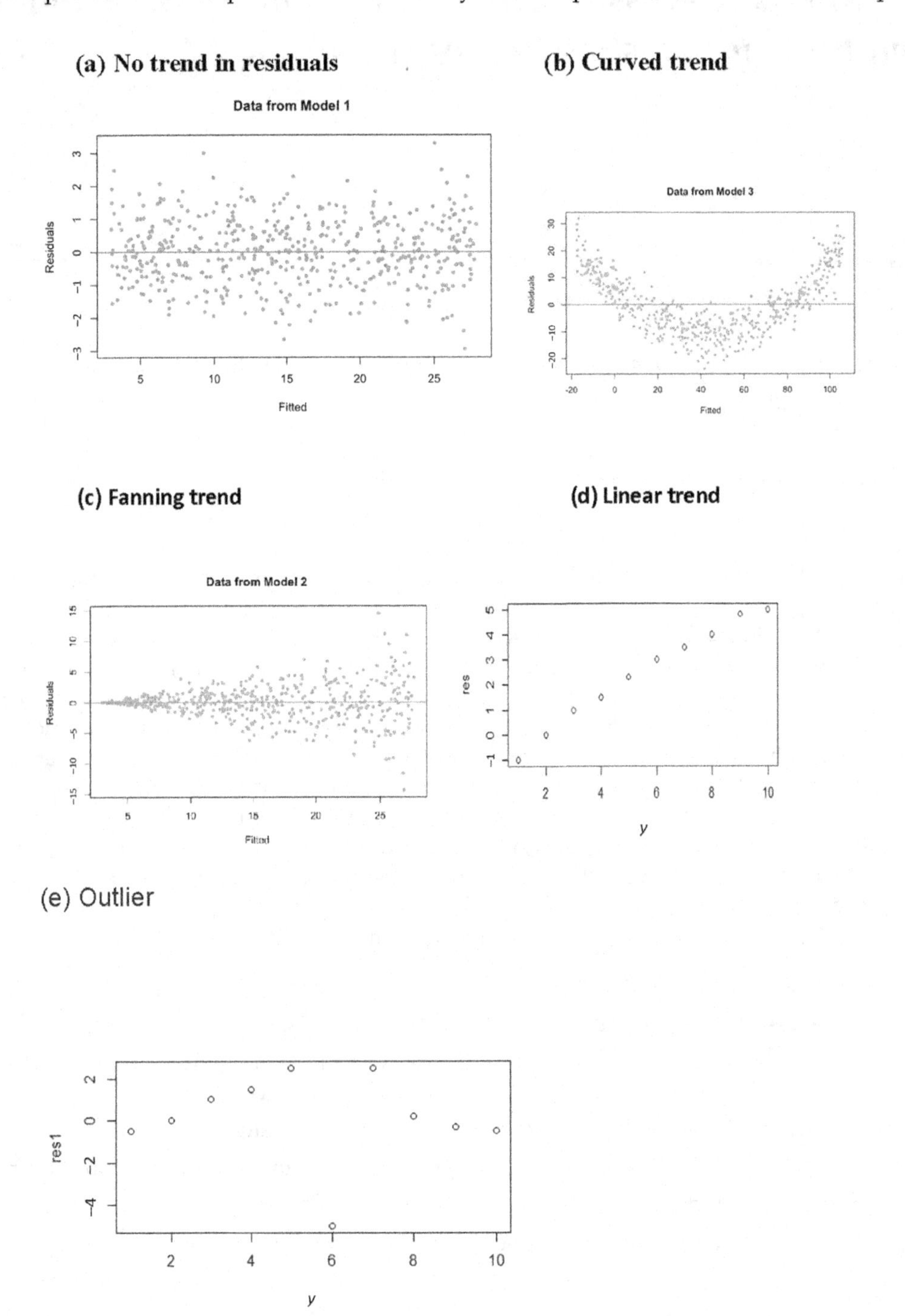

FIGURE 14.1
Patterns for residuals (a) No pattern, (b) Curved pattern, (c) Outliers, (d) Fanning pattern, (e) Linear trend.

from the residual plot perspective. Linear, curve, or fanning trends indicate a problem in the regression model. Afifi and Azen (1979) have a good and useful discussion on corrective action based upon trends found. Percent relative error also provides information about how well the model approximates the original values, and it provides insights into where the model fits well and where it might not fit well. We define percent relative error with Equation 14.1,

$$\%\boldsymbol{RE} = \frac{100|y_a - y_p|}{y_a}. \tag{14.1}$$

14.2 Review of Linear Regression

14.2.1 Correlation of Spring Data

First, let's define correlation. Correlation, ρ, measures the linearity between the data sets X and Y. Mathematically correlation is defined as follows:

The correlation coefficient, equation (2), between X and Y, denoted as ρ_{xy}, is

$$\rho_{xy} = \frac{COV(X,Y)}{\sigma_x \sigma_y} = \frac{E[XY]-\mu_x\mu_y}{\sigma_x \sigma_y}. \tag{14.2}$$

The values of correlation range from –1 to +1. The value of –1 corresponds to a perfect line with a negative slope and a value of +1 corresponds to a perfect line with a positive slope. A value of 0 indicates that that there is no linear relationship.

We present two rules of thumb for correlation from the literature. First, from Devore (2012), for math, science, and engineering data, we have the following:

$0.8 < |\rho| \leq 1.0$ Strong linear relationship
$0.5 < |\rho| \leq 0.8$ Moderate linear relationship
$|\rho| \leq 0.5$ Weak linear relationship

According to Johnson (2012) for nonmath, nonscience, and nonengineering data, we find a more liberal interpretation of ρ:

$0.5 < |\rho| \leq 1.0$ Strong linear relationship
$0.5 < |\rho| \leq 0.3$ Moderate linear relationship
$0.1 < |\rho| \leq 0.3$ Weak linear relationship
$|\rho| \leq 0.1$ No linear relationship

Further, in our modeling efforts, we emphasize the interpretation of $|\rho| \approx 0$. This can be interpreted as either no linear relationship or the existence of a nonlinear relationship. Most students and many researchers fail to pick up on the importance of the nonlinear relationship aspect of the interpretation.

Calculating the correlation between two (or more) variables in R is simple. After loading in the spring data set as an object into R's workspace, we can first visualize the data in tabular format. This lets us be sure that the data is in the proper format and that there are no oddities (missing values, characters entered instead of numbers) that would cause problems.

Using either rule of thumb, the correlation coefficient, $|\rho| = 0.999272$, indicates a strong linear relationship. We obtain this value (look at Figure 14.1), and we see an excellent linear relationship with a positive correlation very close to 1.

```
x y
1  50 0.1000
2  100 0.1875
3  150 0.2750
4  200 0.3250
5  250 0.4375
6  300 0.4875
7  350 0.5675
8  400 0.6500
9  450 0.7250
10  500 0.8000
11  550 0.8750
```

To estimate the correlation between the two columns in this data set, we simply use the cor() command in R on the data table:

```
## Calculate and print correlation matrix
print(cor(spring_data))

## x          y
## x 1.0000000 0.9992718
## y 0.9992718 1.0000000
```

The data's correlation coefficient is 0.9992718, which is very close to 1. Visualizing the data makes this relationship easy to see, and we would expect to see a linear relationship with a positive slope as shown in Figure 14.2.

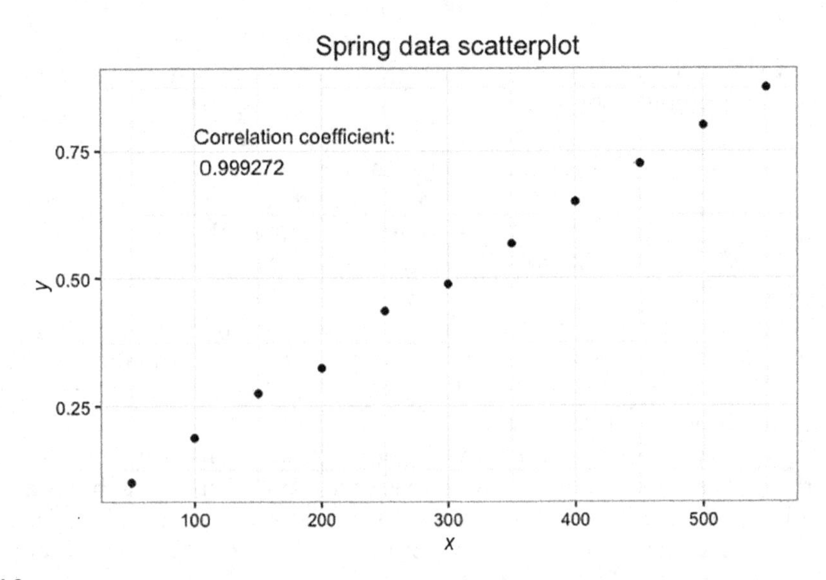

FIGURE 14.2
Plot of spring data with correlation value.

14.2.2 Review of Linear Regression of Spring Data

Fitting an ordinary least-squares (OLS) model with form $y\,x+\epsilon$ to the spring data in R is quite simple. Using the lm() command (short for "linear model") fits the linear model and saves the result as another object in R's workspace.

```
## Fit OLS model to the data
spring_model <- lm(y ~ x, data = spring_data)
```

We can then perform operations on this object to produce tables presenting coefficient estimates and a range of diagnostic statistics to evaluate how well the model fits the data provided.

	Estimate	Std. Error	*t* value	Pr(>|*t*|)
X	0.001537	1.957e–05	78.57	4.437e–14
(Intercept)	0.03245	0.006635	4.891	0.0008579

Fitting linear model: y ~ x

Observations	Residual Std. Error	R^2	Adjusted R^2
11	0.01026	0.9985	0.9984

Analysis of Variance Table

	Df	Sum Sq	Mean Sq	*F* value	Pr(>*F*)
x	1	0.6499	0.6499	6173	4.437e–14
Residuals	9	0.0009475	0.0001053	NA	NA

We visualize this estimated relationship by overlaying the fitted line to the spring data plot. This plot shows that the trend line estimated by the linear model fits the data quite well as shown in Figure 14.3. The relationship between R^2 and ρ is that $R^2 = (\rho)^2$.

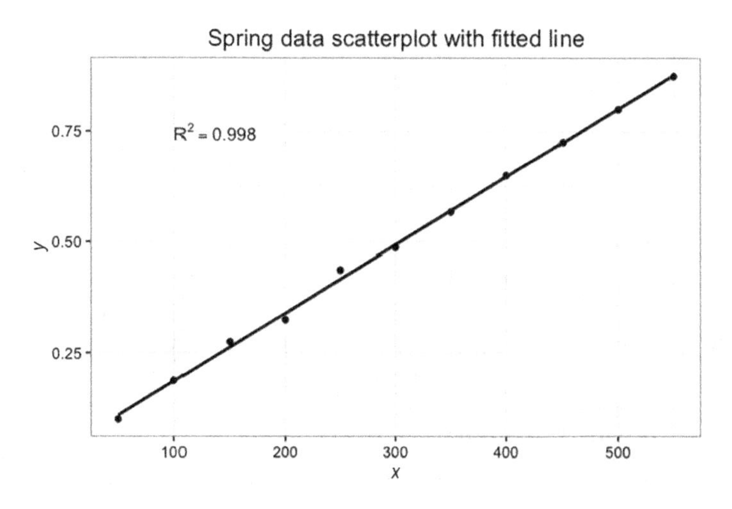

FIGURE 14.3
Regression plot of spring data.

14.3 Review of Polynomial Regression

Exponential decay modeling: Introducing hospital recovery data

```
## # A tibble: 15 × 2
##         T     Y
##     <int> <int>
##  1      2    54
##  2      5    50
##  3      7    45
##  4     10    37
##  5     14    35
##  6     19    25
##  7     26    20
##  8     31    16
##  9     34    18
## 10     38    13
## 11     45     8
## 12     52    11
## 13     53     8
## 14     60     4
## 15     65     6
```

Printing the table of recovery data shows that once again, the structure of the data is amenable to statistical analysis. We have two columns, T (number of days in the hospital) and Y (estimated recovery index), and we want to generate a model that predicts how well a patient will recover as a function of the time they spend in the hospital. Using the cor() command retrieves an initial correlation coefficient of −0.941.

```
## Calculate and print correlation matrix
print(cor(recovery_data))
```

```
##          T          Y
## T 1.0000000 - 0.9410528
## Y -0.9410528 1.0000000
```

Once again, creating a scatterplot, Figure 14.4, of the data helps us visualize how closely the estimated correlation value matches the overall trend in the data.

In this example, we will show linear regression, polynomial regression, and then exponential regression in order to obtain a useful model.

14.3.1 Linear Regression of Hospital Recovery Data

It definitely appears that there is a strong negative relationship: the longer a patient spends in the hospital, the lower their recovery index tends to be. Next, we fit an OLS model to the data to estimate the magnitude of the linear relationship.

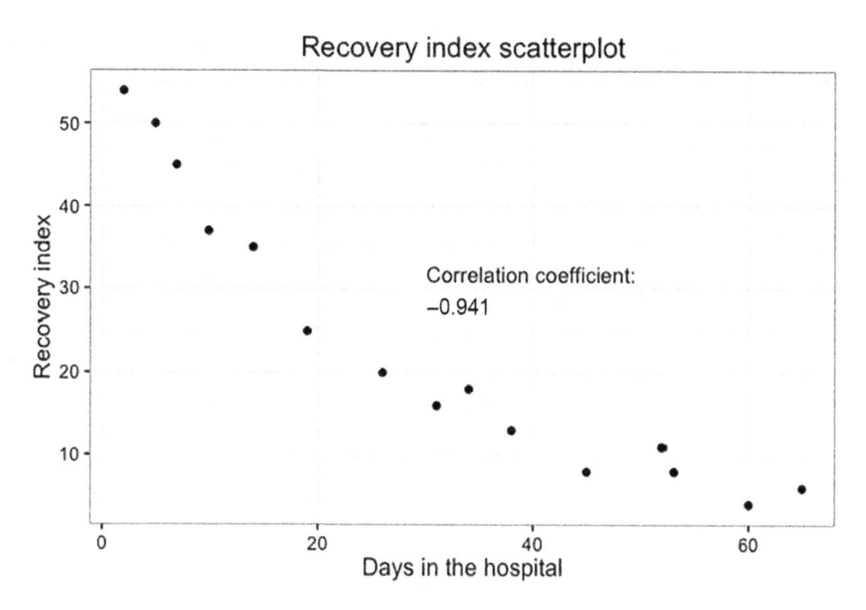

FIGURE 14.4
Scatterplot of days in the hospital and recovery index.

```
## Fit OLS model to the data
recovery_model <- lm(
Y ~ T
, data = recovery_data
)
```

	Estimate	Std. Error	*t* value	Pr(>|*t*|)
T	−0.7525	0.07502	−10.03	1.736e−07
(Intercept)	46.46	2.762	16.82	3.335e−10

Fitting linear model: Y ~ T

Observations	Residual Std. Error	R^2	Adjusted R^2
15	5.891	0.8856	0.8768

Analysis of Variance Table

	Df	Sum Sq	Mean Sq	*F* value	Pr(>*F*)
T	1	3492	3492	100.6	1.736e−07
Residuals	13	451.2	34.71	NA	NA

OLS modeling shows that there is a negative and statistically significant relationship between time spent in the hospital and patient recovery index. However, ordinary least-squares regression may not be the best choice in this case for two reasons. First, we are dealing with real-world data: a model that can produce (for example) negative estimates of recovery index is not applicable to the underlying concepts our model is dealing with. Second, the assumption of OLS, like all linear models, is that the magnitude of the relationship between input and output variables stays constant over the entire range of values in the data. However, visualizing the data suggests that this assumption may not hold—in fact, it appears that the magnitude of the relationship is very high for low values of T and decays somewhat for patients who spend more days in the hospital.

To test for this phenomenon, we examine the residuals of the linear model. Residuals analysis can provide quick visual feedback about model fit and whether the relationships estimated hold over the full range of the data. We calculate residuals as the difference between observed values Y and estimated values $Y&*$, or $Y_i - Y_i^*$. We then normalize residuals as percent relative error between the observed and estimated values, which helps us compare how well the model predicts each individual observation in the data set:

```
## # A tibble: 15 × 6
##        T     Y index  predicted  residuals pct_relative_error
##    <int> <int> <int>      <dbl>      <dbl>              <dbl>
## 1      2    54     1  44.955397  9.0446035          16.749266
## 2      5    50     2  42.697873  7.3021275          14.604255
## 3      7    45     3  41.192857  3.8071435           8.460319
## 4     10    37     4  38.935333 -1.9353325          -5.230628
## 5     14    35     5  35.925301 -0.9253005          -2.643716
## 6     19    25     6  32.162761 -7.1627605         -28.651042
## 7     26    20     7  26.895205 -6.8952045         -34.476023
## 8     31    16     8  23.132665 -7.1326645         -44.579153
## 9     34    18     9  20.875141 -2.8751405         -15.973003
## 10    38    13    10  17.865109 -4.8651085         -37.423912
## 11    45     8    11  12.597553 -4.5975525         -57.469407
## 12    52    11    12   7.329997  3.6700035          33.363668
## 13    53     8    13   6.577489  1.4225115          17.781393
## 14    60     4    14   1.309933  2.6900675          67.251686
## 15    65     6    15  -2.452607  8.4526075         140.876791
```

The data can also be plotted to visualize how well the model fits over the range of our input variable.

The residuals plotted, Figure 14.5, show a curvilinear pattern, decreasing and then increasing in magnitude over the range of the input variable. This means that we can likely improve the fit of the model by allowing for nonlinear effects. Furthermore, the current model can make predictions that are substantively nonsensical, even if they were statistically valid. For example, our model predicts that after 100 days in the hospital, a patient's estimated recovery index value would be −29.79. This has no common sense, as the recovery index variable is always positive in the real world. By allowing for nonlinear terms, we can also guard against these types of nonsense predictions.

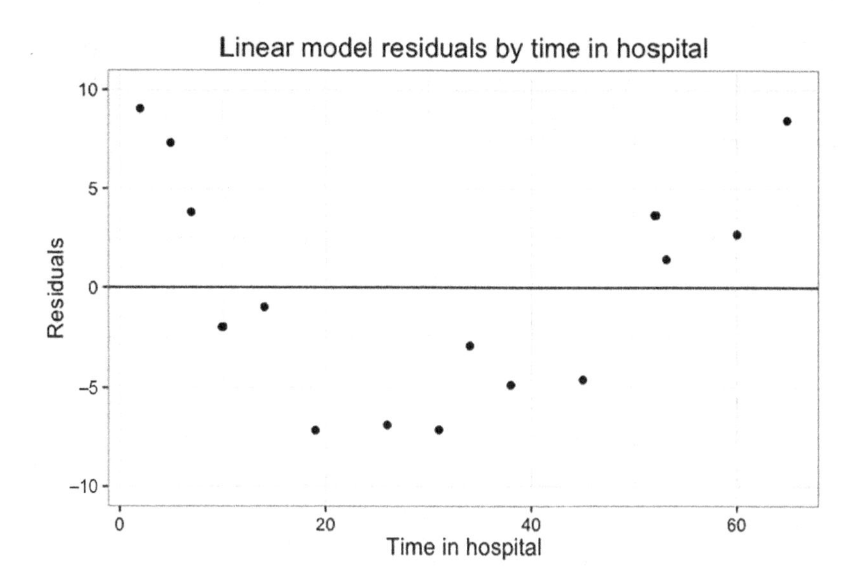

FIGURE 14.5
Residual plot for linear model.

14.4 Review of Quadratic (Polynomial) Regression of Hospital Recovery Data

Including a quadratic term modifies the model formula: $Y = \beta_0 + \beta_1 x + \beta_2 x^2$. Fitting this model to the data produces separate estimates of the effect of T itself as well as the effect of T^2, the quadratic term.

```
## Generate model
recovery_model2 <- lm(Y ~ T + I(T²), data = recovery_data)
```

	Estimate	Std. Error	*t* value	Pr(>|*t*|)
T	−1.71	0.1248	−13.7	1.087e−08
I(T²)	0.01481	0.001868	7.927	4.127e−06
(Intercept)	55.82	1.649	33.85	2.811e−13

Fitting linear model: Y ~ T + I(T²)

Observations	Residual Std. Error	R^2	Adjusted R^2
15	2.455	0.9817	0.9786

Analysis of Variance Table

	Df	Sum Sq	Mean Sq	*F* value	Pr(>*F*)
T	1	3492	3492	579.3	1.59e−11
I(T²)	1	378.9	378.9	62.84	4.127e−06
Residuals	12	72.34	6.029	NA	NA

Including the quadratic term improves model fit as measured by R^2 from 0.88 to 0.98—a sizable increase. To assess whether this new input variable deals with the curvilinear trend we saw in the residuals from the first model, we calculate and visualize the residuals from the quadratic model:

```
## # A tibble: 15 × 6
##        T     Y index predicted  residuals pct_relative_error
##    <int> <int> <int>     <dbl>      <dbl>              <dbl>
## 1      2    54     1 52.460836  1.5391644          2.8503045
## 2      5    50     2 47.640993  2.3590072          4.7180144
## 3      7    45     3 44.575834  0.4241663          0.9425917
## 4     10    37     4 40.200199 -3.2001992         -8.6491871
## 5     14    35     5 34.780614  0.2193857          0.6268164
## 6     19    25     6 28.672445 -3.6724455        -14.6897820
## 7     26    20     7 21.364792 -1.3647924         -6.8239618
## 8     31    16     8 17.033457 -1.0334567         -6.4591042
## 9     34    18     9 14.790022  3.2099781         17.8332119
## 10    38    13    10 12.213370  0.7866302          6.0510012
## 11    45     8    11  8.844363 -0.8443634        -10.5545422
## 12    52    11    12  6.926437  4.0735627         37.0323886
## 13    53     8    13  6.770903  1.2290967         15.3637082
## 14    60     4    14  6.511355 -2.5113548        -62.7838691
## 15    65     6    15  7.214379 -1.2143795        -20.2396576
```

Visually, Figure 14.6, evaluating the residuals from the quadratic model shows that the trend has disappeared. This means that we can assume the same relationship holds

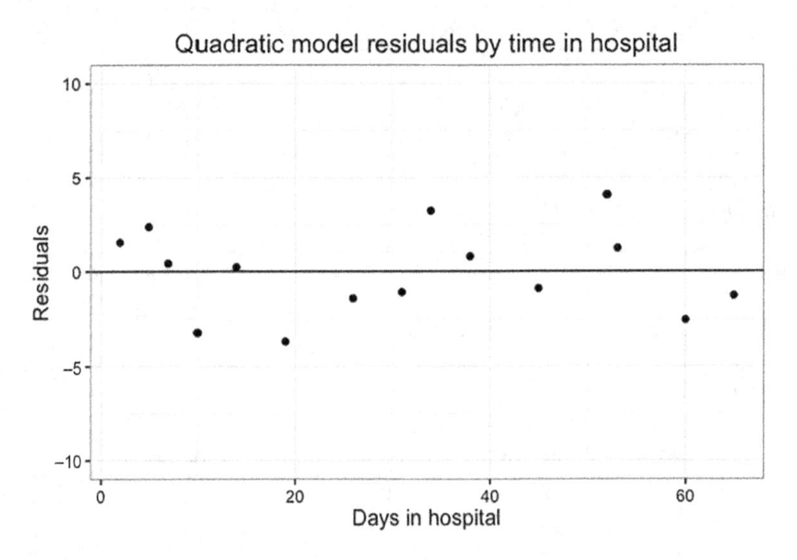

FIGURE 14.6
Residual plot for polynomial regression model.

whether $T = 1$ or $T = 100$. However, we are still not sure if the model produces numerical estimates that pass the common sense test. The simplest way to assess this is to generate predicted values of the recovery index variable using the quadratic model and plot them to see if they make sense.

To generate predicted values in R, we can pass the quadratic model object to the predict() function along with a set of hypothetical input values. In other words, we can ask the model what the recovery index would look like for a set of hypothetical patients who spend anywhere from 0 to 120 days in the hospital.

```
## Create a set of hypothetical patient observations with days in the hospital from 1 to 120
patient_days = tibble(T = 1:120)
## Feed the new data to the model to generate predicted recovery index values
predicted_values = predict(
recovery_model2
, newdata = patient_days)
```

We can then plot these estimates to quickly gauge whether they pass the common sense test for real-world predictive value as shown in Figure 14.7.

```
## # A tibble: 5 × 2
##         T predicted
##   <int>     <dbl>
## 1     1  54.12668
## 2     2  52.46084
## 3     3  50.82461
## 4     4  49.21799
## 5     5  47.64099
```

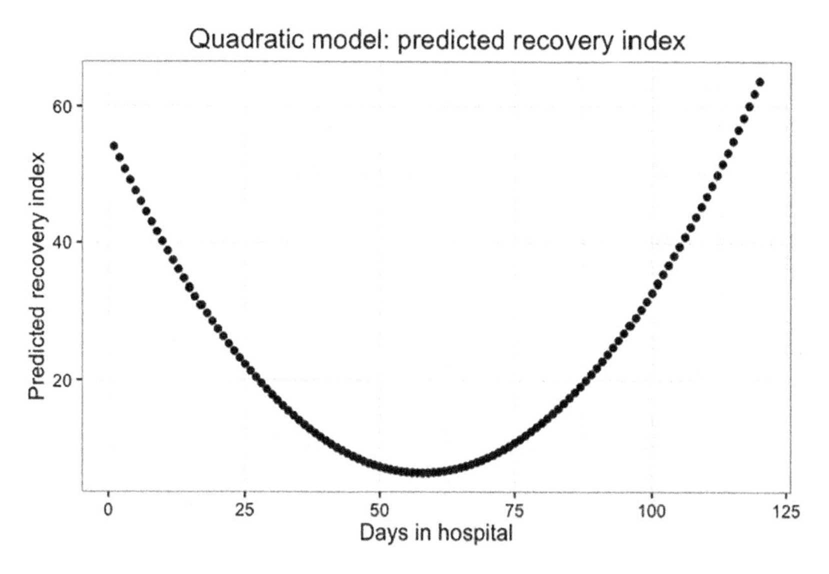

FIGURE 14.7
Polynomial regression plot (quadratic polynomial).

The predicted values curve up toward infinity, Figure 14.7; clearly, this is a problem. The quadratic term we included in the model leads to unrealistic estimates of recovery index at larger values of *T*. Not only is this unacceptable for the context of our model, but it is unrealistic on its face. After all, we understand that people generally spend long periods in the hospital for serious or life-threatening conditions such as severe disease or major bodily injury. As such, we can assess that someone who spends six months in the hospital probably should not have a higher recovery index than someone who was only hospitalized for a day or two.

14.4.1 Exponential Decay Modeling of Hospital Recovery Data

We may be able to build a model that both accurately fits the data and produces estimates that pass the common sense test by using an exponential decay model. This modeling approach lets us model relationships that vary over time in a nonlinear fashion—in this case, we want to accurately capture the strong correlation for lower ranges of T but allow the magnitude of this relationship to decay as T increases, as the data seems to indicate.

Generating nonlinear models in R is done using the nonlinear least-squares or NLS function, appropriately labeled nls(). This function automatically fits a wide range of nonlinear models based on a functional form designated by the user. It is important to note that when fitting an NLS model in R, minimizing the sum of squares $\sum_{i=1}^{n}\left(y_i - a\left(exp\left(bx_i\right)\right)\right)^2$ is done computationally rather than mathematically. That means that the choice of starting values for the optimization function is important—the estimates produced by the model may vary considerably based on the chosen starting values (Fox, 2012)! As such, it is wise to experiment when fitting these nonlinear values to test how robust the resulting estimates are to the choice of starting values. We suggest using a ln-ln transformation of this data to begin with and then transforming back into the original x-y space to obtain "good" estimates. The model $ln(y) = ln(a)+bx$ yields $ln(y) = 4.037159 - 0.03797x$. This translates into the estimated model: $y = 56.66512e^{(-0.03797x)}$. Our starting values for (a, b) should be $(56.66512, -0.03797)$.

```
## Fit NLS model to the data
## Generate model
recovery_model3 <- nls(
Y ~ a * (exp(b * T))
, data = recovery_data
, start = c(
a = 56.66512
, b = −0.03797
)
, trace = T
)
## 11198.06:    1.00        0.05
```

## 9903.899:	10.3397660	−0.1171301
## 5127.568:	1.445986e+01	1.861608e−04
## 3205.214:	22.43186129	−0.01280344
## 1099.753:	37.70528644	−0.02984727
## 68.00072:	57.77928985	−0.04158464
## 49.46803:	58.62130670	−0.03954164
## 49.4593:	58.60566525	−0.03958494
## 49.4593:	58.6065363	−0.0395864

Fitting nonlinear regression model: $Y \sim a * (exp(b * T))$

Parameter Estimates

a	*b*
58.61	−0.03959

residual sum-of-squares: 1.951

The final model is $y = 58.61e^{-0.03959x}$. Overlaying the trend produced by the model on the plot of observed values, Figure 14.8, we see that the NLS modeling approach fits the data very well.

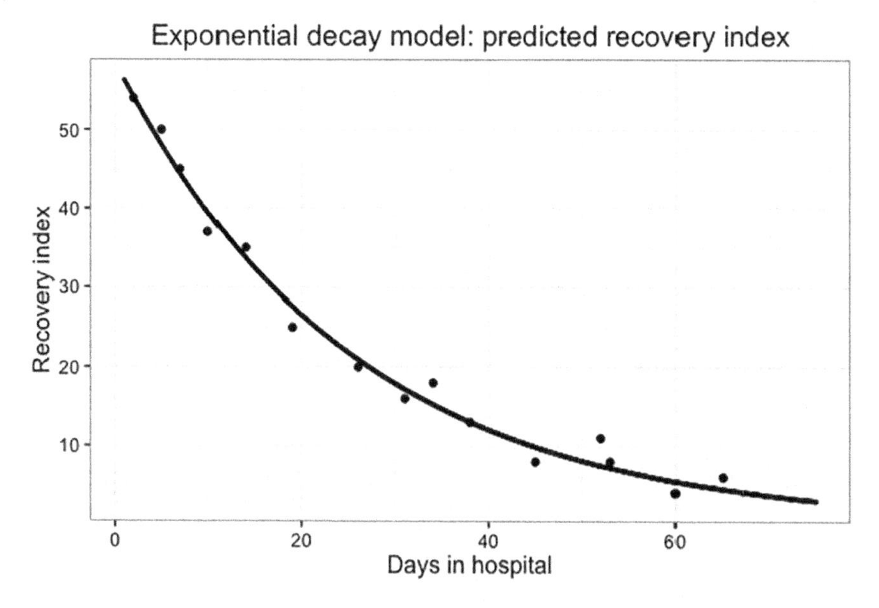

FIGURE 14.8
Exponential regression model and data.

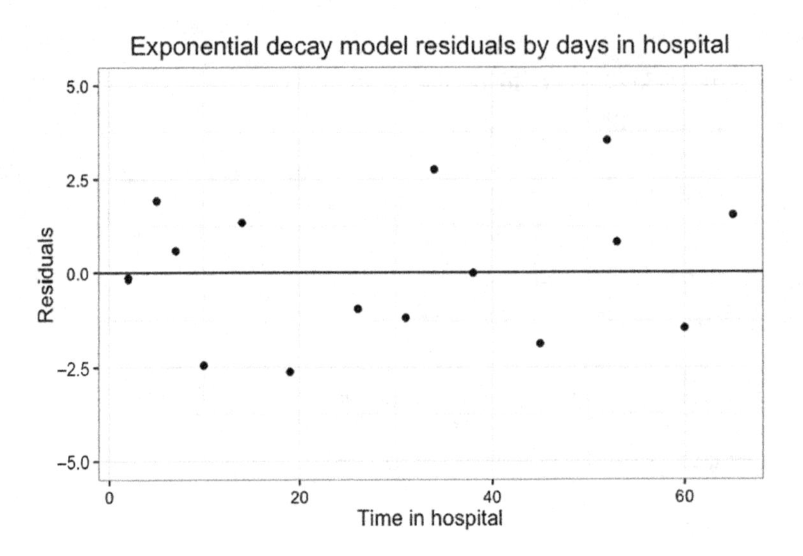

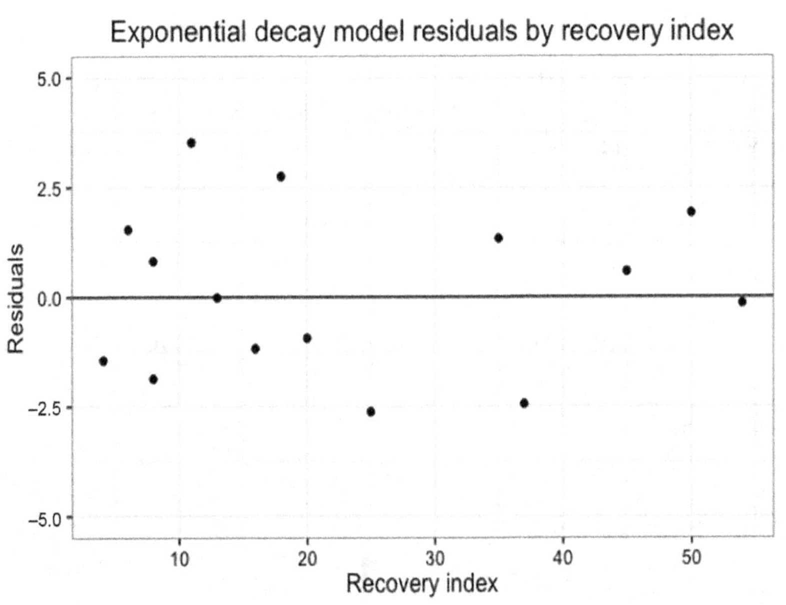

FIGURE 14.9
Residual plot as functions of (a) the time and (b) the model.

Once again, we can visually assess model fit by calculating and plotting the residuals. Figures 14.9 (a) and (b) show the same residuals plotted along both days in the hospital T and recovery index Y.

```
##   # A tibble: 15 × 5
##          T       Y    predicted   residuals  pct_relative_error
##      <int>   <int>        <dbl>       <dbl>               <dbl>
##   1      2      54    54.143916−0.14391597          −0.2665110
```

```
## 2   5  50  48.079006  1.92099420    3.8419884
## 3   7  45  44.418036  0.58196361    1.2932525
## 4  10  37  39.442567 -2.44256693   -6.6015322
## 5  14  35  33.664559  1.33544135    3.8155467
## 6  19  25  27.617389 -2.61738904  -10.4695562
## 7  26  20  20.931299 -0.93129928   -4.6564964
## 8  31  16  17.171407 -1.17140692   -7.3212932
## 9  34  18  15.247958  2.75204170   15.2891206
## 10 38  13  13.014259 -0.01425912   -0.1096856
## 11 45   8   9.863545 -1.86354475  -23.2943094
## 12 52  11   7.475609  3.52439082   32.0399165
## 13 53   8   7.185360  0.81464008   10.1830010
## 14 60   4   5.445805 -1.44580513  -36.1451283
## 15 65   6   4.467574  1.53242564   25.5404274
```

In both cases, we see that there is no easily distinguishable pattern in residuals. Finally, we apply the common sense check by generating and plotting estimated recovery index values for a set of values of ***T*** from 1 to 120.

The predicted values generated by the exponential decay model make intuitive sense. As the number of days a patient spends in the hospital increases, the model predicts that their recovery index will decrease at a decreasing rate. This means that while the recovery index variable will continuously decrease, it will not take on negative values (as predicted by the linear model) or explosively large values (as predicted by the quadratic model). It appears that the exponential decay model not only fits the data best from a purely statistical point of view but also generates values that pass the common sense test to an observer or analyst shown in Figure 14.10.

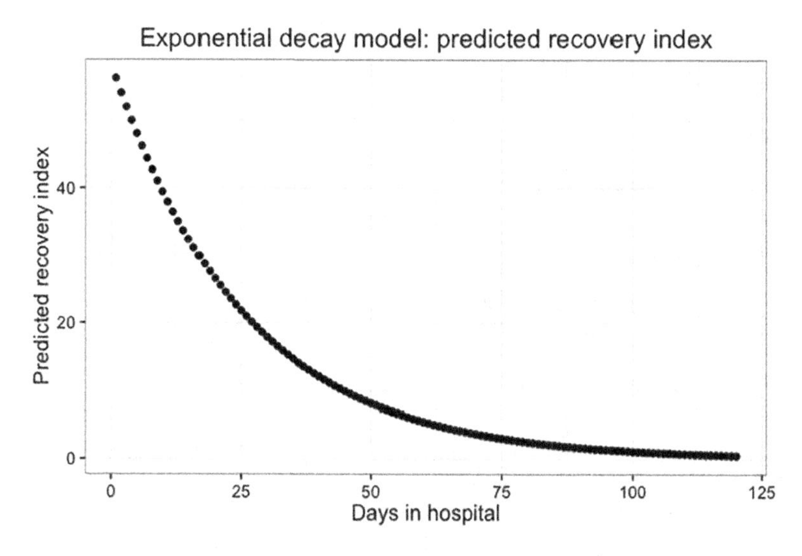

FIGURE 14.10
Plot of exponential regression model.

14.5 Sinusoidal Regression

We introduce some shipping data

```
## Print data as a tibble
print(shipping_data)
##  #A tibble: 20 × 2
##     Month UsageTons
##     <int>     <int>
## 1     1        20
## 2     2        15
## 3     3        10
## 4     4        18
## 5     5        28
## 6     6        18
## 7     7        13
## 8     8        21
## 9     9        28
## 10   10        22
## 11   11        19
## 12   12        25
## 13   13        32
## 14   14        26
## 15   15        21
## 16   16        29
## 17   17        35
## 18   18        28
## 19   19        22
## 20   20        32
## Calculate and print correlation matrix
print(cor(shipping_data))
##                  Month UsageTons
##    Month     1.0000000 0.6725644
##    UsageTons 0.6725644 1.0000000
```

Once again, we can visualize the data in a scatterplot to assess whether this positive correlation is borne out by the overall trend.

Visualizing the data, Figure 14.11, we see that there is a clear positive trend over time in shipping usage. However, examining the data in more detail suggests that a simple linear model may not be best suited to capturing the variation in these data. One way to

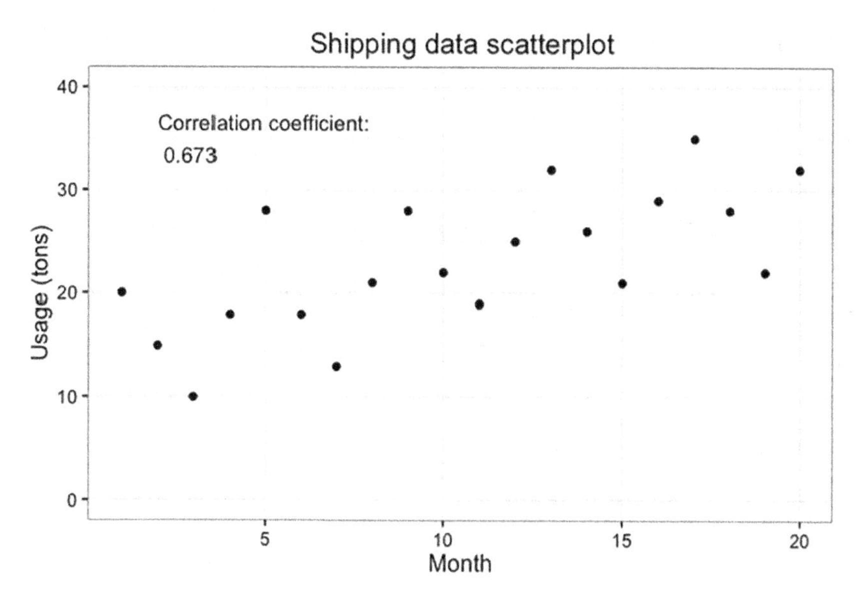

FIGURE 14.11
Scatterplot of shipping data.

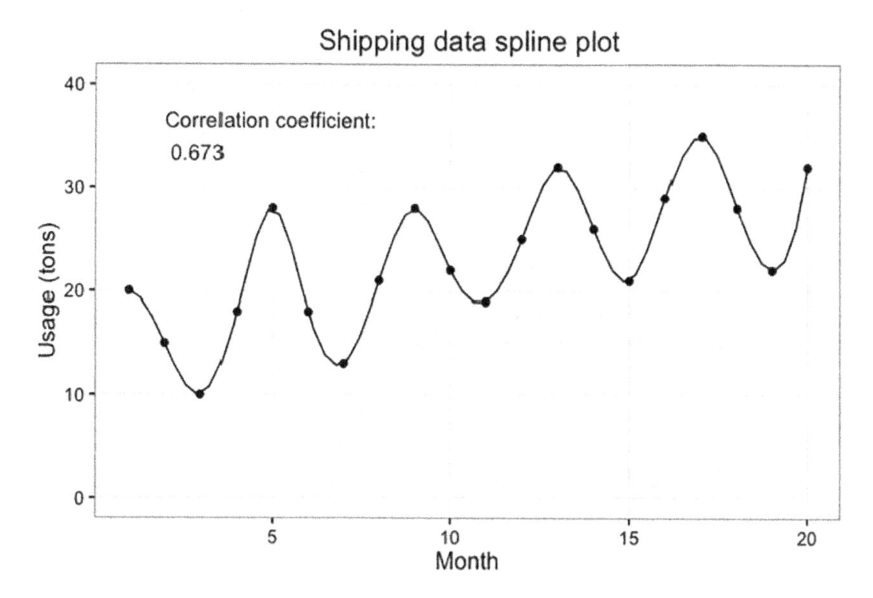

FIGURE 14.12
Shipping data with data points connected show an oscillating trend.

plot more complex patterns in data is through the use of a trend line using polynomial or nonparametric smoothing functions.

Plotting a trend line generated via a spline function shows that there seems to be an oscillating pattern with a steady increase over time in the shipping data, Figure 14.12.

14.5.1 Linear Regression of Shipping Data

As a baseline for comparison, we begin by fitting a standard OLS regression model using the lm() function in R.

```
## Generate model
shipping_model 1 <- lm(UsageTons ~ Month, data = shipping_data)
```

	Estimate	Std. Error	t value	Pr(>|t|)
Month	0.7594	0.1969	3.856	0.001158
(Intercept)	15.13	2.359	6.411	4.907e−06

Fitting linear model: UsageTons ~ Month

Observations	Residual Std. Error	R^2	Adjusted R^2
20	5.079	0.4523	0.4219

Analysis of Variance Table

	Df	Sum Sq	Mean Sq	F value	Pr(>F)
Month	1	383.5	383.5	14.87	0.001158
Residuals	18	464.3	25.79	NA	NA

While the linear model, $y = 15.13 + 0.7954x$, fits the data fairly well, the oscillation identified by the spline visualization suggests that we should apply a model that better fits the seasonal variation in the data.

14.5.2 Sinusoidal Regression of Shipping Data

R treats sinusoidal regression models as part of the larger family of nonlinear least-squares (NLS) regression models. This means that we can fit a sinusoidal model using the same nls() function and syntax as we applied earlier for the exponential decay model. The functional form for the sinusoidal model we use here can be written as:

$$Usage = a * sin(b * time + c) + d * time + e$$

This function can be expanded out trigonometrically as:

$$Usage = a * time + b * sin(c * time) + d * cos(c(time)) + e$$

This equation can be passed to nls(), and R will computationally assess best-fit values for the a, b, c, d, and e terms. It is worth stressing again the importance of selecting good starting values for this process, especially for a model like this one with many parameters to be simultaneously estimated. Here, we set starting values based on pre-analysis of the data. It is also important to note that because the underlying algorithms used to optimize these functions differ between Excel and R, the two methods produce models with different parameters but nearly identical predictive qualities. The model can be specified in R as follows.

```
## Generate model
shipping_model2 <- nls(
UsageTons ~ a * Month + b * sin(c * Month) + d * cos(c * Month) + e
, data = shipping_data
, start = c(
a = 5
, b = 10
, c = 1
, d = 1
, e = 10
)
, trace = T
)
## 45042.53:          5          10          1          1          10
## 663.046:   0.7736951  -1.5386559  0.9616379  4.2289392  15.3202771
## 458.8408:  0.7425778  -0.8555154  0.9595757 -0.1801322  15.3201412
## 380.7509:  0.7687894  -1.5130791  1.3777090  3.7655408  15.3260166
## 126.2519:   0.834506    2.821016   1.487313   4.923127   14.637850
## 99.34237:   0.862460    8.130120   1.583191   2.146993   14.066110
## 22.29435:  0.8478613   6.4959045  1.5747331  0.5860108  14.1975699
## 21.80271:  0.8479764   6.6646276  1.5733725  0.5579265  14.1866924
## 21.80233:  0.8479494   6.6663745  1.5735053  0.5518689  14.1865380
## 21.80233:  0.8479513   6.6663622  1.5735011  0.5520711  14.1865328
```

Fitting nonlinear regression model: UsageTons ~ a * Month + b * sin(c * Month) + d * cos(c * Month) + e

Parameter Estimates

a	*b*	C	*d*	*e*
0.848	6.666	1.574	0.5521	14.19

Residual sum of squares: 1.206

The model found is:

$$\boldsymbol{Usage} = 0.848 * \boldsymbol{time} + 6.666 * \boldsymbol{sin}(1.574 * \boldsymbol{time}) + 0.5521 * \boldsymbol{cos}(\boldsymbol{c}(\boldsymbol{time})) + 14.19.$$

Plotting the trend line, Figure 14.13, produced by the sinusoidal model shows that this modeling approach fits the data much better, accounting for both the short-term seasonal variation and the long-term increase in shipping usage.

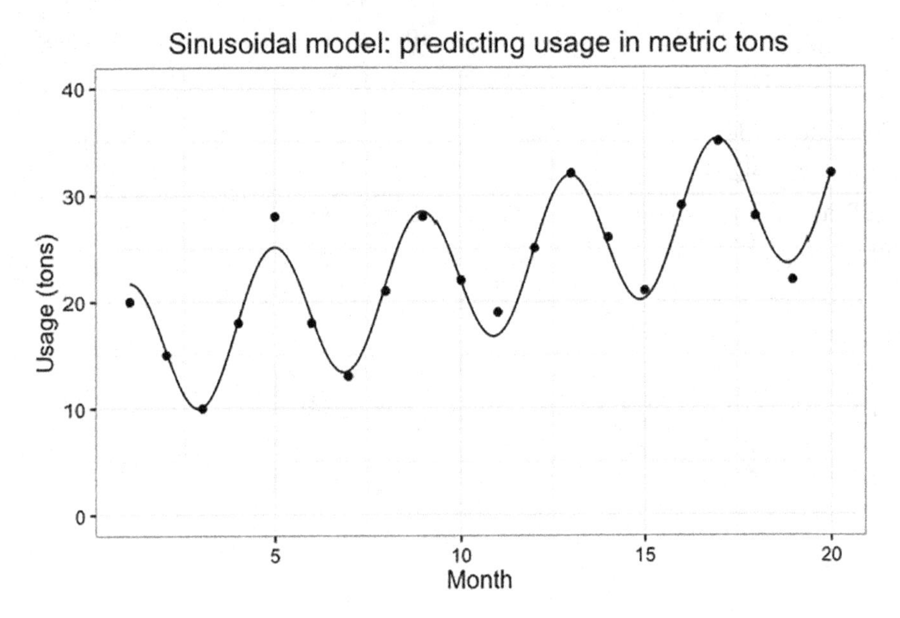

FIGURE 14.13
Overlay of regression model and data.

```
## # A tibble: 20 × 5
##   MonthUsageTonspredicted  residuals pct_relative_error
##    <int> <int>      <dbl>       <dbl>              <dbl>
## 1      1    20   21.69933 −1.69932876         −8.4966438
## 2      2    15   15.29431 −0.29431044         −1.9620696
## 3      3    10   10.06872 −0.06872366         −0.6872366
## 4      4    18   18.20250 −0.20249983         −1.1249991
## 5      5    28   25.08458  2.91542390         10.4122282
## 6      6    18   18.61406 −0.61406056         −3.4114475
## 7      7    13   13.46748 −0.46747667         −3.5959744
## 8      8    21   21.66632 −0.66632268         −3.1729651
## 9      9    28   28.46904 −0.46904406         −1.6751573
## 10    10    22   21.93389  0.06611205          0.3005093
## 11    11    19   16.86701  2.13299137         11.2262704
## 12    12    25   25.13006 −0.13006404         −0.5202562
## 13    13    32   31.85273  0.14726650          0.4602078
## 14    14    26   25.25380  0.74619895          2.8699960
## 15    15    21   20.26732  0.73268136          3.4889588
## 16    16    29   28.59372  0.40628451          1.4009811
## 17    17    35   35.23565 −0.23564539         −0.6732725
```

```
## 18    18    28  28.57381  -0.57380826     -2.0493152
## 19    19    22  23.66841  -1.66840571     -7.5836623
## 20    20    32  32.05727  -0.05726862     -0.1789645
```

Analysis of model residuals bears this out and also highlights the difference in solving methods between Excel and R. The model fitted in R has different parameter estimates and a slightly worse model fit (average percent relative error of 3.26% as opposed to the 3.03% from the Excel-fitted model), but the overall trend identified in the data is virtually identical.

Example 1: Introducing Afghanistan casualty data.

14.5.3 Sinusoidal Regression of Afghanistan Casualties

Visualizing data on casualties in Afghanistan between 2006 through 2008 shows an increasing trend overall and significant seasonal oscillation, Figure 14.14. Once again, we want to fit a nonlinear model that accounts for the oscillation present in the data. We use the same sinusoidal functional form

$$Casualties = a * sin(b * time + c) + d * time + e$$

which, as before, can be expressed as

$$Casualties = a * time + b * sin(c * time) + d * cos(c * time) + e$$

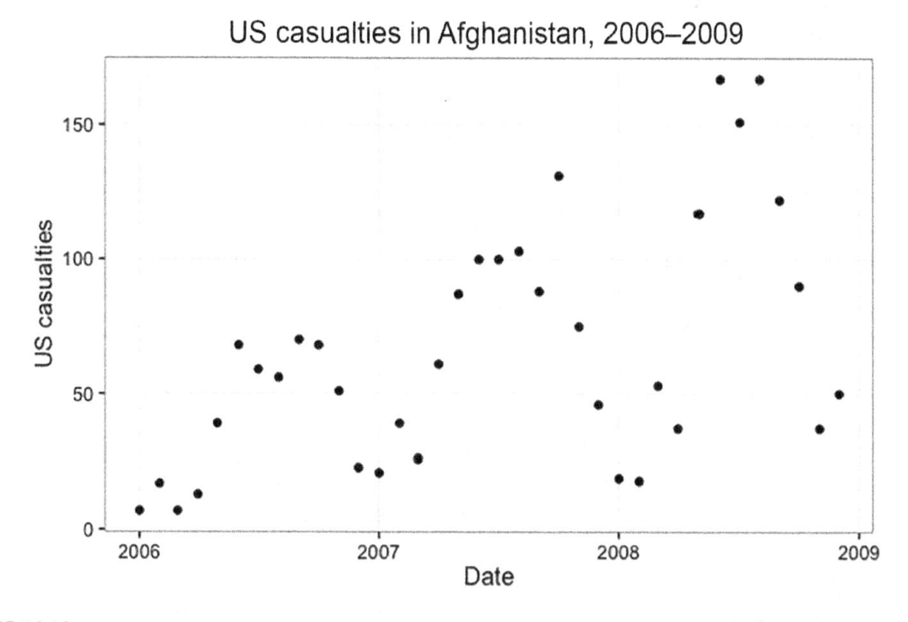

FIGURE 14.14
Casualty data scatterplot.

We fit the model using the nls() function once again:

```
## 138985.7:          1           10           1            10            1
## 49982.63:  2.258671     1.356303    1.029391      1.353397     24.619633
## 35427.2:   2.2068763   -9.4273932   0.6196208    15.1463459    26.0931204
## 24267.23:  2.1318806  -20.5140300   0.5758111    12.0966231    27.9982286
## 18101.34:  1.9724998  -34.2695108   0.5582244     7.4221538    31.3617907
## 14806.36:  1.8863899  -44.3682090   0.5584127    -0.9302451    33.3906528
## 14498.11:  1.8238452  -44.3050251   0.5480953    -9.2653710    34.0511512
## 14423.67:  1.8575261  -43.3800752   0.5492376   -10.3036562    33.5078195
## 14411.39:  1.8442981  -43.0865138   0.5471146   -11.8346191    33.6267612
## 14408.94:  1.851289   -43.005846    0.547460    -11.949005     33.525642
## 14408.47:  1.8484852  -42.9396306   0.5470361   -12.2282504    33.5512463
## 14408.37:  1.8499620  -42.9311529   0.5471265   -12.2338044    33.5310064
## 14408.35:  1.8493503  -42.9186159   0.5470412   -12.2859459    33.5369245
## 14408.35:  1.8496642  -42.9178538   0.5470632   -12.2839399    33.5327759
## 14408.35:  1.8495307  -42.9155167   0.5470459   -12.2938726    33.5341311
## 14408.35:  1.8495976  -42.9155139   0.5470511   -12.2929115    33.5332705
## 14408.35:  1.8495686  -42.9150731   0.5470475   -12.2948380    33.5335758
## 14408.35:  1.8495829  -42.9150984   0.5470487   -12.2945453    33.5333958
## 14408.35:  1.8495765  -42.9150139   0.5470479   -12.2949258    33.5334641
```

Fitting nonlinear regression model: Casualties ~ a * DateIndex + b * sin(c * DateIndex) + d * cos(c * DateIndex) + e

Parameter Estimates

a	*b*	*c*	*d*	*e*
1.85	−42.92	0.547	−12.29	33.53

Residual sum of squares: 21.56

The model found is

$$\textbf{\textit{Casualties}} = 1.85 * \textbf{\textit{time}} \pm 42.92 * \textbf{\textit{sin}}(0.547 * \textbf{\textit{time}}) - 12.19 * \textbf{\textit{cos}}(0.547 * \textbf{\textit{time}}) + 33.53$$

Plotting the trend line, Figure 14.15, identified by the sinusoidal model, shows again that the sinusoidal modeling approach can account for both short-term oscillation and long-term increase. We can now estimate residuals and error metrics and assess how well the model fits over the full range of the data. We see the residual lot in Figure 14.16.

```
## #A tibble: 36 × 8
##      Year  Month  Casualties        Date  DateIndex  predicted    residuals
##     <int>  <int>       <int>      <date>      <int>      <dbl>        <dbl>
## 1    2006      1           7  2006-01-01          1   2.559362   4.44063774
```

```
## 2   2006   2     17  2006-02-01     2  -6.539489   23.53948919
## 3   2006   3      7  2006-03-01     3  -2.862473    9.86247255
## 4   2006   4     13  2006-04-01     4  13.057033   -0.05703292
## 5   2006   5     39  2006-05-01     5  37.112404    1.88759617
## 6   2006   6     68  2006-06-01     6  62.822382    5.17761750
## 7   2006   7     59  2006-07-01     7  83.222777  -24.22277700
## 8   2006   8     56  2006-08-01     8  92.899116  -36.89911626
## 9   2006   9     70  2006-09-01     9  89.566985  -19.56698456
## 10  2006   10    68  2006-10-01    10  74.738779   -6.73877932
## #. . . with 26 more rows, and 1 more variables: pct_relative_error <dbl>
```

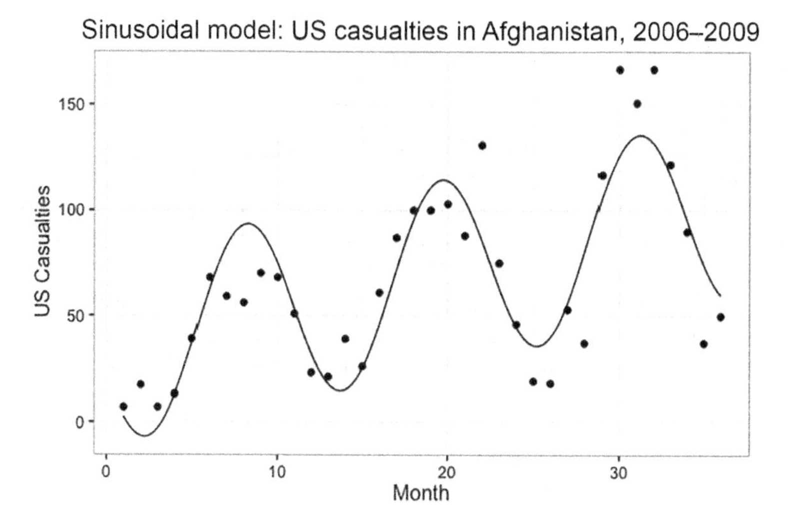

FIGURE 14.15
Model of casualties.

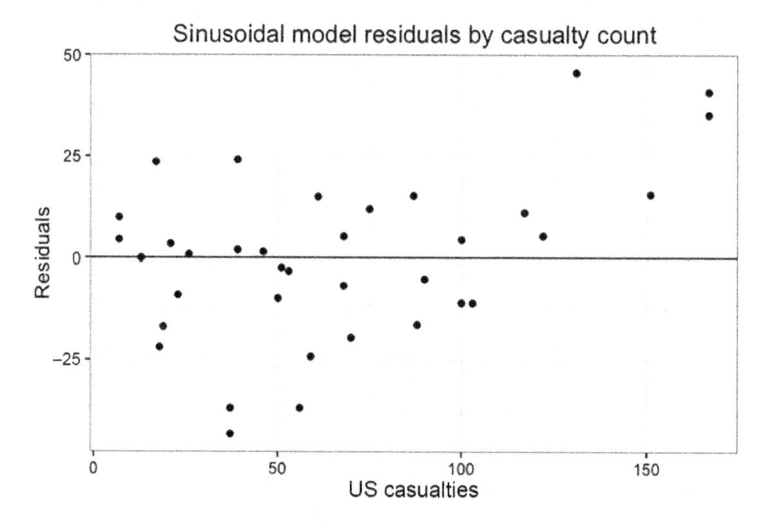

FIGURE 14.16
Residual plot of casualty model.

Again, this highlights both the importance of starting values and the difference in estimation method between *R* and Excel. Despite using different starting values and estimating very different parameters, each model produces very similar estimates of casualties over time: *SSE* for the Excel model is 14,415.2125, almost identical to the *R* model *SS* of 14,408.35.

14.6 Logistic regression

Often our dependent variable has special characteristics. Two such special cases that occur commonly are dependent variables which are binary {0,1} or the dependent variables are counts {0, 1, 2, . . .}. The latter case is typically handled by assuming the counts follow a Poisson distribution with the Poisson regression model. We focus here on the binary case which is often handled with logistic regression models.

14.6.1 A Binary Logistical Regression Analysis of Dehumanization

Example 4: Dehumanization data

We wish to analyze the interrelation of dehumanization's effects (shown through higher percentages of civilian casualties) on the outcome of conflict (shown to be a win "1" or a loss "0"). In this case, the civilian casualty percentages are the independent variable "X" and *Side A's* win/loss outcome from the conflict to be the dependent variable "Y."

Conflict outcomes differ from the data we've examined previously in that the measure of state victory only has two values, 1 and 0. This type of data can be modeled using a binomial logistic (or sometimes "logit") regression. Logistic regression estimates an underlying continuous variable as a linear function using the "logit" function:

$$ln\left(\frac{P}{1-P}\right)=\beta_0+\beta_1 X.$$

Here, P is the probability that Y = 1, or that the binary outcome is a "success". We do not have probabilities, but instead the actual Y = 1 or Y = 0 values. The "logit" function is also called the "log odds" function as the ratio of P to 1-P is the odds of the outcome Y = 1.

The model above is transformed by exponentiating both sides and then solving for P to produce:

$$P=\frac{exp(\beta_0+\beta_1 x)}{1+exp(\beta_0+\beta_1 x)}.$$

This non-linear function is bounded below by 0 and above by 1. This means it is useful for estimating binary (0,1) outcomes with a model:

$$Y=\frac{exp(\beta_0+\beta_1 x)}{1+exp(\beta_0+\beta_1 x)}+\varepsilon.$$

In other words, the model estimates probabilities of observing a 1 versus a 0. Due to the binary nature of the outcome, the errors, ε, are assumed to follow a binomial distribution.

The logistic model in R is treated as one case of a broader range of generalized linear models (GLM), and can be accessed via the conveniently named glm() function. Note that because glm() implements a wide range of generalized linear models based on the inputs provided, it is necessary for the user to specify both the family of model (binomial) and the "link" function (logit). The code below fits the model for the war data:

```
war_model <- glm( side_a ~ cd_pct, data = war_data ,
      family = binomial(link = 'logit'))
```

Output for the model using the summary function includes estimates of the two parameters and tests for their significant. The key portions of the output are:

	Estimate	Std. Error	z value	Pr(>|z|)
cd_pct	1.85	2.556	0.7237	0.4692
(Intercept)	0.004716	1.925	0.00245	0.998

The estimate of the coefficient for the X variable (cd_pct, the percent of civilian casualties) is 1.85. Thus, the model suggests that there is a positive correlation between civilian casualties and state victory; the probability of victory increases with more casualties. However, this relationship is not statistically significant with $p = 0.4692$. This means we cannot reject the null hypothesis that no relationship exists between the input and output variables.

Logistic regression analysis requires much more material than we can provide in this text. For example, model interpretation and assessment are somewhat different than in linear regression models. For a very thorough discussion of these models see Hosmer, Lemeshow, and Sturdivant (2013).

14.7 Conclusions and Summary

We showed some of the common misconceptions by decision-makers concerning correlation and regression. Our purpose of this presentation is to help prepare more competent and confident problem-solvers for the 21st century. Data can be found using part of a sine curve where the correlation is quite poor, close to zero, but the decision-maker can describe the pattern. Decision-makers see the relationship in the data as periodic or oscillating. Examples such as these should dispel the idea that correlation of almost zero implies no relationship. Decision-makers need to see and believe concepts concerning correlation, linear relationships, and nonlinear (or no) relationship.

We recommended the following summary steps.

Step 1. Ensure you understand the problem and what answers are required.

Step 2. Get the data that is available. Identify the dependent and independent variables.

Step 3. Plot the dependent versus an independent variable, and note trends.

Step 4. If the dependent variable is binary {0,1}, then use binary logistics regression. If the dependent variables are counts that follow a Poisson distribution, then use Poisson regression. Otherwise, try linear, multiple, or nonlinear regression as needed.

Step 5. Ensure your model produces results that are acceptable.

14.8 Chapter 14 Exercises

For the following data, (a) plot the data and (b) state the type of regression that should be used to model the data.

14.1 Tire tread

Number	Hours	Tread (cm)
1	2	5.4
2	5	5.0
3	7	4.5
4	10	3.7
5	14	3.5
6	19	2.5
7	26	2.0
8	31	1.6
9	34	1.8
10	38	1.3
11	45	0.8
12	52	1.1
13	53	0.8
14	60	0.4
15	65	0.6

14.2 Let's assume our suspected nonlinear model form is: $Z = a\frac{x^b}{y^c}$ for the data below. If we use our ln-ln transformation, we obtain: $\ln Z = \ln a + b\ln x - c\ln y$. Use regression techniques to estimate the parameters *a*, *b*, and *c*.

ROW	x	y	Z
1	101	15	0.788
2	73	3	304.149
3	122	5	98.245
4	56	20	0.051
5	107	20	0.270
6	77	5	30.485
7	140	15	1.653
8	66	16	0.192
9	109	5	159.918
10	103	14	1.109
11	93	3	699.447
12	98	4	281.184
13	76	14	0.476
14	83	5	54.468
15	113	12	2.810
16	167	6	144.923
17	82	5	79.733
18	85	6	21.821

ROW	x	y	Z
19	103	20	0.223
20	86	11	1.899
21	67	8	5.180
22	104	13	1.334
23	114	5	110.378
24	118	21	0.274
25	94	5	81.304

14.3 Using the basic linear model, $y_i = \beta_0 + \beta_1 x_i$, fit the following data sets. Provide the model, the analysis of variance information, the value of R^2, and a residual plot.

a.

x	y
100	150
125	140
125	180
150	210
150	190
200	320
200	280
250	400
250	430
300	440
300	390
350	600
400	610
400	670

b. The following data represent changes in growth where x = bodyweight and y = normalized metabolic rate for 13 animals.

x	y
110	198
115	173
120	174
230	149
235	124
240	115
360	130
362	102
363	95
500	122
505	112
510	98
515	96

14.4 Ten observations of college acceptances to graduate school.

ADMIT	*GRE*	*TOPNOTCH*	*GPA*
1	380	0	3.61
0	660	1	3.67
0	800	1	4
0	640	0	3.19
1	520	0	2.93
0	760	0	3
0	560	0	2.98
1	400	0	3.08
0	540	0	3.39
1	700	1	3.92

14.5 Data set for lung cancer from E. L. Frome (1983), *Biometrics* 39: 665–674. The number of person years are in parentheses, broken down by age and daily cigarette consumption.

Age	Nonsmokers	Smokes 1–9 Per Day	Smokes 10–14 Per Day	Smokes 15–19 Per Day	Smokes 20–24 Per Day	Smokes 25–34 Per Day	Smokes >35 Per Day
15–20	1 (10366)	0 (3121)	0 (3577)	0 (4319)	0 (5683)	0 (3042)	0 (670)
20–25	0 (8162)	0 (2397)	1 (3286)	0 (4214)	1 (6385)	1 (4050)	0 (1166)
25–30	0 (5969)	0 (2288)	1 (2546)	0 (3185)	1 (5483)	4 (4290)	0 (1482)
30–35	0 (4496)	0 (2015)	2 (2219)	4 (2560)	6 (4687)	9 (4268)	4 (1580)
35–40	0 (3152)	1 (1648)	0 (1826)	0 (1893)	5 (3646)	9 (3529)	6 (1136)
40–45	0 (2201)	2 (1310)	1 (1386)	2 (1334)	12 (2411)	11 (2424)	10 (924)
45–50	0 (1421)	0 (927)	2 (988)	2 (849)	9 (1567)	10 (1409)	7 (556)
50–55	0 (1121)	3 (710)	4 (684)	2 (470)	7 (857)	5 (663)	4 (255)
>55	2 (826)	0 (606)	3 (449)	5 (280)	7 (416	3 (284)	1 (104)

14.6 Modeling absences from class where:

Gender: 1-female 2-male

Ethnicity: 6 categories

School: school 1 or school 2

Math test score: continuous

Language test score: continuous

Bilingual status: 4 bilingual categories

Gender	Ethnicity	School	Math Score	Language Score	Bilingual Status	Days Absent
2	4	1	56.98	42.45	2	4
2	4	1	37.09	46.82	2	4
1	4	1	32.37	43.57	2	2
1	4	1	29.06	43.57	2	3
1	4	1	6.75	27.25	3	3
1	4	1	61.65	48.41	0	13
1	4	1	56.99	40.74	2	11

Gender	Ethnicity	School	Math Score	Language Score	Bilingual Status	Days Absent
2	4	1	10.39	15.36	2	7
2	4	1	50.52	51.12	2	10
2	6	1	49.47	42.45	0	9

14.7 Fit the following nonlinear model with the provided data:

Model: $y = ax^b$

Data:

t	7	14	21	28	35	42
y	8	41	133	250	280	297

14.8 Fit the following model, $y = ax^b$, with the provided data.

Year	0	1	2	3	4	5	6	7	8	9	10
Quantity	15	150	250	275	270	280	290	650	1200	1550	2750

14.9 References and Suggested Reading

Afifi, A. and S. Azen (1979). *Statistical Analysis*, 2nd ed., Academic Press, London, UK, pp. 143–144.

Devore, J. (2012). *Probability and Statistics for Engineering and the Sciences*, 8th ed., Cengage Publisher, Belmont, CA, pp. 211–217.

Fox, W. (2011). Using the EXCEL Solver for Nonlinear Regression, *Computers in Education Journal (COED)*, October–December, **2**(4), pp. 77–86.

Fox, W. (2012a). Issues and Importance of "Good" Starting Points for Nonlinear Regression for Mathematical Modeling with Maple: Basic Model Fitting to Make Predictions with Oscillating Data, *Journal of Computers in Mathematics and Science Teaching*, **31**(1), pp. 1–16.

Fox, W. (2012b). *Mathematical Modeling with Maple*, Cengage Publishers, Boston, MA.

Fox, W. and C. Fowler (1996). Understanding Covariance and Correlation, *Primus*, **VI**(3), pp. 235–244.

Fox, W. and J. Hammond (2019). Advanced Regression Models: Least Squares, Nonlinear, Poisson and Binary Logistics Regression Using R, *Data Science and Digital Business*, Springer, Switzerland, pp. 221–262.

Frome, E. L. (1983). The Analysis of Rates Using Poisson Regression Models, *Biometrics*, **39**, pp. 665–674.

Giordano, F., W. Fox and S. Horton (2013). *A First Course in Mathematical Modeling*, 5th ed., Cengage Publishers, Boston, MA.

Hosmer, D.W., Lemeshow, S., and Sturdivant, R.X. (2013). *Applied Logistic Regression, 3rd Edition*, Wiley, Inc., New Jersey.

Johnson, I. (2012). *An Introductory Handbook on Probability, Statistics, and Excel*, at http://records.viu.ca/~johnstoi/maybe/maybe4.htm (accessed July 11, 2012).

Neter, J., M. Kutner, C. Nachtsheim and W. Wasserman (1996). *Applied Linear Statistical Models*, 4th ed., Irwin Press, Chicago, IL, pp. 531–547.

15

ANOVA in R

Objectives

1. **Know ANOVA procedures.**
2. **Know how to implement the procedures in R.**
3. **Know how to interpret the results of each procedure.**

15.1 Introduction

ANOVA is short for Analysis of Variance. ANOVA is one of the most basic yet powerful statistical models that we can utilize. While it is commonly used for categorical data, because ANOVA is a type of linear model, it can be modified to include continuous data. Although ANOVA is relatively simple as compared to many statistical models, there are still some things we should consider. We explore both the features and functions of ANOVA as handled by R.

Like any statistical routine, ANOVA also comes with its own set of vocabulary. We do not cover it all, but we will discuss two types of ANOVAs that are typically referred to as **one-way** and **two-way**, which is just a way of saying how many factors are being examined in the model. (ANOVAs can be three-way or more, but these are less common or end up with different names.) A **factor** is the categorical variable that you are evaluating, and the different categories within the factor are often called **levels** or **groups**. This language is somewhat flexible. Of course, ANOVAs can have many more than two factors, but as with any model, there are costs and benefits as models increase in complexity. We can add a continuous covariate to our model, which results in a hybrid ANOVA and linear model that is called **ANCOVA**—an analysis of covariance, that we discuss later.

15.1.1 When to Use a One-Way ANOVA

Use a one-way ANOVA when you have collected data about one categorical independent variable and one quantitative dependent variable. The independent variable should have at least three levels (i.e., at least three different groups or categories).

ANOVA tells you if the dependent variable changes according to the level of the independent variable.

DOI: 10.1201/9781003317906-15

15.1.2 Assumptions of ANOVA

The assumptions of the ANOVA test are the same as the general assumptions for any parametric test:

1. **Independence of observations**: The data were collected using statistically valid methods, and there are no hidden relationships among observations. If your data fail to meet this assumption because you have a confounding variable that you need to control for statistically, use an ANOVA with blocking variables.
2. **Normally distributed response variable**: The values of the dependent variable follow a normal distribution.
3. **Homogeneity of variance**: The variation within each group being compared is similar for every group. If the variances are different among the groups, then ANOVA probably isn't the right fit for the data.

The procedure for one-way ANOVA

Let i = the number of treatments compared.

μ_1 = mean of treatment 1

μ_2 = mean of treatment 2

...

μ_i = mean of treatment i

The hypotheses of interest are

H_0: $\mu_1 = m_2 = \ldots = \mu_i$

Hα: at least two means are different

15.2 ANOVA Mechanics

15.2.1 One-Way ANOVA

We start with a broad overview of how ANOVA works. ANOVA analyzes the variance in the data to look for differences. It does this by considering two sources of variance, the **between-group variance** and the **within-group variance**. The between-group variation is calculated by comparing the mean of each group with the overall mean of the data. That is, individual data points don't matter quite as much as just comparing group means. The within-group variation is the variation of each observation from its group mean. For both types of variance, we use a sum of squares (SS). A **sum of squares** (SS) is the numerical metric used to quantify them, and this metric simply sums the distances of each point to the mean. The ratio of these SS (between SS divided by within SS) results in an ***F*-statistic.** This is the test statistic for ANOVA. The *F*-statistic is then combined with the degrees of

TABLE 15.1

ANOVA One-Way Formulas

Source of Variation	Sum of Squares Formula	Degrees of Freedom	Mean Square (MS)	Value of F Ratio
Between samples	$SSB = b\sum_{i=1}^{a}(\overline{x_{.i}} - \overline{x})^2$	(a−1)	MSA = SSA/(a−1)	
Within Samples	$SSW = a\sum_{i=1}^{b}(\overline{x_{j.}} - \overline{x})^2$	a*(b−1)		
Total	$SST = \sum_{i=1}^{a}\sum_{j=1}^{b}(x_{ij} - \overline{x})^2$	ab−1		

freedom (df) to arrive at a ***p*** **-value**. The *p*-value's interpretation is, just as we did in hypothesis testing, discussed in Chapters 10 and 11. Another way to think of it is that small *p*-values come from large *F*-statistics, and large *F*-statistics suggest that the between-group variance is much larger than the within-group variance. So when the differences in group means is larger and yet the groups are not that variable, we tend to have significant factors in our ANOVAs.

ANOVA Table with formulas for one-way ANOVA (see Table 15.1).

First, we will do an example without technology.

Example 1: Our data, A, B, and C are classes, and each class has 10 numeric entries. We want to test if the means of classes A, B, and C are the same versus they are not the same.

Thus, our hypothesis test is

$H_0: \mu_A = \mu_B = \mu_C$

$H_a: \mu_A \neq \mu_B \neq \mu_C$

Number	A	B	C
1	7	4	6
2	9	3	1
3	5	6	3
4	8	2	5
5	6	7	3
6	8	5	4
7	6	5	6
8	10	4	5
9	7	1	7
10	4	3	3

Next we compute both the row means and the column means,

	Number	A	B	C	Column Means
	1	7	4	6	5.666667
	2	9	3	1	4.333333
	3	5	6	3	4.666667
	4	8	2	5	5
	5	6	7	3	5.333333
	6	8	5	4	5.666667
	7	6	5	6	5.666667
	8	10	4	5	6.333333
	9	7	1	7	5
	10	4	3	3	3.333333
Row	Mean	7	4	4.3	5.1

We find the grand mean is 5.1.
We also have 3 classes ($a = 3$) and 10 entries ($b = 10$).
We substitute into our ANOVA table:

Source of Variation	Sum of Squares	Degrees of Freedom	Mean Square MS	F ratio	*p*-Value	F Critical ($\alpha = 0.05$)
SSB	54.6	2	27.3	8.27	0.00169	3.354
SSW	90.1	27	3.3			
Total	144.7	29				

Interpretation: Since 8.27 > 3.354, we reject the null hypothesis that all the means are the same. We note this does not tell us which are different or by how much.

Example 2: ANOVA example repeated using R.

Using R, we first read in the data as shown here. Then we use the commands: aov(dependent variable ~independent variable, data = "name")

```
one.way <- aov(rtimes ~ class, data = data)
> summary(one.way)
data<- read.csv("data.csv", header = TRUE, colClasses = c("factor", "numeric"))
> data
  class times
1   A     7
2   A     9
3   A     5
4   A     8
5   A     6
```

```
6   A   8
7   A   6
8   A   10
9   A   7
10  A   4
11  B   4
12  B   3
13  B   6
14  B   2
15  B   7
16  B   5
17  B   5
18  B   4
19  B   1
20  B   3
21  C   6
22  C   1
23  C   3
24  C   5
25  C   3
26  C   4
27  C   6
28  C   5
29  C   7
30  C   3
> summary(data)
class times
A:10  Min.:     1.00
B:10  1st Qu.:  3.25
C:10  Median:   5.00
      Mean:     5.10
      3rd Qu.:  6.75
> summary(one.way)
          Df  Sum Sq  Mean Sq  F value Pr(>F)
class     2   54.6    27.300   8.181   0.00167 **
Residuals 27  90.1    3.337
—

Signif. codes: 0 '***' 0.001 '**' 0.01 '*' 0.05 '.' 0.1 '' 1
```

Interpretation: We see that our significant codes show the p-value is significant at 0.05 and 0.01, but not 0.001. Therefore, at 0.05 and 0.01, we reject the claim that the means are the same.

Example 3: We have data for the response rate found and the color of the item. Our data is entered into R, and we perform the one-way ANOVA.

```
one.way <- aov(response ~ color, data = data2)
> summary(one.way)
Df Sum Sq Mean Sq F value Pr(>F)
color 1 437.6 437.6 17.13 0.000147 ***
Residuals 46 1174.9 25.5
—
Signif. codes: 0 '***' 0.001 '**' 0.01 '*' 0.05 '.' 0.1 '' 1
```

Interpretation: We find the means are not the same as we reject the null hypothesis, and it is significant at 0.05, 0.01, and 0.001 according to our output from R.

15.2.2 Two-Way ANOVA

The two-way ANOVA compares the mean differences between groups that have been split on two independent variables (called factors). The primary purpose of a two-way ANOVA is to understand if there is an interaction between the two independent variables on the dependent variable. For example, you may want to determine whether there is an interaction between physical activity level (independent variable) and gender (independent variable) on blood cholesterol concentration (dependent variable) in children.

The **interaction term** in a two-way ANOVA informs you whether the effect of one of your independent variables on the dependent variable is the same for all values of your other independent variable (and vice versa).

There are some assumptions to do a two-way ANOVA:

1: Your dependent variable should be measured at the continuous level (i.e., they are interval or ratio variables).

2: Your two independent variables should each consist of two or more categorical, independent groups.

3: You should have independence of observations, which means that there is no relationship between the observations in each group or between the groups themselves.

4: There should be no significant outliers. Outliers are data points within your data that do not follow the usual pattern and most likely fall outside the typical values.

5: Your dependent variable should be approximately normally distributed for each combination of the groups of the two independent variables.

6: There needs to be homogeneity of variances for each combination of the groups of the two independent variables.

15.2.3 Two-Way ANOVA Calculation by Hand

We will do two-way ANOVA with an example.

Example 4: Suppose you want to determine whether the brand of laundry detergent used and the temperature affects the amount of dirt removed from your laundry. To this end, you buy two detergents with different brand ("A" and "B") and choose three different temperature levels ("cold," "warm," and "hot"). Then you divide your laundry randomly into "6 * *r*" pile of equal size and assign each "*r*" pile into the combination of ("A" and "B") and ("cold," "warm," and "hot"). In this example, we are interested in testing the null hypothesis.

$H_0(D)$ = The amount of dirt removed does not depend on the type of detergent.

$H_0(T)$ = The amount of dirt removed does not depend on the temperature.

The example has two factors (factor detergent, factor temperature) at $a = 2$ (Super and Best) and $b = 3$ (cold, warm, and hot) levels. Thus, there are $a * b = 3 * 2 = 6$ different combination of detergent and temperature with each combination. There are $r = 4$ loads. (r is called the number of replicates). This sums up to "$n = a * b * r$" $= 24 = 2 * 3 * 4$ loads in total.

The amounts of $Y(ijk)$ of dirt removed when washing groups k ($k = 1, 2, 3, 4$) with detergent i ($i = 1, 2$) at temperature j ($j = 1, 2, 3$) are recorded in Table 15.2.

We have calculated all the means like detergent mean(M_d), temperature mean(M_t) and mean of every group combination.

TABLE 15.2

Detergent and Temperature Data

	Cold	Warm	Hot
Brand A	4	7	10
	5	9	12
	6	8	11
	5	12	9
Brand B	6	13	12
	6	15	13
	4	12	10
	4	12	13

	Cold	Warm	Hot	M(d) [Y(i)]
A	4	7	10	
	5	9	12	
	6	8	11	
	5	12	9	
	mean(Yij) = 5	mean(Yij) = 9	mean(Yij) = 10.5	8.166
B	6	13	12	
	6	15	13	
	4	12	10	
	4	12	13	
	mean(Yij) = 5	mean(Yij) = 13	mean(Yij) = 12	10
M(t)[Y(j)]	5	11	11.25	9

Now what we only have to do is calculate the sum of squares (ss) and degree of freedom (df) for temperature, detergent, and interaction between factor and levels.

First calculate the SS(within)/df(within); we have already know how to calculate SS(within)/df(within) in one-way ANOVA; we calculated this, but in two-way ANOVA, the formula is different:

Step 1. Formula for calculation of SS(within) is:

Y_{ijk} are the elements in the groups.
Y¯(ij) is mean of combinations

When we put the values and do calculations with this formula, we will get SS(within) is

$$SS_{within} = \sum_{i=1}^{2}\sum_{j=1}^{3}\sum_{k=1}^{4}\left(Y_{ijk} - \bar{Y}_{ij}.\right)^2 = (4-5)^2 + (5-5)^2 + (6-5)^2 + (5-5)^2$$

$$+ (7-9)^2 + (9-9)^2 + (8-9)^2 + (12-9)^2$$

......

$$+ (12-12)^2 + (13-12)^2 + (10-12)^2 + (13-12$$

$$= 37$$

Calculate the df(within):

df(within) = $(r - 1) * a * b$ = 3 * 2 * 3 = 18

Calculate MS(within):

MS(within) = SS(within)/df(within) = 38/18 = 2.1111

Step 2. Calculate SS(detergent) and df(detergent) and MS(detergent)

Y¯(i) is the mean of detergent

Y¯ is the total mean detergent and temperature $SS_{detergent} = r \cdot b \cdot \sum_{i=1}^{2}\left(\bar{Y}_{i..} - \bar{Y}_{...}\right)^2$

= 4 * 3 [(8.16666 – 9.08333)² + (10 – 9.0833)²]
= 20.17

Calculate df(detergent):

df(detergent) = a – 1 = 2 – 1 = 1

Calculate MS(detergent):

MS(detergent) = SS(detergent)/df(detergent)
= 20.17/1 = 20.17

Step 3. Calculate the SS(temperature), df(temperature) and MS(temperature)

Y¯(i) is the mean of detergent
Y¯ is the total mean detergent and temperature

$$SS_{temperature} = r \cdot a \cdot \sum_{j=1}^{3} \left(\bar{Y}_{\cdot j \cdot} - \bar{Y}_{\ldots} \right)^2$$

$= 4 * 2 * [(5 - 9.0833)^2 + (11 - 9.0833)^2 + (11 - 9.0833)^2]$
$= 200.33$

Calculate df(temperature):

df(temperature) = b – 1 = 3 – 1 = 2

Calculate MS(temperature):

MS(temperature) = SS(temperature)/df(temperature)
= 200.33/2 = 100.165

Step 4. Calculate SS(interaction), df(interaction) and MS(interaction)
Y¯(ij) is mean of combinations
Y¯(i) is the mean of detergent
Y¯(j) is the mean of temperature
Y¯ is the total mean detergent and temperature

Calculate SS(interaction):

$$SS_{interaction} = r \times \sum_{i=1}^{2} \sum_{j=1}^{3} \left(\bar{Y}_{ij\cdot} - \bar{Y}_{i\ldots} - \bar{Y}_{\cdot j \cdot} + \bar{Y}_{\ldots} \right)^2$$

$= 4 \times (5 - 8 - 5 + 9.0833)^2 + (9 - 8 - 11 + 9.0833)^2 +$
$(10 - 8 - 11 + 9.0833)^2 + \cdots + (12 - 11 - 10 + 9.0833)^2$
$= 16.33$

Calculate df(interaction):

df(interaction) = (a – 1) * (b – 1) = (2 – 1) * (3 – 1) = 2

Calculate MS(interaction):

MS(interaction) = SS(interaction)/df(interaction)
= 16.33/2
= 8.165

It is time to calculate the F-test: Calculate critical F-value
MS(detergent)/MS(within) ~ F(df(detergent), df(within))
MS(temperature)/MS(within) ~ F(df(temperature), df(within))
MS(interaction)/MS(within) ~ F(df(interaction), df(within))

Let's put this into a table to see the results more clearly:

	Df	*SS*	*MS*	*F Value*	*P(Fstat > F)*	**Significant at $\alpha = 0.05$**
Brand	1	20.17	20.17	9.811	0.00576	Yes
Temp	2	200.33	100.17	48.730	0.0000000544	Yes
Interaction	2	16.333	8.17	3.973	0.03722	Yes
Residuals	18	37.00	2.06			
Total	23	273.83				

We can do this in R. We enter the data from Table 15.2.

```
> two.way <- aov(y ~ temp + Brand+ Brand*temp, data = data)
> summary(two.way)
             Df   Sum Sq   Mean Sq   F value       Pr(>F)
Brand         1    20.17     20.17     9.811   0.00576 **
temp          2   200.33    100.17    48.730  5.44e-08 ***
Brand:temp    2    16.33      8.17     3.973   0.03722 *
Residuals    18    37.00      2.06
—
Signif. codes: 0 '***' 0.001 '**' 0.01 '*' 0.05 '.' 0.1 '' 1
```

Interpretation: We find all independent variables and the interaction terms are all significant at $\alpha = 0.05$ in explaining the dependent variable.

Example 5: One- and two-way ANOVA on crops with R.

```
crop.data <- read.csv("crop.csv", header = TRUE, colClasses = c("factor", "factor", "factor",
  "numeric"))
> crop.data
  density block fertilizer  yield
1    1      1       1      177.2287
2    2      2       1      177.5500
3    1      3       1      176.4085
4    2      4       1      177.7036
5    1      1       1      177.1255
6    2      2       1      176.7783
7    1      3       1      176.7463
8    2      4       1      177.0612
9    1      1       1      176.2749
```

10	2	2	1	177.9672
11	1	3	1	176.6013
12	2	4	1	177.0305
13	1	1	1	177.4795
14	2	2	1	176.8741
15	1	3	1	176.1144
16	2	4	1	176.0084
17	1	1	1	176.1083
18	2	2	1	178.3574
19	1	3	1	177.2624
20	2	4	1	176.9188
21	1	1	1	176.2390
22	2	2	1	176.5731
23	1	3	1	176.0393
24	2	4	1	176.8179
25	1	1	1	176.1606
26	2	2	1	177.2264
27	1	3	1	175.9385
28	2	4	1	177.1649
29	1	1	1	175.3608
30	2	2	1	177.2770
31	1	3	1	175.9454
32	2	4	1	175.8828
33	1	1	2	176.4793
34	2	2	2	176.0443
35	1	3	2	177.4125
36	2	4	2	177.3608
37	1	1	2	177.3855
38	2	2	2	176.9758
39	1	3	2	177.3798
40	2	4	2	177.9980
41	1	1	2	176.4349
42	2	2	2	176.9333
43	1	3	2	175.9835
44	2	4	2	177.0341
45	1	1	2	176.4368
46	2	2	2	176.0677
47	1	3	2	177.1210
48	2	4	2	177.1977
49	1	1	2	176.6037
50	2	2	2	177.2082

51	1	3	2	177.1488
52	2	4	2	176.8191
53	1	1	2	176.9991
54	2	2	2	178.1346
55	1	3	2	176.4292
56	2	4	2	176.6683
57	1	1	2	176.8959
58	2	2	2	177.7795
59	1	3	2	176.4145
60	2	4	2	176.8789
61	1	1	2	177.5807
62	2	2	2	176.9573
63	1	3	2	175.7475
64	2	4	2	177.3526
65	1	1	3	177.1042
66	2	2	3	178.0796
67	1	3	3	176.9034
68	2	4	3	177.5403
69	1	1	3	177.0327
70	2	2	3	178.2860
71	1	3	3	176.4054
72	2	4	3	176.4308
73	1	1	3	177.3963
74	2	2	3	176.9256
75	1	3	3	177.0550
76	2	4	3	177.3442
77	1	1	3	177.1284
78	2	2	3	177.1683
79	1	3	3	176.3539
80	2	4	3	179.0609
81	1	1	3	176.3005
82	2	2	3	177.5934
83	1	3	3	177.1152
84	2	4	3	177.7945
85	1	1	3	177.0040
86	2	2	3	178.0369
87	1	3	3	177.7014
88	2	4	3	177.6328
89	1	1	3	177.6523
90	2	2	3	177.1004
91	1	3	3	177.1880
92	2	4	3	177.4053

```
93  1   1   3   178.1416
94  2   2   3   177.7106
95  1   3   3   177.6873
96  2   4   3   177.1182
> summary(crop.data)
density  block  fertilizer  yield
1:48     1:24   1:32        Min.: 175.4
2:48     2:24   2:32        1st Qu.: 176.5
         3:24   3:32        Median: 177.1
         4:24               Mean: 177.0
                            3rd Qu.: 177.4
                            Max.: 179.1
> one.way <- aov(yield ~ fertilizer, data = crop.data)
> summary(one.way)
            Df  Sum Sq  Mean Sq  F value  Pr(>F)
fertilizer   2  6.07    3.0340   7.863    7e−04 ***
Residuals   93  35.89   0.3859
—
Signif. codes: 0 '***' 0.001 '**' 0.01 '*' 0.05 '.' 0.1 '' 1
```

Interpretation: The one way ANOVA shows that the fertilizer is significant.

```
> two.way <- aov(yield ~ fertilizer + density, data = crop.data)
> summary(two.way)
            Df  Sum Sq  Mean Sq  F value    Pr(>F)
fertilizer   2   6.068    3.034    9.073  0.000253 ***
density      1   5.122    5.122   15.316  0.000174 ***
Residuals   92  30.765    0.334
—
Signif. codes: 0 '***' 0.001 '**' 0.01 '*' 0.05 '.' 0.1 '' 1
```

Interpretation: Both the fertilizer and the density are significant factors to the yield.

```
> interaction <- aov(yield ~ fertilizer*density, data = crop.data)
> summary(interaction)
                    Df  Sum Sq  Mean Sq  F value    Pr(>F)
fertilizer           2   6.068    3.034    9.001  0.000273 ***
density              1   5.122    5.122   15.195  0.000186 ***
fertilizer:density   2   0.428    0.214    0.635  0.532500
Residuals           90  30.337    0.337
—
Signif. codes: 0 '***' 0.001 '**' 0.01 '*' 0.05 '.' 0.1 '' 1
```

Interpretation: The interaction term is not significant.

15.3 ANOVA Using lm()

We can run our ANOVA in R using different functions. The most basic and common functions in R that we can use are aov() and lm(). Note that there are other ANOVA functions available, but aov() and lm() are built into R and will be the functions we start with.

Because ANOVA is a type of linear model, we can use the lm() function. Let's see what lm() produces for a fish-size analysis.

```
size <- c(30, 40, 50, 60, 40, 50, 60, 70, 70, 80, 90, 100)
> pop <- c("A", "A", "A", "A", "B", "B", "B", "B", "C", "C", "C", "C")
> lm.model <- lm(size ~ pop)
> summary(lm.model)
```

Call:

```
lm(formula = size ~ pop)
```

Residuals:

```
Min   1Q  Median 3Q Max
-15.0 -7.5 0.0    7.5 15.0
```

Coefficients:

	Estimate	Std. Error	t value	Pr(>|t|)
(Intercept)	45.000	6.455	6.971	6.53e–05 ***
popB	10.000	9.129	1.095	0.30177
popC	40.000	9.129	4.382	0.00177 **

—

Signif. codes: 0 '***' 0.001 '**' 0.01 '*' 0.05 '.' 0.1 ' ' 1

Residual standard error: 12.91 on 9 degrees of freedom

Multiple R-squared: 0.698, Adjusted R-squared: 0.6309

F-statistic: 10.4 on 2 and 9 DF, p-value: 0.004572

First, we see our model call simply reminds us of the model that we fit. Next, we see our residuals that we may want to evaluate for model fit. Assuming the model fit is reasonable, we can now look at some of the statistics from the ANOVA. Under coefficients, we see three categories, although popA is missing and (Intercept) is listed as the first coefficient. This is because the function defaults to an effects parameterization, whereby the first categorical group, popA, is the reference or baseline group and is called the intercept. The coefficient estimate for that group is the first group's mean, but the coefficients for the other groups represent the effect of being in that group—hence, the effect of parameterization. The way to get the group coefficients for all groups after the first group is to add the estimate of the intercept (first group) to each of the other groups. So in the example above, popA has a mean maximum size of 45, while the mean maximum size of popB is 55, and popC has a mean maximum size of 85. It is advisable to ignore the test statistics and p-values for the individual coefficients, as there are better ways to examine them.

With all that being said about the coefficients, you may want to skip immediately looking at the coefficients and look to the bottom of the model summary where the F-statistic is reported along with the p-value on that F-statistic. The F-statistic is the test statistic for ANOVA and is a combination of the sums of squares described above. The associated p-value can help provide a significance interpretation on the F-statistic. That is, a p-value < 0.05 tells us that at least one group mean differs from another at the $\alpha = 0.05$ level of significance and a p-value <0.01 tells us the group means differ at an a-level of 0.01. What does all this mean? It means if you run your ANOVA and skip to the p-value, a p-value >0.05 suggests no group means differ from each other and you may be done with that model. If the p-value < 0.05, then you have at least one group that differs from the other(s), and additional steps need to be taken to quantify those differences.

15.4 ANOVA Using aov()

A simple and perhaps preferred way to do an ANOVA in R is to use the aov() function. Let's try that function on the same model we examined above with the lm() function.

```
aov.model <- aov(size ~ pop)
summary(aov.model)
aov.model <- aov(size ~ pop)
> summary(aov.model)
            Df  Sum Sq  Mean Sq  F value      Pr(>F)
pop          2    3467   1733.3     10.4  0.00457 **
Residuals    9    1500    166.7
—
Signif. codes: 0 '***' 0.001 '**' 0.01 '*' 0.05 '.' 0.1 ' ' 1
>
```

It is worth noting that your categorical variable in the aov() needs to be a factor. For example, you may have categorical groups labeled 1–10, but if those labels are numeric or integer in the eyes of R, then they won't work in aov(). Fortunately, the as.factor() wrapper usually does the trick.

Back to the aov() output. First, notice it is greatly reduced compared to the lm() output. This is because aov() just reports on the ANOVA-specific information, which again is somewhat basic to start. We can clearly see the F-statistic along with the p-value—and we are happy to see that they match the F-statistic and p-value from the lm output. We get a little different information in the aov() function, however. We get the degrees of freedom and the sum of squares and mean squares, which you may want for your model reporting.

We may want the ANOVA coefficients, which are not included in the summary. Fortunately, those coefficients can easily be done by creating subsets the model object.

```
aov.model$coefficients
## (Intercept)  popB  popC
##       45      10     40
```

And again, they match the lm output, as expected. Also note that despite the relatively basic model summary for aov, there is much information in the model object, much of it beyond what the basic modeler will need.

We have an ANOVA that has detected a significant effect of the factor, which in this case is population. We know this because the p-value <0.05. We could simply report this and call it done, but chances are you or your audience will want to know *which* groups differ from each other. Recall that we cannot just infer this from a visual of the data, but fortunately, there are statistical tests to help us understand the group differences.

15.5 References and Further Readings

Devore, J. (2012). *Probability and Statistics for Engineering and the Sciences*, 8th ed., Cengage Publisher, Belmont, CA, pp. 211–217.

Neter, J., W. Wasserman and M. Kutner (1996). *Applied Linear Statistical Models*, 4th ed., Irwin Press, Chicago, IL, pp. 531–547.

Zelazo, P., N. Zelazo and S. Kolb (1972). Walking in the Newborns, *Science*, **176**, pp. 314–315.

15.6 Chapter 15 Exercises

15.1 Zelazo et al. (1972) investigated the variability in age at first walking in infants. Study infants were grouped into four groups, according to reinforcement of walking and placement: (1) active, (2) passive, (3) no exercise, and (4) 8-week control. Sample sizes were 6 per group, for a total of $n = 24$. For each infant, study data included group assignment and age at first walking, in months. Perform a one-way ANOVA. The following are the data and consist of recorded values of age (months) by group:

Active Group	Passive Group	N-Exercise Group	8-Week Control
9.00	11.0	11.5	13.25
9.50	10.0	12.0	11.5
9.75	10.0	9.0	12
10.0	11.75	11.5	13.5
13.0	10.50	13.25	11.5
9.50	15.0	13.0	12.35

Source: Zelazo et al. (1972) "Walking in the newborn." *Science* 176: 314–315.

15.2 Fish growth as a function of light and temperature. Perform both a one-way and two-way ANOVA.

Light (light)	Water Temp (temp)	Fish Growth (grwoth)
1=low	1=cold	4.55
1=low	1=cold	4.24
1=low	2=lukewarm	4.89
1=low	2=lukewarm	4.88
1=low	3=warm	5.01
1=low	3=warm	5.11
2=high	1=cold	5.55
2=high	1=cold	4.08
2=high	2=lukewarm	6.09
2=high	2=lukewarm	5.01
2=high	3=warm	7.01
2=high	3=warm	6.92

15.3 Suppose the National Transportation Safety Board (NTSB) wants to examine the safety of compact cars, midsize cars, and full-size cars. It collects a sample of three for each of the treatments (cars types). Using the hypothetical data provided below, test whether the mean pressure applied to the driver's head during a crash test is equal for each types of car. Use $\alpha = 5\%$.

Compact	Midsize	Full-Size
643	469	484
655	427	456
702	525	402

15.4 Times that three workers required to perform an assembly-line task were recorded of five randomly selected occasions. Times were rounded to the nearest minute.

Tom	Dick	Susan
8	8	10
10	9	9
9	9	10
11	8	11
10	10	9

15.5 Patient response times versus various new drugs for 15 patients across three drugs. Perform a one-way ANOVA at $\alpha = 0.05$.

Drug 1	5.9	5.92	5.91	5.89	5.88
Drug 2	5.5	5.50	5.50	5.49	5.50
Drug 3	5.01	5	4.99	4.98	5.02

15.6 Perform a one-way ANOVA at $\alpha = 0.05$.

Group 1	Group 2	Group 3
51	23	56
45	43	76
33	23	74
45	43	87
67	45	56

15.7 Three different traffic routes are tested for mean driving time. The entries in the table are the driving times in minutes on the three different routes. Conduct a one-way ANOVA, and interpret.

Route 1	Route 2	Route 3
30	27	16
32	29	41
27	28	22
35	36	31

15.8 Suppose a consumer group is interested in determining the age at which teenagers obtain their drivers licenses. The following data are collected from five regions of the country on the average age a teenager in their region gets their license.

Northeast	South	West	Midwest	East
16.3	16.9	16.4	16.2	17.1
16.1	16.5	16.5	16.6	17.2
16.4	16.4	16.6	16.5	16.6
16.5	16.2	16.1	16.4	16.8

15.9 Five basketball teams took a random sample of players on how high each player can jump vertically. Perform an ANOVA, and interpret.

Team 1	Team 2	Team 3	Team 4	Team 5
36	32	48	38	41
42	35	50	44	39
51	38	39	46	40

16

Two-Way ANCOVA Using R

Objectives

1. **Know ANCOVA procedures.**
2. **Know how to implement the procedures in R.**
3. **Know how to interpret the results of each procedure.**

What is ANCOVA?

Analysis of covariance (ANCOVA) is a general linear model that blends ANOVA and regression. According to Neter et al. (1996), it is used for either observational studies or designed experiments. Basically, it augments the ANOVA model containing the factor effects of one or more additional quantitative variables that are related to the dependent variable. The purpose is to reduce the variance in the error term of the model. Thus, this will make the model more precise. We treat covariance models as a special case of regression model.

In this chapter, we will introduce single-factor as well as multifactor analysis.

ANCOVA evaluates whether the means of a dependent variable (DV) are equal across levels of a categorical independent variable (IV), often called a treatment, while statistically controlling for the effects of other continuous variables that are not of primary interest, known as covariates (CV) or nuisance variables. Mathematically, ANCOVA decomposes the variance in the DV into variance explained by the CV(s), variance explained by the categorical IV, and residual variance. Intuitively, ANCOVA can be thought of as "adjusting" the DV by the group means of the CV(s).

The ANCOVA model assumes a linear relationship between the response (DV) and covariate (CV) shown in Equation 16.1:

$$y_{ij} = \mu + \tau_i + \mathrm{B}\left(x_{ij} - \bar{x}\right) + \varepsilon_{ij}. \quad (16.1)$$

where:

y_{ij} is the jth observation under the ith categorical group;

τ_I are the fixed treatment effects subject to the restriction $\Sigma\, \tau_I = 0$

B is a regression coefficient for the relationship of Y and X.

X_{ij} are constants

$\bar{x}$ is the mean of the X_{ij}

ε_{ij} are the error terms as independent $N(0, \sigma^2)$

$i = 1, 2 \ldots r, j = 1, 2, \ldots, n$

Under this specification, the categorical treatment effects sum to zero. The standard assumptions of the linear regression model are also assumed to hold.

DOI: 10.1201/9781003317906-16

16.1 ANCOVA vs. Regression

Both ANCOVA and regression are statistical techniques and tools. ANCOVA and regression share many similarities but also have some distinguishing characteristics. Both ANCOVA and regression are based on a covariate, which is a continuous predictor variable.

ANCOVA stands for analysis of covariance. It is a combination of one-way ANOVA (analysis of variance) and linear regression, a variant of regression. It deals with both categorical and continuous variables. It is a specific statistical method for determining the extent of the variance of one variable that is due to the variability in some other variable.

ANCOVA is basically ANOVA with more sophistication and the addition of a continuous variable to an existing ANOVA model. Another form of ANCOVA is MANCOVA (multivariate analysis of covariance). Moreover, ANCOVA is a general linear model that has a continuous outcome variable and two or more predictor variables. The two predictor variables are both continuous and categorical variables.

In a continuous variable, the data is quantitative and scaled, while categorical data is characterized as nominal and non-scaled. ANCOVA is mainly used to control factors that cannot be randomized but can still be calculated on an interval scale in experimental designs, while on the observational designs, it is used to erase the variable effects that change the relationship between categorical independents and interval dependents. MANCOVA also has some use in regression models, where its main function is to fit the regressions in both categorical and interval independents. In summary:

1. ANCOVA is a specific linear model in statistics. Regression is also a statistical tool, but it is an umbrella term for a multitude of regression models. Regression is also the name from the state of relations.
2. ANCOVA deals with both continuous and categorical variables, while regression deals only with continuous variables.
3. ANCOVA and regression share one particular model: the linear regression model.
4. Both ANCOVA and regression can be done using specialized software to perform the actual calculations.
5. ANCOVA came from the field of agriculture, while regression originated from the study of geography.
6. Thus, this technique answers the question: are mean differences or interactive effects likely to have occurred by chance after scores have been adjusted on the dependent variable because of the effect of the covariate?

16.2 Introduction to the ANCOVA Process by Example

We'll use a sea otter predation example to walk our way through how to carry out an ANCOVA in R. The data gives the year and the approximate number of sea otters in the bay and the lagoon.

Year	1	2	3	4	5	6	7	8	9	10	11	12
Bay	235	220	211	225	234	199	220	210	199	202	201	183
Lagoon	178	175	162	174	161	172	168	178	170	155	149	168

We will convert specified columns to add codes for the bay (1) and lagoon (2).

The data are laid out with the response variable in one column and two additional columns for the predictor variables: one contains the codes for the categorical variables; the other contains the values of the numeric variable. Thus, the first column (Otters) contains the sea otter abundances, the second column (Location) contains the codes for the study population (levels: "Lagoon" and "Bay"), and the third column (Year) contains the observation year (1993–2004).

16.2.1 Visualizing the Data

As always, we should start by visualizing our data. Here, we use the ggplot command.

```
ggplot(seaotters, aes(x = Year, y = Otters, colour = Location)) +
geom _ point()
```

Figure 16.1 suggests that sea otter abundances have declined in both locations, with a greater decline where sea otters were exposed to being hunted by killer whales (the bay location).

We should also consider the assumptions of the ANCOVA. The scatterplot suggests that, within each location, the relationship between x and y is linear. The numeric predictor variable (study year) is measured on an interval scale, and the response variable (otter abundance) is measured on ratio scale. Year is obviously measured without error. What about the independence assumption?

16.2.2 Independence

Can you think of any reasons why the independence assumption may be problematic in this example? Think about how the data have been collected—they are a time series of abundances in a pair of adjacent populations.

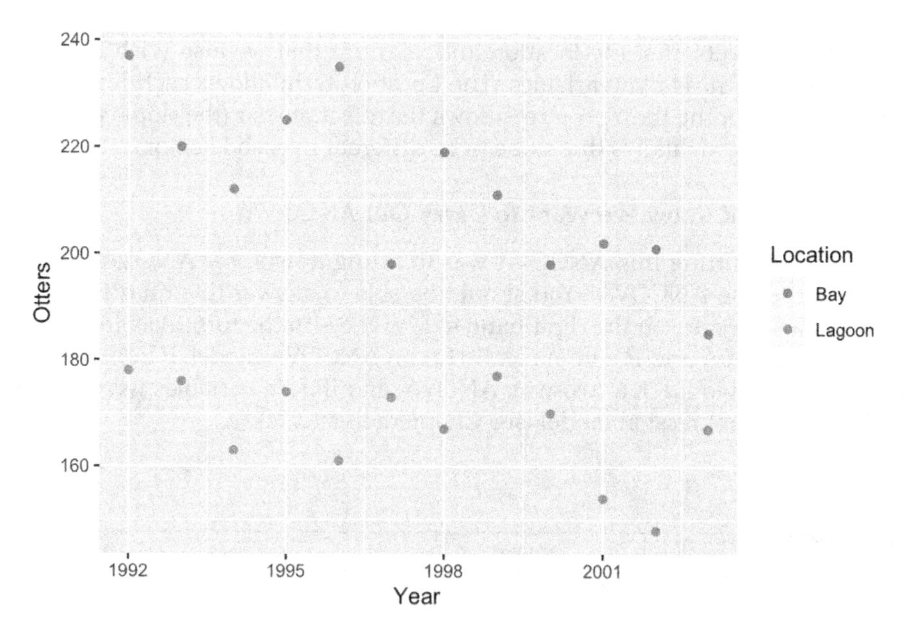

FIGURE 16.1
Plot of sea otters in the bay and lagoon from 1992 to 2004.

We'll assume the independence assumption has been met in these data. As with regression, the remaining assumptions are probably best measured using regression diagnostics after we've fitted the model.

16.3 Fitting an ANCOVA

As you should expect by this point, carrying out ANCOVA in R is a two-step process. The first step is the model-fitting step. This is where R calculates the best-fit intercepts and slopes for each group (i.e., each location in this example), along with additional information needed to carry out the evaluation of significance in step 2.

We carry out the model-fitting step using the lm function:

```
otters.model <- lm(Otters ~ Location + Year + Location:Year, data = Otter)
```

This is just more of the same model fitting with lm as we did in the regression chapters (Chapters 13 and 14). We assigned two arguments:

1. The first argument is a **formula**. The variable name on the left of the ~ must be the response variable (Otters), and the terms on the right must only include the two predictor variables (Location and Year).
2. The second argument is the name of the data frame that contains the variables listed in the formula:

   Otters ~ Location + Year + Location: Year

There are three terms, each separated by a + symbol: the two main effects (Location and Year) and their interaction (Location: Year). This tells R that we want to fit a model accounting for the main effects of study location and year but that we also wish to include the interaction between these two variables. The Location term allows each line to cross the y-axis at a different point, the Year term allows the effect of year (the slope) to be nonzero, and the interaction term allows this slope to be different in each location.

16.3.1 How Does R Know We Want to Carry Out ANCOVA?

Notice how similar fitting this ANCOVA was to fitting a two-way ANOVA. How does R know we want to use ANCOVA? You should be able to answer this question. R looks at what type of variables are on the right-hand side of the ~ in the formula. Since Location is a factor and Year is numeric, R automatically fits an ANCOVA model. If both variables had been factors, then R would fit a two-way ANOVA, and if both variables were numeric, we would fit a multiple regression model (see Chapters 13 and 14).

16.4 Diagnostics

Before we go on to look at the *p*-values, we should check the remaining assumptions using the diagnostics. We'll make the same plots as if we'd fitted a linear regression.

First, we'll evaluate the linearity assumption by constructing a **residuals vs. fitted values** plot, Figure 16.2.

```
plot(otters.model, add.smooth = FALSE, which = 1)
```

There's no evidence of a systematic trend (as in our chapter on regression with linear, curve, etc.) here, so the linearity assumption is fine. We'll move on to the normality assumption next, by making a normal probability plot, Figure 16.3. Here we are looking for linearity.

```
plot(otters.model, which = 2)
```

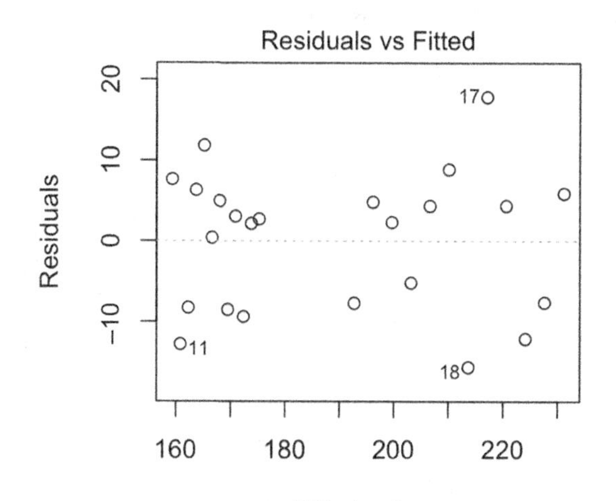

FIGURE 16.2
Residual plot of errors versus fitted model.

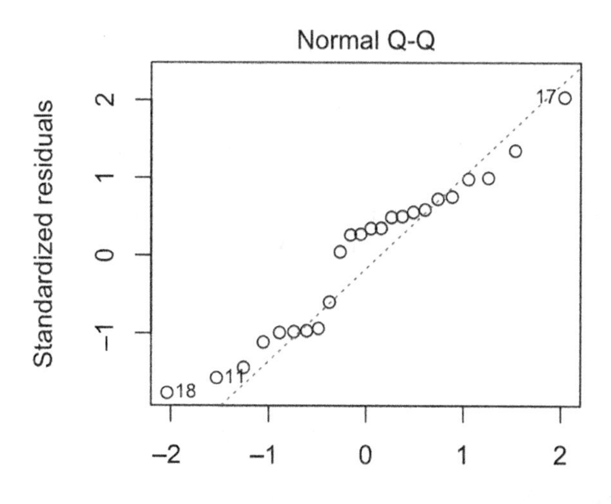

FIGURE 16.3
Normal Q–Q plot.

This doesn't look great, very few of the points are on the dashed line, and there appears to be a systematic trend away from the line. We'll carry on for now, as we're just using this as an example of how to carry out an ANCOVA. If we were really interested in the results of this analysis, we should consider transforming our response variable.

16.4.1 Normality Assumption

Can you think of any reason that we might expect the residuals in these data not to be normally distributed? What kind of transformation might help?

Finally, we'll consider the constant variance assumption using the **scale location plot**.

```
plot(otters.model, add.smooth = FALSE, which = 3)
```

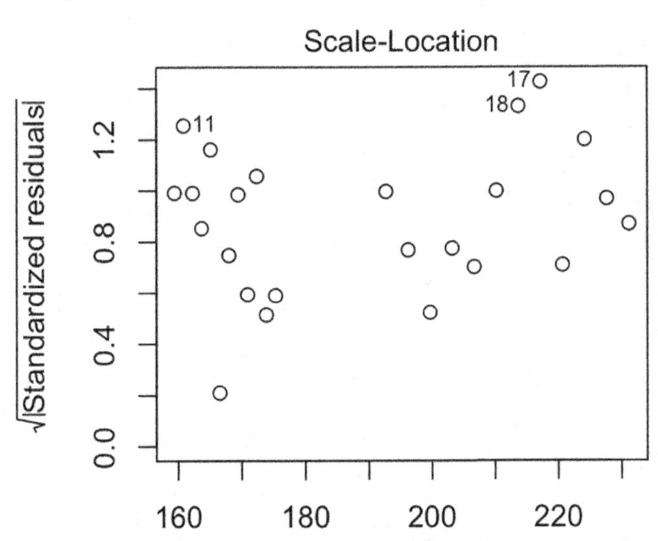

Fitted values
lm(Otters ~ Location + Year + Location:Year)

Here, we're on the lookout for a systematic pattern in the size of the residuals and the fitted values—does the variability go up or down with the fitted values? There doesn't appear to be a strong pattern here.

16.5 Interpreting the Results

Next, we use the ANOVA function in R to determine whether the main effects and the interaction are significant, by passing it the name of the fitted regression model object (otters.model):

```
anova(otters.model)
result <- aov(Otters~year + location + year * location, data = input)
> print(summary(result))
               Df  Sum Sq  Mean Sq  F value      Pr(>F)
year            1    1558     1558   17.762  0.000426 ***
location        1   11660    11660  132.940  2.75e-10 ***
year:location   1     318      318    3.624    0.071450.
Residuals      20    1754       88
—
Signif. codes: 0 '***' 0.001 '**' 0.01 '*' 0.05 '.' 0.1 '' 1
```

The first line reminds us that we are looking at an ANOVA table. Remember, this doesn't necessarily mean we are dealing with an ANOVA model—we are definitely examining an ANCOVA because of the type of independent variables in the model. The second line reminds us what variable we analyzed (i.e., the response variable). The critical part of the output is the table at the end:

```
result <- aov(Otters~year + location + year * location, data = input)
> print(summary(result))
               Df  Sum Sq  Mean Sq  F value      Pr(>F)
year            1    1558     1558   17.762  0.000426 ***
location        1   11660    11660  132.940  2.75e-10 ***
year:location   1     318      318    3.624    0.071450.
Residuals      20    1754       88
—
Signif. codes: 0 '***' 0.001 '**' 0.01 '*' 0.05 '.' 0.1 '' 1
```

This summarizes the parts of the analysis of variance calculations, as they apply to ANCOVA. These are Df, degrees of freedom; Sum Sq, the sum of squares; Mean Sq, the mean square; *F* value, the *F*-statistic (i.e., variance ratio), Pr(>*F*) (i.e., the *p*-value).

The *F*-statistics (variance ratios) are the key terms. When working with an ANCOVA, these relate to how much variability in the data is explained when we include each term in the model, taking into account the degrees of freedom it "uses up." Larger values indicate a stronger effect. The *p*-value gives the probability that the relationship could have arisen through sampling variation, if in fact there were no real association: a *p*-value of less than 0.05 indicates a less than 1 in 20 chance of the result being due to chance, and we take this as evidence that the relationship is real.

We need to interpret these *p*-values. The two main effects are significant ($p < 0.001$), but the interaction is not ($p = 0.0714$). An ANOVA table tells us nothing about the direction of the effects; we have to plot the data to be able to do this. If we look back at the scatterplot, it is apparent that the significant main effects are supporting the observation that otter abundances are higher in the bay area and that, in general, otter abundances have declined over the course of the study. The interaction term is insignificant—though we only just missed the conventional $p < 0.05$ cutoff. We are forced to conclude that the data do not support the hypothesis that the population abundances in each location have declined by different amounts.

16.6 Presenting the Results

We will need to provide a succinct factual summary of the analysis in the results section of the report:

There were significant effects of location (ANCOVA: $F = 132.94$, df = 1,20, $p < 0.001$) and year ($F = 17.762$, df = 1,20, $p < 0.001$) on sea otter abundance. The interaction between location and year was not significant ($F = 3.5$, df = 1,20, $p = 0.0714$). Sea otter abundances were generally higher in the bay but declined by a similar amount in both locations during the study (Figure 16.4).

Notice that we never referred to "treatments" in this summary. It does not make any sense to describe the variables in this data set as treatments, as we are describing the results from an observational study. Of course, there is nothing to stop us using ANCOVA to analyze experimental data if it is appropriate.

For presentation, it is best to present the results as a figure such as in Figure 16.4. We can produce publication-quality figures to summarize ANCOVA in much the same way as we summarize a fitted regression model. We are aiming to produce a figure that shows two pieces of information: a scatterplot and lines of best fit. We also want to differentiate the data and best-fit lines for each location. We know how to produce a scatterplot, so the main challenge is to add the lines of best fit. We use the predict function to do this.

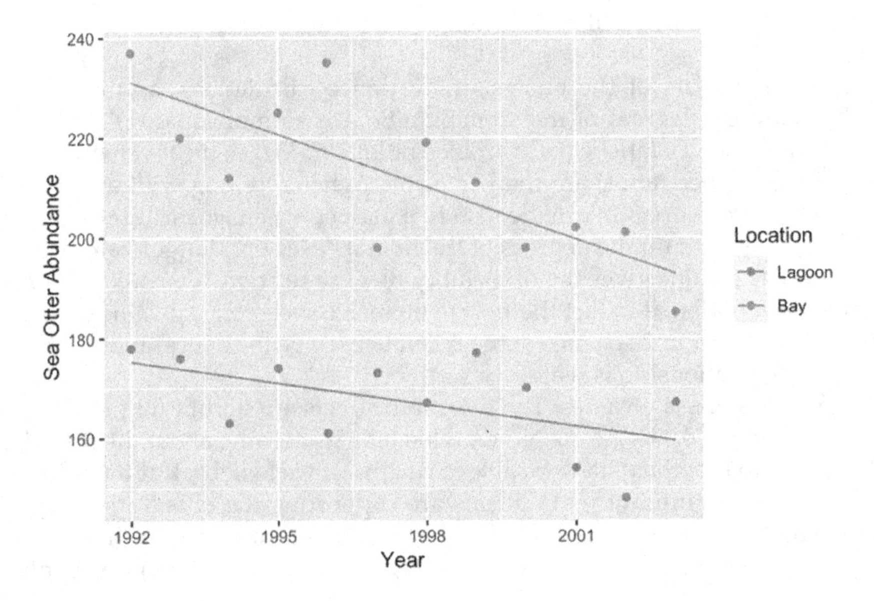

FIGURE 16.4
Data with regression lines added showing downward trends.

16.7 Illustrious Example

We begin with data on temperature and pulse rates.
The data:

Temperature (C°)	Pulses per Second	Temperature (C°)	Pulses per Second
20.8	67.9	17.2	44.3
20.8	65.1	18.3	47.2
24	77.3	18.3	47.6
24	78.7	18.3	49.6
24	79.4	18.9	50.3
24	80.4	18.9	51.8
26.2	85.8	20.4	60
26.2	86.6	21	58.5
26.2	87.5	21	58.9
26.2	89.1	22.1	60.7
28.4	98.6	23.5	69.8
29	100.8	24.2	70.9
30.4	99.3	25.9	76.2
30.4	101.7	26.5	76.1
30.5	100.5	26.5	77
31	102	26.5	77.7
32	103	28.6	84.7

The means are

mean 88.452941 mean 62.4

```
mypdata = read.csv("pulse.csv") # read csv file
> mypdata
location temp pulse
 1 1 20.8  67.9
 2 1 20.8  65.1
 3 1 24.0  77.3
 4 1 24.0  78.7
 5 1 24.0  79.4
 6 1 24.0  80.4
 7 1 26.2  85.8
 8 1 26.2  86.6
 9 1 26.2  87.5
10 1 26.2  89.1
11 1 28.4  98.6
12 1 29.0 100.8
13 1 30.4  99.3
14 1 30.4 101.7
```

```
15  1   30.5 100.5
16  1   31.0 102.0
17  1   32.0 103.0
18  2   17.2  44.3
19  2   18.3  47.2
20  2   18.3  47.6
21  2   18.3  49.6
22  2   18.9  50.3
23  2   18.9  51.8
24  2   20.4  60.0
25  2   21.0  58.5
26  2   21.0  58.9
27  2   22.1  60.7
28  2   23.5  69.8
29  2   24.2  70.9
30  2   25.9  76.2
31  2   26.5  76.1
32  2   26.5  77.0
33  2   26.5  77.7
34  2   28.6  84.7
>

input1<-mypdata
> r2<-aov(pulse~temp + location + temp * location, data = input1)
> print(summary(r2))
               Df  Sum Sq  Mean Sq   F value      Pr(>F)
temp            1   10169    10169  2652.371  <2e-16 ***
location        1     592      592   154.524  2.3e-13 ***
temp:location   1       0        0     0.079      0.781
Residuals      30     115        4
—
Signif. codes: 0 '***' 0.001 '**' 0.01 '*' 0.05 '.' 0.1 '' 1
```

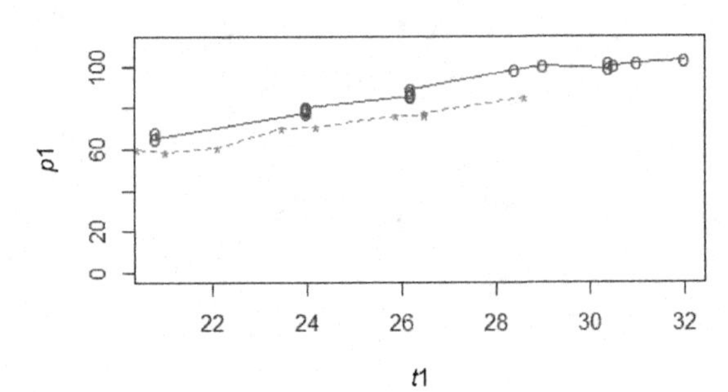

```
> input1<-mypdata
> r2<-aov(pulse~temp + location + temp * location, data = input1)
> print(summary(r2))
               Df  Sum Sq  Mean Sq   F value     Pr(>F)
temp            1   10169    10169  2652.371  < 2e-16 ***
location        1     592      592   154.524  2.3e-13 ***
temp:location   1       0        0     0.079      0.781
Residuals      30     115        4
—
Signif. codes: 0 '***' 0.001 '**' 0.01 '*' 0.05 '.' 0.1 '' 1
>
```

Interpretation: We find here that only location and temperature are significant, and the interaction term is not significant.

16.8 Chapter 16 Exercises

16.1 Given the data on the effects of special diets in pre- and posttreatments for cholesterol in the table below, perform and interpret an ANCOVA.

		Pretreatment	Posttreatment	Cholesterol Levels	
Treatments					
A	A	B	B	C	C
PRE	POST	PRE	POST	PRE	POST
175	135	205	165	210	185
175	145	175	195	170	170
235	205	230	160	235	210
215	175	190	155	185	210
195	140	155	150	180	200
195	190	185	170	220	240

16.2 Three methods of instruction for elementary computer programming.

To assess the relative merits of three methods of instruction for elementary computer programming, a curriculum researcher randomly selected 12 fifth graders from each of three elementary schools in a certain school district. Each group, within the setting of its home school, then received a six-week course of instruction in one or another of the three methods. The following table shows the measure of how well each of the 36 subjects, 12 per group, learned the prescribed elements of the subject matter. Perform and interpret an ANCOVA.

	Method A	Method B	Method C
	29 24 14	15 28 13	32 27 15
	27 27 28	36 29 27	23 26 17
	27 32 13	31 33 32	25 14 29
	35 32 17	15 30 26	22 30 25
means	25.4	26.3	23.8

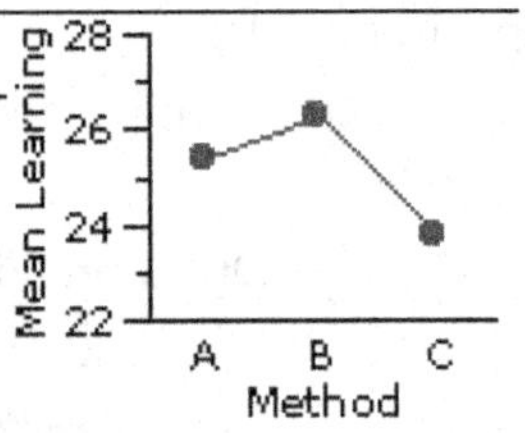

Mean Learning
28
26
24
22
A B C
Method

16.3 Given data for the use of drugs and placebo for treatment,

Placebo	3	4
	2	1
	5	5
	2	1
	2	2
	2	2
	7	7
	2	4
	4	5
Low Dose	7	5
	5	3
	3	1
	4	2
	4	2
	7	6
	5	4
	4	2
High Dose	9	1
	2	3
	6	5
	3	4
	4	3
	4	3
	4	2
	6	0
	4	1
	6	3
	2	0
	8	1
	5	0

16.4 Suppose an investigator is investigating the effects on two treatments to improve achievement test scores. He has 24 students and decides to separate them into two groups of 12 to compare treatment A and treatment B on scores. IQ scores for each are also available and provided. Perform and interpret an ANCOVA model.

Group A		Group B	
IQ	Score	IQ	Score
100	23	96	19
113	31	108	26
98	35	122	31
110	28	103	22
124	40	132	36
135	42	120	38
118	37	111	31
93	29	93	25
120	34	115	29
127	45	125	41
115	33	102	27
104	25	107	21

16.5 Three groups of six people are being compared by treatment as to how the variable x affects y. Perform and interpret an ANCOVA model.

Group 1		Group 2		Group 3	
x	y	x	y	x	y
12	26	11	32	6	23
10	22	12	31	13	35
7	20	6	20	15	44
14	34	18	41	15	41
12	28	10	29	7	28
11	26	11	31	9	30

16.6 Perform and interpret an ANCOVA model. Let gender 1 be a female and gender 2 a male.

Gender	Years	Interact	Salary
1	5	5	80
1	3	3	50
1	2	2	30
1	1	1	20
1	4	4	60
2	3	6	78
2	1	2	43
2	5	10	103
2	2	4	48
2	4	8	80

16.9 References and Further Readings

Devore, J. (2012). *Probability and Statistics for Engineering and the Sciences*, 8th ed., Cengage Publisher, Belmont, CA, pp. 211–217.

Neter, J., W. Wasserman and M. Kutner (1996). *Applied Linear Statistical Models*, 4th ed., Irwin Press, Chicago, IL, pp. 531–547.

Appendix A: Labs/Projects

Lab 1 Descriptive Statistics and Displays

This lab consists of data analysis and uses the following data.

Student Count	Gender (male 0, female 1	Age	Height	Weight	Class (Fr 1, Soph 2, Jr 3,sr 4 N/A)	Time	Family Size
1	1	18	64	125	1	10	4
2	0	20	71	163	3	10	6
3	1	18	64	125	1	7	3
4	0	17	74	150	1	10.5	7
5	1	19	70	200	3	10	6
6	0	19	72	175	1	12.5	4
7	1	19	70	140	2	8	6
8	1	18	65	135	1	8	4
9	0	20	73.5	170	3	8	6
10	0	18	72	159	1	7	4
11	0	19	72	150	1	15	5
12	1	19	62	145	1	12	6
13	1	18	67	135	1	10	3
14	1	18	66	210	1	11	6
15	0	19	74	175	2	10	4
16	0	19	71	160	1	15	4
17	0	19	67	177	1	10	3
18	0	18	73	175	1	9	4
19	1	18	65	120	1	10	4
20	1	18	70	145	1	14	4
21	1	19	67	140	1	15	6
22	0	19	70	150	1	8	5
23	0	19	69	140	1	8	4
24	1	18	65	120	1	10	5
25	1	18	70	135	1	5	6
26	1	19	61	150	2	10	5
27	1	18	67	150	1	6	5
28	0	19	69.5	140	2	5	6
29	0	22	68	230	4	6	4
30	0	19	70	135	1	10	3
31	1	19	63	150	1	18	4
32	1	18	65	115	1	10	5
33	0	19	72	162	2	10	4
34	0	18	70	145	1	11	4
35	0	18	72	190	1	20	7
36	1	20	64	160	1	16	6

Using the data for gender and the data for class (freshman-other), obtain a frequency and a relative frequency table, and then produce either by hand or by technology using either the frequency or relative frequency information the following displays.

a) Pie chart

b) Bar chart

1. Which graph, in your opinion better describes the qualitative data for each data set?
2. Using the data for weight produce (a) a histogram using Excel for weight and also a stem and leaf plot by hand. Comment about the shape for each plot in terms of symmetry.

3. Using the data as requested, answer the following:
   a) Use the data for gender, find the appropriate descriptive statistics: mean, median, mode, variance, standard deviation, range, Q1, Q2, and Q3. Which measure of location best describes this data?
   b) Using the weight data, find the descriptive statistics: mean, median, mode, variance, standard deviation, range, Q1, Q2, and Q3. Which measure of location best describes this data?
   c) Draw, by hand, a boxplot of the weight data.
   d) Separate the weight data by gender. Keep the weights of males and weights of females separate. Find the five-number summary of the male and female weights, and then make, by hand, a side by side boxplot of each on the same axis. Make a comment about what you see in these boxplots.
   e) For the original weight data, find the coefficient of skewness using the formula below, and use the results to comment about the skewness of the weight data. Does it follow your result of question 2, in Part I of this lab.

$$S_k = \frac{3(\bar{x} - \tilde{x})}{s}$$

Lab 2 Probability and Distributions

Assume you are working at a major medical complex. You are asked to examine the patient results to a new drug treatment that is given with their medical care. You will advise management on the results of studies submitted by patients. Recently, you randomly sampled 35 of the medical complex's 200,000 patients over the past year from the same clinic. The patient's final ratings were required after taking the new drug for four days. Consider a poor rating any value greater than or equal to 200.

200 242 216 178 194 198 160 182 182 198

182 238 198 188 166 204 182 178 212 164

230 186 162 182 218 170 200 176 175 202

205 199 210 250 172

REQUIREMENTS:

Part I (Descriptive Statistics).

Write a short descriptive paragraph summarizing what you learned from analyzing (describing) this data. Ensure you explain in terms of patient cholesterol level with this new drug. Include references to the following in your discussion. Include all these computations and graphs as attachments to your paragraph(s).

a) Compute the mean, median, mode, variance, standard deviation, and coefficient of skewness for this new drug testing data.

b) Create a stem and leaf plot of this data.

c) Describe the data set as completely as you can. Use any of your calculated statistics that help in your analysis (you can calculate the coefficient of skewness). Include a brief discussion of any symmetry or lack of symmetry (skewness) seen in this data.

Part II.

d) Show all solution setups and your work. If you use the calculator, tell me what values you put into the calculator. Round only to three decimal places.

e) Using only the data, find the chance (percent) that a randomly selected patient has poor rating (values greater than or equal to 200).

f) Using only the data, find the chance (percent) that a randomly selected patient does not have a poor rating (less than 200).

g) Using your probability found above for a randomly selected patient with medical care, and assuming this probability is the same from trial to trial, compute the probability that 8 out of the next 20 patients will having a poor rating (greater than or equal to 200). We assume success is a poor rating. Assume trials are independent and identical.

(1) State which discrete distribution you are using to solve this problem.

(2) For values of $x = 0, 1, 2, \ldots, 20$, compute $P(X = x)$.

(3) Compute the probability that at most 7 will have a poor rating?

(4) Compute the probability that at least 9 will have a poor rating satisfied?

(5) Compute the probability that exactly 12 have a good rating.

(6) Compute the mean and variance for this discrete distribution.

Part III.

We now want to assume that this data is *continuous data*. Assume this data approximately follows a normal distribution with the mean and variance that you calculated in your descriptive statistics (you may round to two decimal places) Use your answers from Part I (a).

For each probability, show all work, and draw a sketch of a normal distribution indicating the region (shade the region you are finding), find the following probabilities using a normal distribution:

a) That a randomly selected patient has a poor rating.

b) That a randomly selected patient does not have a poor rating.

c) That a randomly selected patient has a reading above 225.

e) That a randomly selected patient has a reading below 210.

f) That a randomly selected patient has a reading between 185 and 210.

g) That a randomly selected patient has a reading between 165 and 185.

h) Find the value of the random variable X and the standard normal random variable Z that corresponds to the 70th percentile for this normal distribution.

i) Find the value of the random variable X and the standard normal random variable Z that corresponds to the middle 90th percentile for this normal distribution.

Lab 3 Central Limit Theorem

One of the most important theorems in all of statistics is called the *Central Limit Theorem* or the *Law of Large Numbers*. The introduction of the Central Limit Theorem requires examining a number of new concepts as well as introducing a number of new commands in the R programming language. Consequently, we will break our introduction of the Central Limit Theorem into several parts.

In this first part of the introduction to the Central Limit Theorem, we will show how to draw and visualize a sample of random numbers from a distribution. From there, we will examine the mean and standard deviation of the sample and then examine the distribution of the sample means.

We begin by learning how to draw random numbers from a distribution.

The Letter r: Drawing Random Numbers

We introduced the use of the letters "d," "p," and "q" in relation to the various distributions (e.g., normal, uniform, and exponential). A reminder of their use follows:

- "d" is for "density." It is used to find values of the probability density function.
- "p" is for "probability." It is used to find the probability that the random variable *lies to the left* of a given number.
- "q" is for "quantile." It is used to find the quantiles of a given distribution.

There is a fourth letter, namely "r," that is used to draw random numbers from a distribution. So, for example, **runif** and **rexp** would be used to draw random numbers from the uniform and exponential distributions, respectively.

Let's use the **rnorm** command to draw 500 numbers at random from a normal distribution having mean 100 and standard deviation 10.

1) Create 500 numbers at random from an exponential distribution having a mean of 10 and a standard deviation of 10. Here is the command in R:

```
> x = rexp(500,rate = 1/10)
```

2) Obtain a histogram, and comment about the shape. It should be skewed, isn't it?

```
> hist(x, prob = TRUE)
```

The Distribution of Sample Means

3) "What is the mean of our sample?"

```
> mean(x)
```

4) What is the standard deviation of our sample?

```
> sd(x)
```

5) Of course, if we take another sample of 500 random numbers from an exponential distribution with mean 10, we get a new sample that has a different mean. Try the following commands, and obtain the mean. Is it different?

```
> x = rexp(500,rate = 1/10)
> mean(x)
```

Producing a Vector of Sample Means

In the next activity, we will repeatedly sample from the normal distribution. Each sample will select five random numbers from the normal distribution having mean 100 and standard deviation 10. We will then find the mean of the five numbers in our sample. We will repeat this experiment 500 times, collecting the sample means in a vector **xbar** as we go.

We begin by declaring the mean and standard deviation of the distribution from which we will draw random numbers. Then we declare the *sample size* (the number of random numbers drawn).

```
> mu = 10;
> n = 30
> xbar = rep(0,500)
```

The **rep** command "repeats" the entry 0 500 times. As a result, the vector **xbar** now contains 500 entries, each of which is 0.

It is easy to draw a sample of size $n = 30$ from the exponential distribution having mean $\mu = 1/10$. We simply issue the command **rexp(n, ate = 1/mu)**. To find the mean of this result, we simply add the adjustment **mean(rexp(n, ate = 1/mu)**. The final step is to store this result in the vector **xbar**. Then we must repeat this same process an additional 499 times for a total of 500 sample means. This requires the use of a **for** loop.

```
> for (i in 1:500) { xbar[i] = mean(rexp(n, mean = 1/mu)) }
```

The **for** construct used by R is similar to the "for loops" used in many programming languages.

- The **i** in **for (i in 1:500)** is called the *index* of the "for loop."
- The index **i** is first set equal to 1; then the "body" of the "for loop" (the part between the curly braces) is executed. On the next iteration, **i** is set equal to 2, and the body of the loop is executed again. The loop continues in this manner, incrementing the index **i** by 1, finally setting the index **i** to 500, upon which the body of the loop executes one last time. Then the "for loop" is terminated.
- In the body of the "for loop," we have **xbar[i] = mean(rexp(n, rate = 1/mu))**. This draws a sample of size $n = 30$ from the exponential distribution; calculates the mean of the sample; and stores the result in **xbar[i]**, the *i*th entry of **xbar**.
- When the "for loop" completes 500 iterations, the vector **xbar** contains the means of 500 samples of size $n = 30$ drawn from the exponential distribution having mean $\mu = 1/10$.

6) Obtain a histogram, and check its symmetry. It is a simple task to sketch the histogram of the sample means contained in the vector **xbar**.

```
> hist(xbar, prob = TRUE, breaks = 12,xlim = c(0,20),ylim = c(0,1))
```

7) Obtain the mean and standard deviation of our **xbar** sample.

Lab 4 Hypothesis Testing

In the March 4, 2018, edition of the *Daily Press,* they ran an article about homicides in Hampton and Newport News. Assume you are working at a major complex. You are asked to examine this data and make some critical observations and test claims by those cities.

Part I.

1) For each set of data: homicides in Hampton and homicides in Newport News obtain the sample mean and sample standard deviation.
2) Since there are 48 data elements in each, state the distribution for the average number of homicides for each city. What theorem are you using?
3) Provide the appropriate mean and standard deviation to use to compute probabilities and hypothesis testing concerning the averages.
4) Find the following probabilities for Hampton:
   a. $P(\overline{X} > 13)$
   b. $P(10 < \overline{X} < 15)$
5) Find the following probabilities for Newport News.
   a. $P(\overline{X} > 21)$
   b. $P(20 < \overline{X} < 25)$
6) Build a 95% confidence interval for each city's average homicides. Interpret the confidence intervals.

Part II. Hypothesis Testing Claims.

7) Hampton claims that their average number of homicides is less than 11. Test this claim using an $\alpha = 0.05$ level of significance. List all steps, and state your conclusion in terms of the context of the problem.
8) Newport News claims that that their average number of homicides is less than 20. Test this claim using an $\alpha = 0.05$ level of significance. List all steps, and state your conclusion in terms of the context of the problem.
9) Local VA legislatures have learned that you know how to compare the means of two samples. They want to know if Newport News has on average twice the number of homicides as Hampton. Use the "matched pairs" differences using $d_i = NN_i - 2 * HH_i$ for each data point. For this project, use d_i as described. You have to modify the correct data to be used to find the differences and not just Hampton-Newport News. Test the claim at both a 95% and 99% level of significance. State the conclusions in the context of the data.

Lab 5 Regression in R

Given the following data

t	1	2	3	4	5	6	7	8	9	10	11	12	13	14	15
y	121	93	82	70	53	50	45	42	35	25	21	13	15	5	6

Using R

(1) Plot the data and discuss the trends.
(2) Obtain the correlation coefficient, and interpret its value in terms of the linear relationship.
(3) Build two models, a) $y = b_0 + b_1x$ and b) $y = b_0 + b_1x + b_2x^2$
Obtain the model and all diagnostics.
(4) Use each to predict the value when $x = 25$, $x = 32$, $x = 100$

Lab 6 Reliability

Consider a stereo system with CD player, AM-FM radio turner, speakers (dual), and power supply as displayed with their reliabilities in the figure below. What assumptions are required by your model? Determine the reliability of the system as drawn.

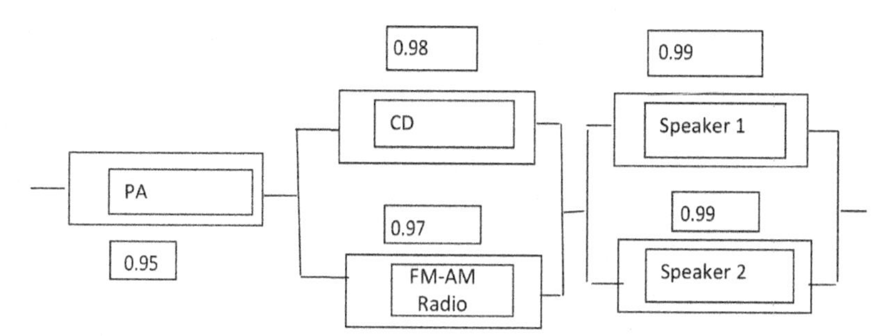

Appendix B: Answers to Selected Exercises

Chapter 2 Exercises

2.1 a) Flip a dime that lands as a "head" or "tail": Categorical
b) The color of peanut M&M's: Categorical
c) The number of calories in the local fast food selections from McDonald's: Quantitative
d) The life expectancy for males in the United States: Quantitative
e) The life expectancy for females in the United States: Quantitative
f) The number of babies born on New Year's eve: Quantitative
g) The dollars spent each month out of the allocated supply budget: Quantitative
h) The number of hours that a woman works per week: Quantitative
i) Amount of car insurance paid per year: Quantitative
j) Whether the bride is older, younger, or the same age as the groom: Categorical
k) The difference in ages of a couple at a wedding: Quantitative
l) Average low temperature in your hometown in January: Quantitative
m) The eye color of a student: Categorical
n) The gender of a student: Categorical
o) The number of intramural sports a person plays per year: Quantitative
p) The distance a bullet travels from a specific weapon: Quantitative
q) The number of roommates in three years: Quantitative
r) The size of your immediate family: Quantitative

2.2 The largest raw number of injuries could be taken as worst, and then when a ratio is formed of injuries/participants, a different largest ratio appears.

2.22 The decimal point is 1 digit(s) to the right of the |

10 | 0515
12 | 1911357
14 | 669
16 | 0
18 | 0

Skewed right

2.23 The decimal point is 1 digit(s) to the left of the |

0 | 05
2 | 23
4 | 0
6 | 225692

Skewed left

2.24 The decimal point is 1 digit(s) to the right of the |

6 | 352
8 | 13234
10 | 56
12 | 15

Slightly skewed right

2.25 Skewed right

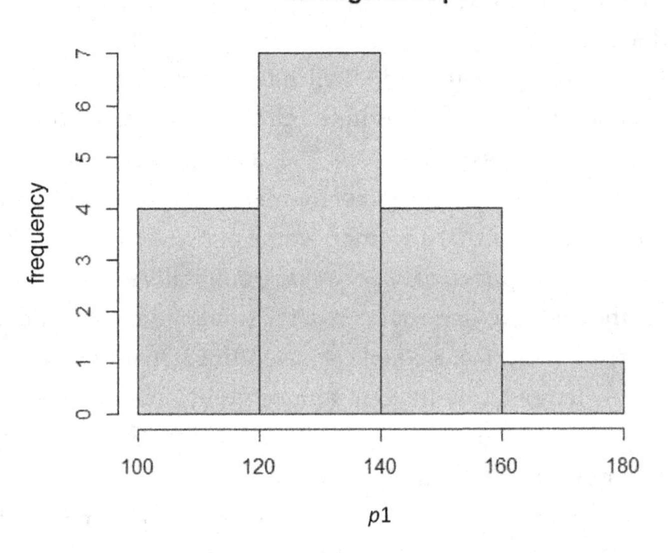

2.26 Skewed left

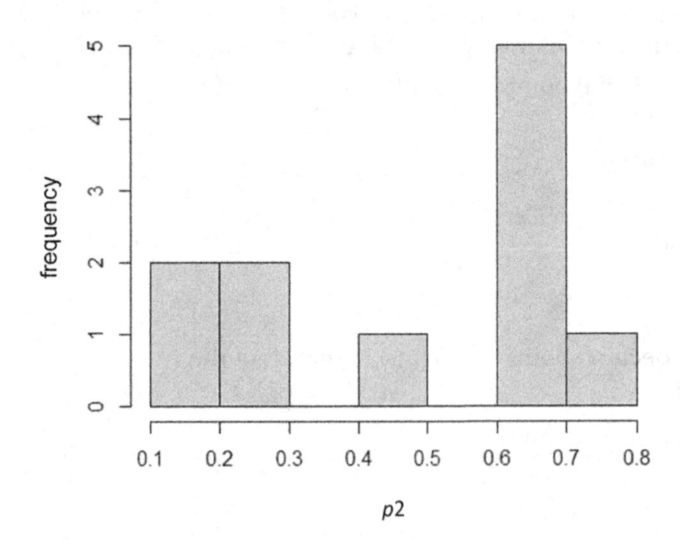

2.27 Slightly skewed right

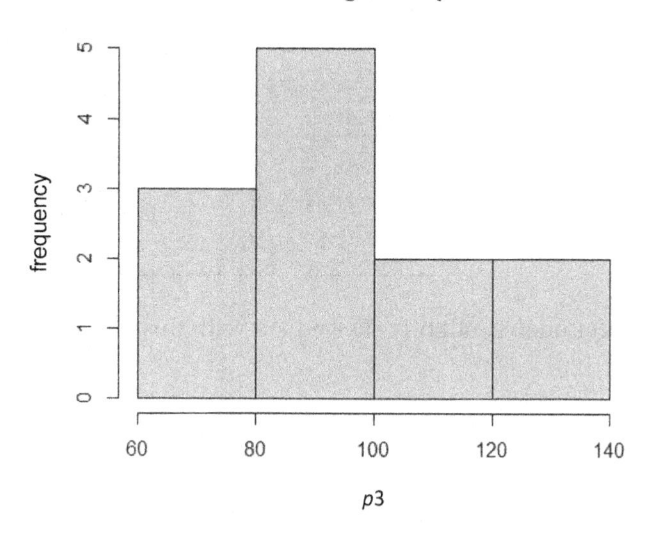

2.28 Skewed right

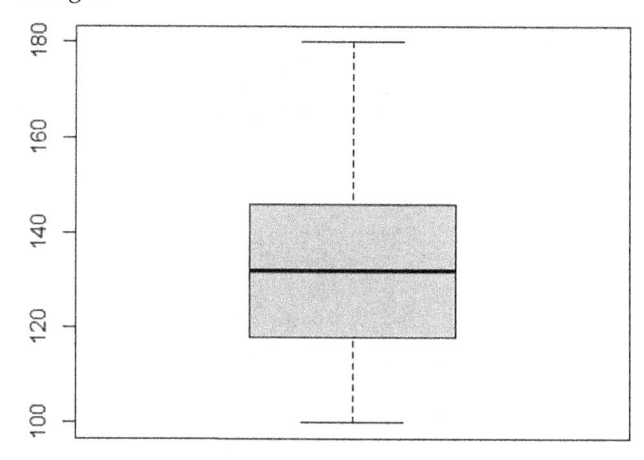

2.29 Skewed left

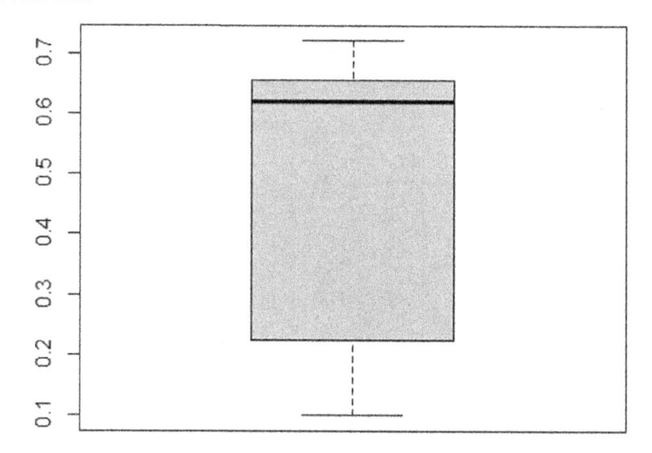

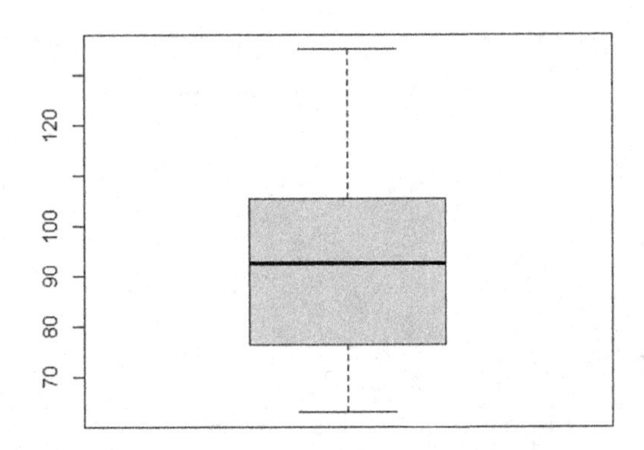

2.30 Almost symmetric, slightly skewed left with this plot.

2.31

```
# Pie Chart with Percentages
slices <- c(43.9,116.7,13.4,17.6)
lbls <- c("Never Married", "Married", "Widowed", "Divorced")
pct <- round(slices/sum(slices)*100)
lbls <- paste(lbls, pct) # add percents to labels
lbls <- paste(lbls, "%", sep = "") # ad % to labels
pie(slices, labels = lbls, col = rainbow(length(lbls)),
  main = "Pie Chart of Women's Status")
```

Pie Chart of Women's Status

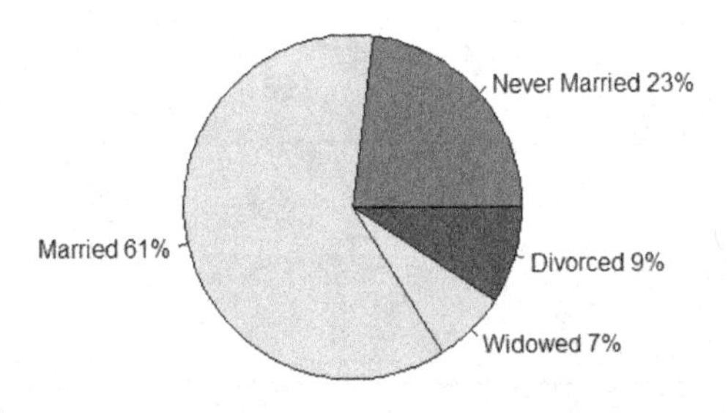

2.32 Symmetric

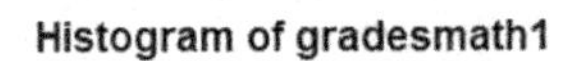

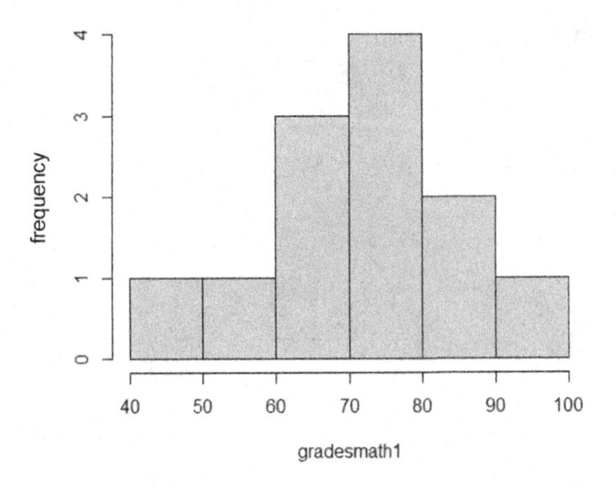

Chapter 3 Exercises

3.1 a. 9 b. 41 c, 0 d. 14

3.3 mean = 2.5 median 2.5 mode {1, 2, 3, 4} or none

3.5 mean = 3, median = 2 mode = 2

3.7 mean = [(1)(1) + (2)*(2) + (7)*(1)]/4 = 3 median = 2 mode = 2

3.13 132, 133, 139, 145, 147, 148, 150,153, 162, 162 (put data in order first)
mean = 147.1, median = 147.5, mode = 162

3.15 mode = 2 (18 times), median = 2, mean = 2.5333 = (152/60)

3.19 (a) cannot find mean as 450* and 500* are not exact. (b) median is the middle value, so 421 is the median.

3.21 n = 800 a. mean = 2.04875 (1639/800) median = 1.5

3.25 Assume this is a sample: range 4–1, variance = s^2 = 1.6666, standard deviation = s = 1.2909

3.27 Range = 7–1, s^2 = 7.3333, s = 2.708

3.29 Data from table is 1, 2, 2, 3, 5, so range = 5–1, s^2 = 2.3000, s = 1.5166

3.31 mean = 146.9, median = 147.5, mode = all or none (each number is seen once), range = 162–132, s = 10.1592, s^2 = 103.209

3.33 n = 20, mean = $(8+20+18)\ /\ 20 = 2.3$ s = 2.2734

3.35 mean = 26.25, median = 22.5, mode = 21, range 50–10, s = 12.35437

3.42 mean = 15.314, s = 1.6767, s^2 = 2.811 all in thousands

3.43 mean = 12, s = 2.7386, s^2 = 7.4999

Chapter 4 Exercises

4.1 a) 0! = 1 b) 1! = 1 c) 11! = 39916800 d) 52! = $8.06517517 \times 10^{67}$
e) 11! – 7! = 39911760 f) 20! = $2.432902008 \times 10^{18}$ g) ${}_{10}P_2 = 90$
h) ${}_{20}P_5 = 1860480$ i) ${}_{10}C_2 = 45$ j) ${}_{20}C_5 = 15504$

4.2 $P(X = 9) = 40/5^{10}$. Here is why.
Denominator is 5^{10} = 9,765,625
Numerator 9 of 10 right 1 answer for 9 questions and (5–1) for 1 = 4
P(guess 9 correct) = 4/9,765,625 = 4.096×10^{-7}
Now, the 9 correct might be any position
WRRRRRRRRR
RWRRRRRRRR
RRWRRRRRRR
RRRWRRRRRR

. . .

RRRRRRRRRW

Thus, we have 10 positions, so the P(9 correct guesses) is 10 * 4/(5^{10}) = 4.096×10^{-6}

4.3 Three aces and two kings in a five card deal. $({}_4C_3)({}_4C_2)/({}_{52}C_5) = (4 * 6) / ({}_{52}C_5) = 24 / ({}_{52}C_5) = 9.23 \times 10^{-6}$

4.4 $P(X \geq 1) = 0.1169$ (see birthday problem in text)

4.5 and 4.6 are tree diagrams

4.7 (4)(5)(6) = 120 ways.

4.8 $({}_{12}P_1 * {}_8P_1) / {}_{20}P_2 = 96/380 = 0.2526$

4.9 (a) 10! / 7! = 10 * 9 * 8 = 720, (b) 12! / 10! 2! = 12 * 11/2 = 66

4.10 4 * 3 * 3 * 4 = 144 ways

4.11

Flip a coin

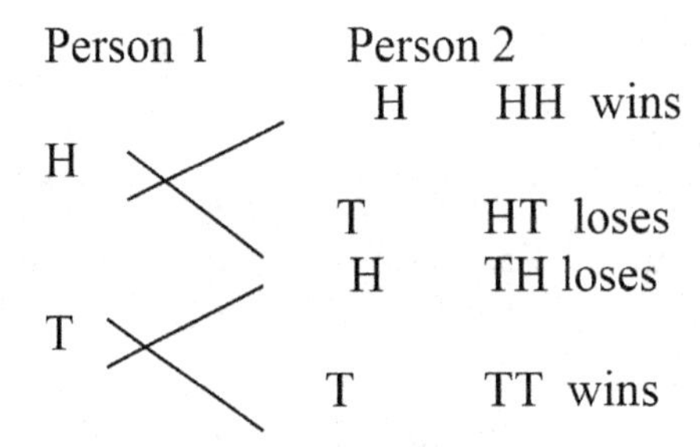

Play 1 times: win P(win) = (1/2) and P(loses) = (1/2)

Three times still P(win) = (1/2)

4.12 2^{20} = 1048576

4.13 a) Order is important, so ${}_{11}P_4 = 7920$

b) order not important, ${}_{11}C_4 = 330$

c) 2 men and 2 females, ${}_5C_2 * {}_6C_2 = 150$

d) 150/330 = 0.4545

4.14 a) 7! = 5040

b) 25! / 20! = 6375600

c) ${}_{12}P_4 = 11880$

d) ${}_7P_7 = 5040 = 7!$

e) 0! = 1

4.15 There are 25 members in the math club, we want to select four of the members.

a) ${}_{25}P_4 = 303600$

b) ${}_{25}C_4 = 12650$

c) ${}_{25}C_6 = 177100$

d) ${}_{15}C_3 * {}_{10}C_3 = 54600$

e) ${}_{15}C_3 * {}_{10}C_3 / {}_{25}C_6$ = 54,600/177,100 = 0.3083

4.16 a) ${}_{10}P_4 = 5040$ b) ${}_{10}C_4 = 210$ c) $\binom{8}{4} = {}_8C_4 = 70$ d) ${}_5P_5 = 5! = 120$
e) $5! = 120$ f) $11!/(11-6)! = 332640$

4.17 a) T b) T c) T

4.18 a) $P(\text{Event } A) = 6/16$ b) $P(\text{Event } B) = 7/16$ c) $P(\text{Event } C) = 8/16$
d) $P(\text{Event } D) = 0$ e) $P(\text{Event } E) = 1/16$

4.19 b d, f, j, k cannot be probability values (some are <0 and others are >1)

4.20 In each situation a–p given below, first clearly circle the correct response if there is replacement (R) or no replacement (NR). Next circle the method you will use to solve the problem: general multiplication rules (M), permutation (P), or combination (C). Finally, calculate the number of ways.

a. NR, C, ${}_{12}C_4 = 495$
b. You flip a fair coin 10 times. R, M, $2^{10} = 1024$
c. You roll a die 5 consecutive times. R, M, $6^5 = 7776$
d. You flip a coin twice and then roll a die. R, M, $2 * 6 = 12$
e. R, M, 480 ways
f. NR, C, ${}_8C_3 = 56$
g. NR, P, ${}_8P_4 = 1680$
h. NR, P, use the multinomial theorem: $11!/(4!)(4!)(2!) = 34650$
i. NR, P, ${}_9P_9 = 9!$ 362880
j. R, M, $9^9 = 387420489$
k. NR, C, ${}_{21}C_{15} = 54264$
l. NR, C, ${}_8C_4 = 70$
m. NR, C, ${}_9C_5 = 126$
n. NR, C, ${}_7C_3 * {}_{11}C_4 * {}_5C_2 = 115500$
o. R, M, $2^{15} = 32768$
p. R, M, $10^4 = 10000$

4.21

a) ${}_{15}C_2 = 105$
b) Both blue is ${}_9C_2 = 36$
c) $P(\text{two blue marbles}) = {}_9C_2 / {}_{15}C_2 = 36/105 = 0.3428$, which also equals $9/15 * 8/14$
d) One marble of each color ${}_6C_1\ {}_9C_1 = 54$; probability then is $54/105 = 0.5143$, since order does not matter.

4.22 (a) How many ways can this be done? ${}_9P_2 = 9 * 8 = 72$ ways
(b) How many ways can the first marble drawn be green? ${}_5P_1$
(c) How many ways can the second marble be green (think)? ${}_4P_1$
(d) What is the probability that both marbles drawn were green? 20/72

4.23 ${}_{14}C_3 * {}_{21}C_2 / {}_{35}C_5 = 0.2354$

4.24 $({}_4C_3 * {}_4C_2) / {}_{52}C_5 = 0.0000092333$

4.25 $3 * 5 * 4 = 60$ ways

4.26 Replacement, so multiplication rule of counting: 5^{10}

4.27 ${}_{10}P_5$

4.28 6!

4.29 9!

4.30 $6 * 6 * 6 = 216$ ways

4.31 144 ways

4.32 4 ways {HH, HT, TH, TT}

4.33 ${}_{10}P_3 = 720$

4.34 $6! = 720$

4.35 $2^8 = 25643$

4.36 (Replication) $2 * 24^3$ and (No replication) $2 * {}_{23}P_3$

4.37 Tree diagram—like three flips of a fair coin

4.40 Let A be McDonald's and B be the movies. Draw the Venn diagram.

$P(A \cap B) = 0.55$

$P(A) = 0.655$

$P(B) = 0.738$

$P(A \cup B) = P(A) + P(B) - P(A \cap B) = 0.655 + 0.738 - 0.55 = 0.843$

$P(\text{Only } A) = 6 = 0.655 - 0.55 = 0.105$

$P(\text{Only } B) = 0.738 - 0.55 = 0.188$

$P(A \cup B)' = 1 - 0.843 = 0.157$

$P(A \cap B)' = 1 - 0.55 = 0.45$

$P(A \cap B') = 0.105$

4.41
a) At least one procedure is done, 0.68
b) Only teeth cleaning is done, 0.05
c) Only a cavity is filled, 0.15
d) Teeth cleaned and cavity filled, no tooth pulled, 0.12
e) No cleaning, filling cavities, or pulling teeth is done at the dentist, 0.32

4.44
a) P(drink Diet Coke) = 190/300
b) P(drink Diet Coke from a can) = 125/300
c) P(drink Pepsi One from a can) = 75/300
d) P(drink a 20 oz. bottle | drink Pepsi One) = 35/110
e) P(drink Diet Coke | drink a 20 oz. bottle) = 125/190
f) Are the events 12 oz. cans and 20 oz. bottles independent? Show work. No they are mutually exclusive.

4.45

4.46 $P(E \mid A) = 0.05$

4.47 $P(E \mid B) = 0.06$

4.48 $P(E' \mid A) = 0.95$

4.49 $P(E' \mid B) = 0.94$

4.50 $P(E \mid C) = 0.07$

4.51 $P(E' \mid C) = 0.93$

4.52 $P(E) = 0.0623$

4.53 $P(E') = 1 - 0.0623 = 0.9377$

4.54 $P(A \mid E) = 0.2006$

4.55 $P(A \mid E') = 0.25327$

4.56 $P(C \mid E) = 0.53932$

4.57 $P(B \mid E') = 0.27066$

4.58 $P(B \mid E) = 0.2600$

4.59 $P(C \mid E') = 0.47605$

4.60 $P(E) = 0.076$

4.61 $P(E) = 0.17508$

4.62 $P(E) = 0.689$ $P(E') = 1 - 0.689 = 0.311$

4.63 $P(E) = 0.1475$, $P(E') = 0.8525$

4.64 (a) What is the probability that a randomly selected American is color-blind? 0.030254

(b) What is the probability that a randomly selected American who is color-blind is female? 0.481 * 0.004/0.0329254 = 0.5029

P(Color-blind) = 0.0355 + 0.001924 = 0.037424

P(Female|Color blind) =
0.001924/0.037424 = 0.051411

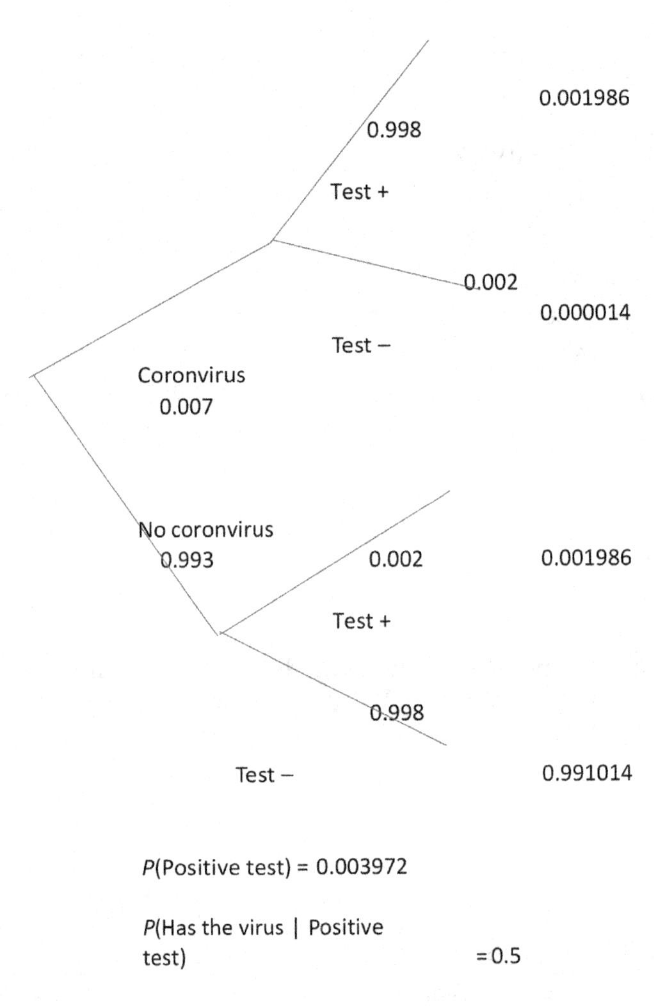

4.66 $P(M) = 0.475848$
$P(\text{advanced degree} \mid M) = 0.0898$
$P(\text{not a high school graduate} \mid M) = 0.16539$

4.68 $P(D) = 0.56312$
$P(\text{Graduated from college} \mid D) = 0.21043$
$P(\text{grade school education } D) = 0.21419$

4.70 $P(\text{polygraph indicated guilty} \mid \text{suspect innocent}) = 0.00346$

Chapter 5 Exercises

5.1 a. $P(X = 5) = 0.058399$
b. $P(X \leq 5) = 0.078126$
c. $\mu = np = 7.5$ and $\sigma^2 = np(1 - p) = 0.25(7.5) = 1.875$

5.2 a. $P(X = 1) = 0.35429$

b. $P(X > 2) = 1 - P(X \leq 1) = 0.114265$

c. mean = 0.6 and variance = 0.54

5.3 PMF

	Lose 3 lbs.	Lose 5 lbs.	Lose 6 lbs.	Lose 8 lbs.	Lose 10 lbs.
$P(X = x)$	0.17	0.34	0.28	0.12	0.09
$P(X \leq x)$	0.17	0.51	0.79	0.91	1.0

$P(X \leq 6) = 0.79$ $P(X \geq 6) = 0.49$

$P(6 \leq X \leq 10) = 1.0 - 0.51 = 0.49$

$m = E[X] = 5.75$

$s^2 = E[X^2] - (E[X])^2 = 36.79 - (5.75)^2 = 3.7275$

5.5 a) We should hire one more additional teller with approximately Ws = 5.96 minutes

b) We should hire one more additional teller with approximately Ws = 4.28 minutes

5.7 $P(X = 7) = 0.2013$

5.9 P(4 candies in a box are pink out of 10) = 0.04009?

5.11 a. All five people are still living. $P(X = 5) = 0.13168$

b. At least three people are still living. $P(X \geq 3) = 1 - P(X \leq 2)$ 0.79

c. Exactly two people are still living. $P(X = 2) = 0.16467$

5.13 $P(X = 3) = 0.2429$ $P(X \geq 1) = 1 - P(X \leq 0) = 0.94368$

5.15 a) $P(X = 3) = 0.024$

b) $1 - P(X = 0) = 0.5563$

5.17 a. No one buys a hot dog. $P(X = 0) = 0$

b. Between 23 and 35 buy hot dogs = 0.55305

c. Between 23 and 35 do not buy hot dogs. = 0.1227

d. mean = (0.7) * 50 = 35 variance (0.7)(0.50)(0.3) = 10.5.

5.19 Given Y is a binomial random variable with $N = 20$ trials and $P = 0.65$, compute the following:

$P(Y = 12)$ = binompdf(20,0.65,12) = 0.16135

$P(Y \leq 12)$ = binomcdf(20,0.65,12) = 0.3989733

$P(Y < 12) = P(Y \leq 11)$ = binomcdf(20,.65,11) = 0.237622

$P(Y > 12)$ = Convert to CDF format with complements = $1 - P(Y \leq 12)$ = $1 - 0.3989733 = 0.6010267$

$P(Y \geq 12)$ = Convert to CDF format with complements = $1 - P(Y < 12)$ = $1 - P(Y \leq 11)$ $1 - 0.237622 = 0.762376$

$P(Y \leq 6)$ = binomcdf(20, 0.65, 6) = 0.0015

$P(Y < 6) = P(Y \leq 5)$ = binomcdf(20, 0.65, 5) = 0.00003

$P(6 < Y < 12) = F(12\text{-}) - F(6) = 0.237622$ – binomcdf(20, 0.65, 6) = 0.237622 – 0.0015 = 0.236122

$P(6< Y \leq 12) = F(12) - F(6) = 0.3989733 - 0.0015 = 0.3974733$

$P(6 \leq Y < 12) = F(12\text{-}) - F(6\text{-}) = 0.237622 - 0.00003 = 0.237619$

$P(6 \leq Y \leq 12) = F(12) - F(6\text{-}) = 0.3989733 - 0.00003 = 0.3989433$

5.20 A shipment of 35 Apple 11 iPhones contains seven that are found defective. Twelve of these 35 are selected at random. Compute the expected value and the variance. What is the probability that exactly 3 are defective? What is the probability that at most 3 are defective? What is the probability that at least 3 are defective? What is the probability that between 3 and 7 are defective (inclusive)?

E[X] = (7/35) 12 = 2.4

V[X] = $[12(7)(35 - 7)(35 - 12)]/35^2(34) = 1.298$

$P(X = 3)$

```
> dhyper(3, 7, 28, 12)
[1] 0.289701
```

$P(X \leq 3)$

```
> phyper(3, 7, 28, 12)
[1] 0.8365586
```

$P(X \geq 3)\ 1 - P(X \leq 2)$

```
> 1 – phyper(2, 7, 28, 12)
[1] 0.4531424
```

$P(3 \leq X \leq 7) = P(X \leq 7) - P(X \leq 2)$

```
> phyper(7, 7, 28, 12) – phyper(2, 7, 28, 12)
[1] 0.4531424
```

5.22 Consider rolling a fair die until our favorite number appears. What is the probability that the first 3 appears in the sixth roll?

$P(X = 3) = (1/6)(5/6)^5 = 0.06697$

5.23 Consider rolling a pair of fair die until our favorite number appears. What is the probability that the first 7 appears in the seventh roll?

$P(X = 7) = (1/6)(5/6)^6 = 0.0558$

5.24 Consider the COVID-19 pandemic. Consider observing patients with COVID-19 until one recovers. What is the probability that here are 100 patients until we see the first recovery if the $P(s) = 1/100$?

$P(X = 10) = (0.01)(0.99)^{99} = 0.003697$

5.25 An oil company has a probability of success of 35% of striking oil when drilling a new oil well on JR Ewing's Texas property. Find the probability that a company drills $x = 6$ *oil* wells to strike oil three times?

$r = 3$

```
> p = 0.35
> n = 6 – r
> #exact
> dnbinom(x = n, size = r, prob = p)
[1] 0.1177455
```

5.26 An oil company has a probability of success of 45% of striking oil when drilling a new oil well on JR Ewing's Texas property. Find the probability that a company drills $x = 6$ *oil* wells to strike oil four times?

$r = 4$

$p = 0.45$

$n = 6 - r$

dnbinom($x = n$, size $= r$, prob $= p$)

[1] 0.1240439

5.31 What % falls between 7.5 and 8.5?

Mean-end point = half interval size

$8 - 7.5 = 0.5$

half interval size/standard deviation = k

$0.5/0.025 = 20$

Since, $k > 1$, we may proceed.

$(1 - 1/k^2) = 99.75\%$

So, at least 99.75% of the data falls in this interval.

5.32 mean = 6.824, $s = 2.8556$

At least 82% of false alarms must fall between _ *a*__ and _*b*__ false calls.

$(1 - 1/k^2) = 0.75$

$-0.25 = -1/k^2$

$k^2 = 1/0.25 = 4$

$k = 2$

So, 75% must lie between two standard deviations of the mean. From above, Sx = 2.856, so two standard deviations is equivalent to (2.856×2) or 5.712.

The mean is 6.824. Thus, 75% fall between $(6.824 - 5.712)$ and $(6.824 + 5.712)$, which is [1.112, 12.536].

5.33 What % of our data in Exercise 2 falls between 3.5 and 10.324?

Mean-end point = half interval size

$6.824 - 3.5 = 3.324$

half interval size/standard deviation = k

$3.324/2/2.8556 = 0.4684$

Since, $k < 1$, we may not proceed.

Chapter 6 Exercises

6.1 (a) $X \sim N(\mu = 10, \sigma = 2)$, $P(X > 6) = 0.97724$ (b) $X \sim N(\mu = 10, \sigma = 2)$, $P(6 < x < 14) = 0.95449$

6.3 $P(X > 0.10) = 0.1586$

6.4 X ~ N(0,1). Compute P (X > 2) =0.0227

6.5 X ~ N(0,1). Compute P (X > 1) = 0.1586

6.6 X ~ N(0,1). Compute P (X > 3) = 0.00134

6.7 X ~ N(0,1). Compute P (0 < X < 2) = 0.4772

6.8 (95% percentile) 14.65 (95% percentile) 13.289

6.9 (95% percentile) 26.59

6.10 (99% percentile) 29.3054

6.11 90% is 1.28155, 95% is 1.64485, 99% is 2.3263

6.12 1.6953

6.13 2.5758

6.14 a) P(x < 40) =0.9937
 b) P(x > 21) = 0.9877
 c) P(30 < x < 35) = 0.39435

6.15 P(X > 100) = 0.1586

6.16 P(50 < X < 70) = 0.4087887

6.18 a) P(4.98 < X < 5.02) = 0.68268
 b) P(4.986 < X < 5.024) = 0.64296

6.20 a) P(X < 19.5) = 0.40129
 b) p(20 < X < 22) = 0.34134

6.21 (a) 0.1586, (b) 0.86444 (c). 0.13567

6.22 a) P(X < $40,000) = 0.3085
 b) P($45,000 < X < $65,000) = 0.37207
 c) P(X > $70,000) = .1586 or 15.86%

6.26 $\int_4^5 f(x)dx = 0.0813$

6.27 0.4965, about 10.39 days

6.28 (a) 0.4965

6.29 0.3934

6.30 P(X < 85) = 0.6867.

6.31 (a) 0.6246 (b) 0.13566

6.32 P(X < 15) = 0.131 so 80 times 001381 = 1.05 calls. P(X < 20) = 0.13326 so 80 * .13326 = 10.66 calls.

6.33 728.155

6.34 (a) 0.3857, (b) 0.0834, (c) 0.41696

6.37 0.3943, 0.8111

6.38 P(X < 85) = 0.4479987

6.40 103.35

Chapter 7 Exercises

7.1 E[X] = $\alpha/(\alpha + \beta) = 1/(1+4) = 1/5$. Since we have 1000 customers, we would expect 1000 * 0.2 = 200 customers to purchase premium gasoline

7.2 $P(X > 0.30)$ = 1-pbeta(0.30, 2, 5) = [1] 0.420175

7.3 $P(T > 10) = e^{-105} = e^{-2} = 0.1353$

$P(X \geq 3) = 1 - P(X < 3) = 1 - P(X \leq 2) = 0.0089$

7.4 $P(T > 9) = e^{-9/6} = 0.2231$

$P(X \geq 2) = 1 - P(X \leq 1) = 1 - 0.7827 = 0.21723$

7.7 (a) less than 110 hours $P(X < 110/365) = 0.0454$

(b) between 65 and 105 hours 0.0255.

(c) Find the expected value of the vacuum. E[X] = 1

7.9

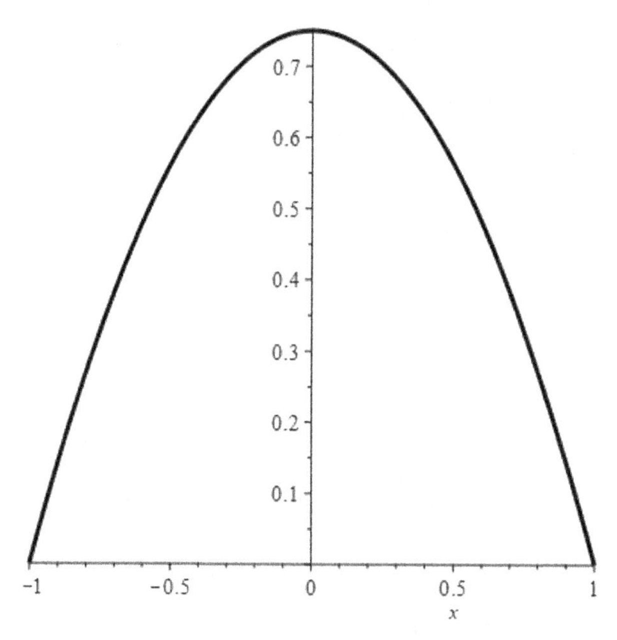

a) sketch the graph of $f(x)$

b) Compute $P(X > 0) = 0.5$

c) Compute $P(-0.5 < X < 0.5) = 0.6875$

d) Compute $P(-0.25 < X \text{ or } X > 0.25) = 0.6328$

Chapter 8 Exercises

8.1 $\bar{X} \sim N(\mu = 10, \sigma = 6)$, $n = 64$ $P(\bar{X} > 9) = 0.9087$

8.2 $\bar{X} \sim N(\mu = 10, \sigma = 6)$, $n = 64$ $P(9 < \bar{X} < 11) = 0.81757$

8.3 $P(X > 0.06) = 0.0227$

8.4 a) $P(\bar{x} < 30.1) = 0.5987$ b) $P(\bar{x} > 30.4) = 0.1587$ c) $P(30.4 < \bar{x} < 30.85)$

8.5 0.01222

8.6 0.4967

8.7 0.1586

8.9 a) less than 10.7 months. 0 b) between 10.7 and 12.1 months. 0.6178

8.11 The probability is 0.0359

Chapter 9 Exercises

9.1 95% CI [7.307, 7.993] does not include 8
99% CI [7.1992, 8.1008] does include 8.

9.2 95% CI [0.5002,-.5198]
99% CI [.49712, 0.52288]

9.3 95% CI [29162, 30830]
99% CI [28899,31101]
$30,000 is in both intervals.

9.4 95% CI [-.1859, 2.5859] 99% [-0.6214, 3.0214]

9.5 8.6 does not falls in either interval
99% [8.6342, 6.6858] 95 % [8.6404, 8.6796]

9.6 Using t-distribution
95% CI [15.574, 16.426]
99% CI [15.411, 16.589]

9.7 $p = 0.092609$, $1 - p = 0.907391$, $n = 1123$ so $[n * P * (1 -- p)] = 94/36 > 10$.
95% CI [0.07565, 0.1096] 99% CI [0.0703,0.1149]

9.8 $E = 0.016955$

9.9 $n = 3228.19$ or 3229 rounded

9.10 Find a 95% confidence interval for the true population mean given the following data from a normal distribution:

```
x
[1]20.0 14.0 29.0 10.0 10.4 21.0 16.0 14.0 6.0 6.8 19.5
> mean(x)
[1]15.15455
> sd(x)
[1]6.887578
> qt(0.025,9)
[1]–2.262157
20.093448 13.934039 29.179828 9.948202 10.434492 21.144037
16.182559 14.360259 6.681212 19.543834
```

9.11 [67.203, 72.797]

9.13 [62.348,67.652]

9.12 ($1000 – 1.96*$200/sqrt(60), $1000 + 1.96*$200/sqrt(60)) ($949.39, $1050.61) We are 95% confident that the interval ($949.39, $1050.61) covers the true mean monthly rent of Durham apartments listed on Craig's List. 2. To what population of apartments can you appropriately infer from your sample in #1? We can most accurately infer to Durham apartments listed on Craig's List. We should not infer to all apartments in Durham because we do not know if apartments on Craig's List are representative of all apartments in Durham.

9.13 We want $z * \sigma$/sqrt(n) to equal $50. $**50** = **50** = $\boldsymbol{z * \sigma n}$ = **1.645** * **200*n*** = **329*n*** Solve for n. sqrt(n) = 329/50 = 6.58. Square both sides, and you get an n equal to 43.3. To calculate a margin of error +/- $50, you would need to randomly sample 44 rental apartments on Craigslist.

9.14 (23.4 – 1.96 * 0.9/sqrt(100), 23.4 + 1.96 * 0.9/sqrt(100)) We are 95% confident that the interval (23.22 mpg, 23.58 mpg) includes the true mean mpg of Duncan's automobile.

9.15 The 95% CI is [5.222, 50.778]

9.16 99% CI is [55.73, 60.27]

9.17 99% CI [9.3791, 22.921]

9.18 The 95% CI is [3.6622, 4.2178]

9.19 95% CI is [22.211, 24.989]

9.21 a) Z→[1.951, 2.049] b) using t and s = 0.12 is [1.9361, 2.0639] c) z versus t.

9.22 The 95% CI is [6.5896, 9.2104], 99% is [6.1088,9.6912].

9.24 x-bar = 69821

9.26 Since $n*p*(1 - p) < 10$, we cannot do the problem.

9.28 [7.1155, 8.8845]

9.29 [6.8272, 9.1728]

9.30 [2.2667, 2.7555]

Chapter 10 Exercises

10.1 One-sided right-tail test because more than 4 out of 5 is good. p-value = 0.13, we fail to reject H_0

10.2 Two-sided test, p-value is 0.089 so we fail to reject,

10.3 p-Value is 0.0146 so if $\alpha = 0.05$ we reject H_0.

10.4 p-Value is approximately 0.002799. Since this is less than either 0.01 or 0.05, we reject H_0 at either level of significance.

10.5 p-Value > α, so we fail to reject H_0.

10.6 p-Value is 0.001, so we reject at both 0.05 and 0.01 using a left-tail test.

10.7 We do conform to safety requirements.

$Z = 1.169$, $p = 0.05264$, Fail to reject H_0.

10.8 phat = 0.383, $n = 1000$, $Z = 4.347$ $P(Z > 4.347) = 1.3785 \times 10^{-5}$, reject the null hypothesis

10.10 H_0: $\mu = 60$, H_a $\mu > 60$, $t = 4.32$, $p = 0.00127$, we reject H_0.

10.11–10.20 *p*-values

10.11 One-sided right-tail test because more than 4 out of 5 is good. *p*-value = 0.13, we fall to reject H_0.

10.12 Two-sided test, *p*-value is 0.089, so we fail to reject.

10.13 *p*-Value is 0.0146, so if $\alpha = 0.05$, we reject H_0, and if $\alpha = 0.01$, we fail to reject.

10.14 *p*-Value is approximately 0.002799. Since this is less than either 0.01 or 0.05, we reject H_0 at either level of significance.

10.15 *p*-Value > α, so we fail to reject H_0.

10.16 *p*-Value is 0.001, so we reject at both 0.05 and 0.01 using a left-tail test.

10.21 The circumference of a steel screw is only acceptable if the standard deviation is at most 0.06 mm. Use a 0.05 level of significance to test the suitability of the company's steel screws. Our sample has 36 rods and the sample standard deviation is 0.065 mm.

H_0: $\sigma^2 = \sigma_0^2 = 0.0036$

Hα = : σ2 > σ02, σ2 > 0.0036

α = 0.05

Test statistic

$$\chi^2 = \frac{(n-1)s^2}{\sigma_0^2} = \frac{(36-1).0.004225}{0.0036} = 41.07$$

Determine rejection regions.

$\chi^2 \geq \chi^2_{\alpha,n-1} \chi^2 > \chi^2_{0.05,35} = 49.8108$

Using R

```
qchisq(0.05,35,lower.tail = FALSE)
[1] 49.8108
```

Since 41.07 < 49.8018, we fail to reject H_0.

10.22 A random sample of 25 college student heights comes from a normal distributed population. We want a 95% confidence interval on the variance. We compute the standard deviation as 5.61.

95% CI is [11.6885, 38.07563]

```
chiupper = qchisq((1 – 0.95)/2, 23, lower.tail = FALSE)
> chiupper
[1]  38.07563
> chilower = qchisq((1 – 0.95)/2, 23)
> chilower
[1]11.68855
```

Chapter 11 Exercises

11.1 a. $Z = 3.4006$, $P = 3.313 \times 10^{-4}$, since $P < 0.01$, we reject H_0. Test, at the 1% level of significance.

b. p-Value = 3.313×10^{-4}

11.3 $Z = [(8.6 - 13.8) - 6.3]/\ 02259 = -4.869$, p-value = 5.616×10^{-7}. We reject H_0.

11.5 $Z = -1.7159$. p-value = 0.08617. Since p-value > α, we fail to reject the null hypothesis.

11.7 $Z = -25.48$, p-value = 0, Reject H_0.

11.9 Assumptions: The data are simple random values from both the populations.

Both populations are follows a binomial distribution.

When both mean (np) and variance($n(1 - p)$) values are greater than 10, the binomial distribution can be approximated by the normal distribution.

11.10 The population must follow a normal distribution.

11.11 p-value = 0.436, so we fail to reject H_0.

Chapter 12 Exercises

12.1 a) Find the reliability for 1.2 time periods, $R(1.2) = 0.8703$

12.3 P(1 helicopter) = 0.714 sending a backup yields 0.9185.

12.7 You are in charge of stage lighting for an outdoor concert. There is some concern about the reliability of the lighting system for the stage. The lights are powered by a 1.5 KW generator that has a MTTF of 7.5 hours.

a. Find the reliability of the generator for 10 hours if the generator's reliability is exponential. $1 - F(7.5) = 1 - \exp(-10/7.5) = 0.7364$.

b. Series = 0.542.

Parallel = 0.9305 (better).

c. Five: four has probability = 0.988 and $n = 5$ is 0.997 > 0.99

N	Poisson
0	0.263606
1	0.615072
2	0.849376
3	0.953509
4	0.988219
5	0.997475

12.8 Find the system reliability for six months when $x = 0.96$ is 0.9002

a. Find the system reliability for six months when $x = 0.939$ is 0.8994

Chapter 13 Exercises

13.1 a)

```
x = c(−1, 0, 1, 2)
> y = c(0, 2, 4, 5)
> fit<-lm(y~x)
> summary(fit)
Call:
lm(formula = y ~ x)
Residuals:
1 2 3 4
−0.2 0.1 0.4 −0.3
Coefficients:
```

	Estimate	Std. Error	t value	Pr(>|t|)
(Intercept)	1.9000	0.2121	8.957	0.0122 *
x	1.7000	0.1732	9.815	0.0102 *

—

```
Signif. codes: 0 '***' 0.001 '**' 0.01 '*' 0.05 '.' 0.1 ' ' 1
Residual standard error: 0.3873 on 2 degrees of freedom
Multiple R-squared: 0.9797,    Adjusted R-squared: 0.9695
F-statistic: 96.33 on 1 and 2 DF, p-value: 0.01022
```

Find the least square regression line for the following set of data $\{(-1, 0),(0, 2),(1, 4),(2, 5)\}$

b) Plot the given points and the regression line.

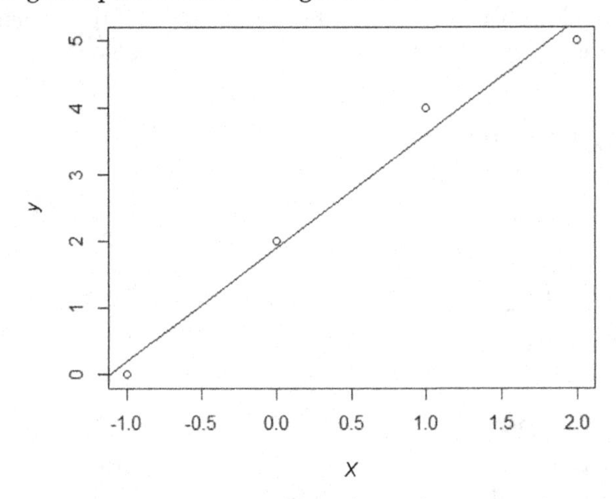

13.3 The sales of a company (in million dollars) for each year are shown in the table below.

x (year)	2005	2006	2007	2008	2009
y (sales)	12	19	29	37	45

a) sales = 8.4*year – 16830.4

b) U sales(2012) = 8.4(2012) – 16830.4 = 70.4 se the least-squares regression line as a model to estimate the sales of the company in 2012.

13.5 a. cor(adv, sales)

[1]0.9804402

b.
```
> sales = c(368, 340, 665, 954, 331, 556, 376)
> adv = c(1.7, 1.5, 2.8, 5, 1.3, 2.2, 1.3)
> mfit<-lm(adv~sales)
> summary(mfit)
Call:
lm(formula = adv ~ sales)
Residuals:
      1         2         3         4         5         6         7
0.25495   0.21192  −0.31008   0.26974   0.06238  −0.29901  −0.18990
Coefficients:
              Estimate   Std. Error   t value   Pr(>|t|)
(Intercept)  −0.6180149   0.2797988    −2.209   0.078224.
sales         0.0056062   0.0005033    11.139   0.000102 ***
—
Signif. codes: 0 '***' 0.001 '**' 0.01 '*' 0.05 '.' 0.1 '' 1
Residual standard error: 0.2857 on 5 degrees of freedom
Multiple R-squared: 0.9613, Adjusted R-squared: 0.9535
F-statistic: 124.1 on 1 and 5 DF, p-value: 0.0001017
```

For the data on student and books.

c.
```
> sales_model2 <- lm(adv ~ sales + I((sales)^2)
+)
> summary(sales_model2)
Call:
lm(formula = adv ~ sales + I((sales)^2))
Residuals:
        1           2           3            4            5            6
0.2130074   0.0861153   0.0053090   −0.0009715   −0.0922219   −0.0017698
7
−0.2094686
Coefficients:
Estimate   Std. Error   t value   Pr(>|t|)
(Intercept)  1.216e+00  5.632e−01   2.159 0.0970.
sales −1.293e−03 2.053e−03 −0.630 0.5630
I((sales)^2) 5.514e−06 1.625e−06 3.394 0.0274 *
```

—

Signif. codes: 0 '***' 0.001 '**' 0.01 '*' 0.05 '.' 0.1 '' 1

Residual standard error: 0.1622 on 4 degrees of freedom

Multiple R-squared: 0.99, Adjusted R-squared: 0.985

F-statistic: 198.3 on 2 and 4 DF, p-value: 9.968e−05

Note: adding terms to the model always makes the model better,

13.7 The time x in years that an employee spent at a company and the employee's hourly

pay, y, for 5 employees are listed in the table below.

a) 0.9689

b) $y = 16.0711 + 1.5498x$

c) $R^2 = 0.9388$

Chapter 14 Exercises

14.1 Scatterplot

Scatterplot of Tread (cm) vs Hours

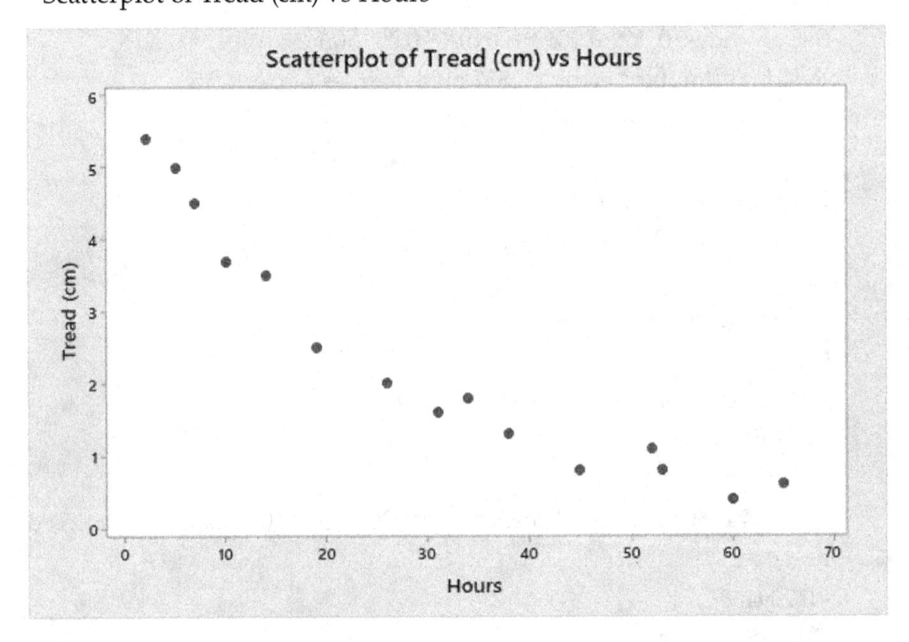

Nonlinear exponential regression yields

tread = 5.86066 $e^{(-0.03959\ hours)}$

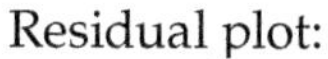

Residual plot:

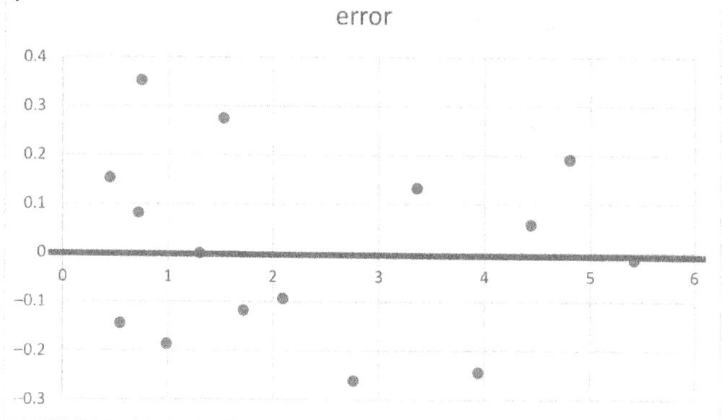

Appears random. SSE = 0.43 and SST = 39.43333

R^2 is 0.987457 as 1 – SSE/SST.

14.2 Regression Analysis: ln(*Z*) versus ln(*x*), ln(*y*)

```
The regression equation is
ln(Z) = - 0.045 + 2.50 ln(x) - 4.34 ln(y)
which converts to Z = 0.955997 x^2.5/y^4.34

Predictor    Coef   SE Coef       T      P
Constant  -0.0455    0.7534   -0.06  0.952
ln(x)      2.4972    0.1630   15.32  0.000
ln(y)    -4.33833   0.06268  -69.21  0.000

S = 0.196990 R-Sq = 99.6% R-Sq(adj) = 99.5%

Analysis of Variance

Source            DF       SS      MS        F      P
Regression         2  194.121  97.061  2501.24  0.000
Residual Error    22    0.854   0.039
Total             24  194.975
```

14.3 a. graph

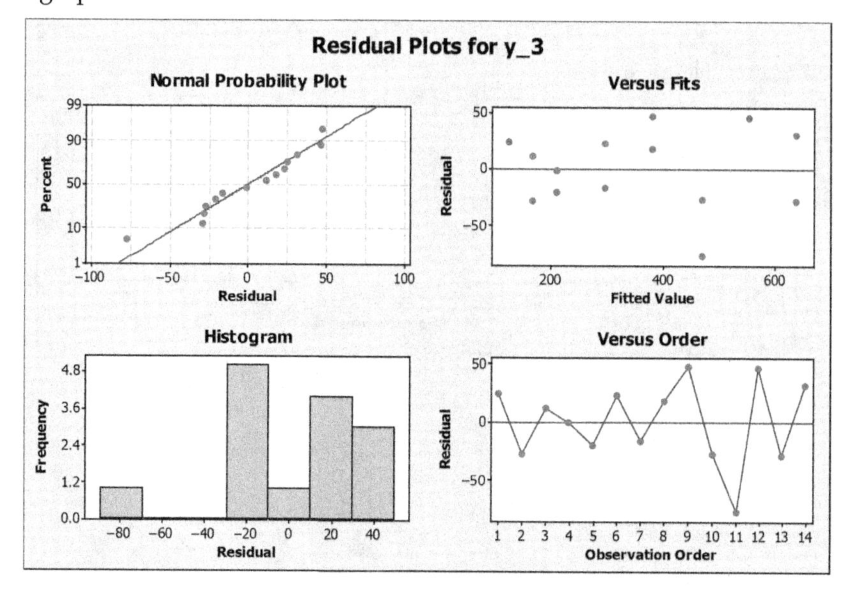

Regression analysis: y versus x

```
The regression equation is
y = - 45.6 + 1.71 x

Predictor   Coef   SE Coef      T       P
Constant  -45.55     25.47  -1.79   0.099
x _ 4     1.71143  0.09969  17.17   0.000

S = 36.7485 R-Sq = 96.1% R-Sq(adj) = 95.8%

Analysis of Variance

Source          DF      SS       MS       F       P
Regression       1  398030   398030  294.74   0.000
Residual Error  12   16205     1350
Total           13  414236

Unusual Observations

Obs  x _ 4   y _ 3    Fit  SE Fit  Residual  St Resid
11     300  390.00  467.88   11.73    -77.88    -2.24R

R denotes an observation with a large standardized residual.
```

14.3 b.

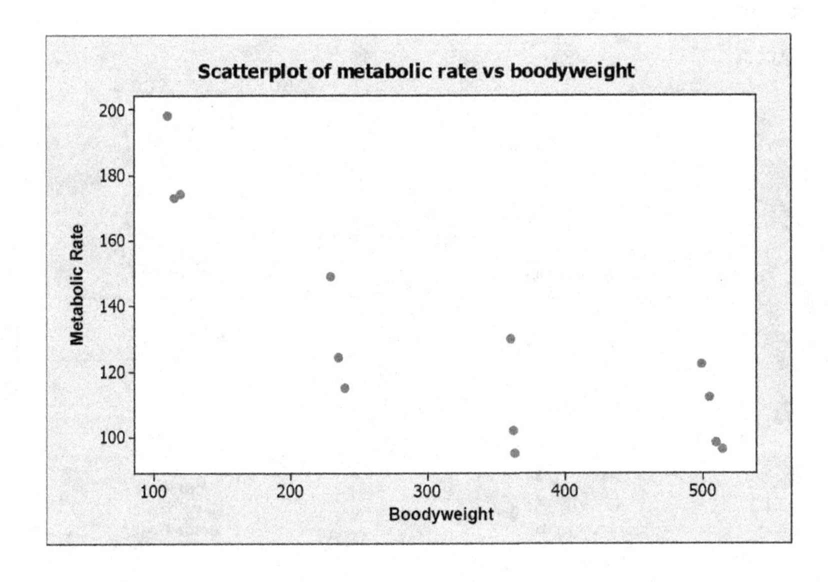

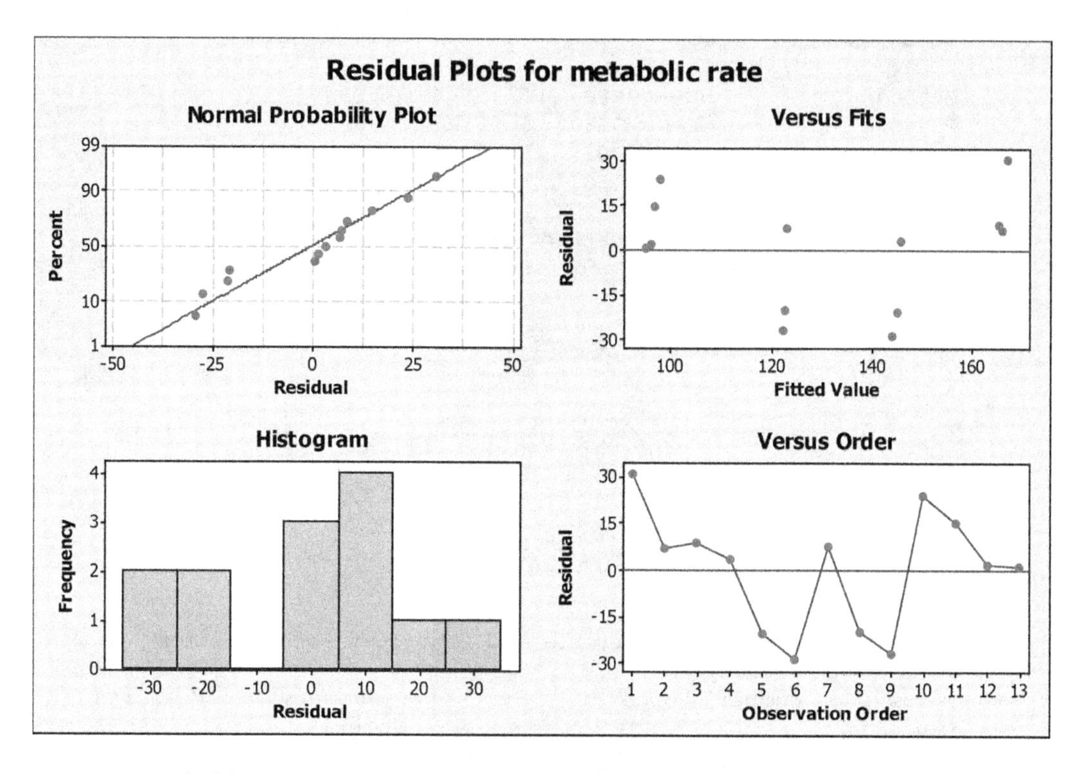

metabolic rate = 187 - 0.177 boodyweight

Predictor	Coef	SE Coef	T	P
Constant	186.61	13.04	14.32	0.000
boodyweight	-0.17717	0.03683	-4.81	0.001

S = 19.9657 R-Sq = 67.8% R-Sq(adj) = 64.8%

Analysis of Variance

Source	DF	SS	MS	F	P
Regression	1	9222.8	9222.8	23.14	0.001
Residual Error	11	4384.9	398.6		
Total	12	13607.7			

3. Logistic Regression Table

						95% CI	
Predictor	Coef	SE Coef	Z	P	Odds Ratio	Lower	Upper
Constant	15.6645	18.7682	0.83	0.404			
GRE	-0.0307517	0.0243202	-1.26	0.206	0.97	0.92	1.02
TOP	4.90852	5.02915	0.98	0.329	135.44	0.01	2585632.77
GPA	0.0755136	5.06531	0.01	0.988	1.08	0.00	22100.12

Log-Likelihood = -3.561
Test that all slopes are zero: G = 6.339, DF = 3, p-value = 0.096

Goodness-of-Fit Tests

Method	Chi-Square	DF	P
Pearson	5.63758	6	0.465
Deviance	7.12107	6	0.310
Hosmer-Lemeshow	5.63758	8	0.688

Table of Observed and Expected Frequencies:
(See Hosmer-Lemeshow Test for the Pearson Chi-Square Statistic)

Group Value	1	2	3	4	5	6	7	8	9	10	Total
1											
Obs	0	0	0	0	0	1	1	0	1	1	4
Exp	0.0	0.0	0.0	0.2	0.3	0.3	0.5	0.6	1.0	1.0	
0											
Obs	1	1	1	1	1	0	0	1	0	0	6
Exp	1.0	1.0	1.0	0.8	0.7	0.7	0.5	0.4	0.0	0.0	
Total	1	1	1	1	1	1	1	1	1	1	10

Measures of Association:
(Between the Response Variable and Predicted Probabilities)

Pairs	Number	Percent	Summary Measures	
Concordant	22	91.7	Somers' D	0.83
Discordant	2	8.3	Goodman-Kruskal Gamma	0.83
Ties	0	0.0	Kendall's Tau-a	0.44
Total	24	100.0		

Answer Chapters 15 and 16

Chapter 15 Exercises

15.1 > summary(one.way)

	Df	Sum Sq	Mean Sq	F value	Pr(>F)
class	3	15.74	5.247	2.402	0.0979.
Residuals	20	43.69	2.184		

—

Signif. codes: 0 '***' 0.001 '**' 0.01 '*' 0.05 '.' 0.1 ' ' 1

>

With a p-value of 0.0979, we fail to reject the null hypothesis that means are the same.

15.2

One-Way Analysis of Variance

Analysis of Variance for growth

Source	DF	SS	MS	F	P
light	1	2.980	2.980	4.27	0.066

```
Error       10    6.973    0.697
Total       11    9.953
                                 Individual 95% CIs For Mean
                                 Based on Pooled StDev
Level        N     Mean   StDev---+---------+---------+---------+---
1            6   4.7800   0.3251 (----------*----------)
2            6   5.7767   1.1353 (----------*---------)
                                 ---+---------+---------+---------+---
Pooled StDev = 0.8350               4.20      4.90      5.60      6.30
```

One-Way Analysis of Variance

```
Analysis of Variance for growth
Source     DF        SS       MS       F       P
temp        2     3.984    1.992    3.00   0.100
Error       9     5.968    0.663
Total      11     9.953
                                 Individual 95% CIs For Mean
                                 Based on Pooled StDev
Level        N     Mean   StDev----+---------+---------+---------+--
1            4   4.6050   0.6595 (--------*--------)
2            4   5.2175   0.5847 (--------*--------)
3            4   6.0125   1.1012 (--------*--------)
                                 ----+---------+---------+---------+--
Pooled StDev = 0.8143              4.0       5.0       6.0       7.0
```

Interpretation: neither one-way analysis is significant.

Two-Way Analysis of Variance

```
Analysis of Variance for growth
Source       DF      SS      MS       F       P
light         1   2.980   2.980   10.39   0.018
temp          2   3.984   1.992    6.95   0.027
Interaction   2   1.268   0.634    2.21   0.191
Error         6   1.721   0.287
Total        11   9.953
```

Interpretation: Both factors, light and temp, are significant by themselves but not the interaction term.

15.3 **One-Way Analysis of Variance**

```
Analysis of Variance for car
Source  DF      SS      MS       F        P
type     2   86050   43025   25.18    0.001
Error    6   10254    1709
Total    8   96304
                                      Individual 95% CIs For Mean
                                      Based on ooled StDev

Level  N      Mean    StDev --+-------+-------+-------+----
1                3        666.67   31.18  (-----*-----)
2                3        473.67   49.17  (----*-----)
3                3        447.33   41.68  (-----*-----)
                                          --+-------+-------+-------+----
Pooled     StDev =        41.34            400     500     600     700
```

Interpretation: results are significant at 0.05, 0.01.

Chapter 16 Exercises

16.1 Let A = 1, B = 2, C = 3

SUMMARY OUTPUT	
Regression Statistics	
Multiple R	0.716994
R Square	0.514081
Adjusted R Square	0.409956
Standard Error	21.97348
Observations	18

ANOVA

	df	*SS*	*MS*	*F*	*Significance F*
Regression	3	7151.439	2383.813	4.93713	0.015222452
Residual	14	6759.673	482.8338		
Total	17	13911.11			

Interpretation: Model appears significant with p-value of 0.015 at alpha = 0.05 but the individuals terms are not significant.

```
The regression model equation is
POST = - 50 + 61.0 treat + 0.967 PRE - 0.215 interact

Predictor    Coef     StDev       T      P
Constant    -49.7     125.1   -0.40  0.697
treat       61.04     57.05    1.07  0.303
PRE        0.9671    0.6271    1.54  0.145
interact  -0.2146    0.2847   -0.75  0.464

S = 21.97 R-Sq = 51.4% R-Sq(adj) = 41.0%

Analysis of Variance

Source          DF      SS      MS      F      P
Regression       3  7151.4  2383.8   4.94  0.015
Residual    Error       14  6759.7  482.8
Total               17 13911.1

Source    DF    Seq   SS
treat      1  4218.7
PRE        1  2658.4
interact   1   274.3
```

Index

Printed in the United States
by Baker & Taylor Publisher Services